T0134611

Culture, Environment and Health in the Yucatan Peninsula

Hugo Azcorra • Federico Dickinson
Editors

Culture, Environment and Health in the Yucatan Peninsula

A Human Ecology Perspective

Editors
Hugo Azcorra
Departamento de Ecología Humana
Cinvestav-Mérida
Mérida, Yucatán, México

Centro de Investigaciones Silvio Zavala
Universidad Modelo
Mérida, Yucatán, México

Federico Dickinson
Departamento de Ecología Humana
Cinvestav-Mérida
Mérida, Yucatán, México

ISBN 978-3-030-27003-2 ISBN 978-3-030-27001-8 (eBook)
https://doi.org/10.1007/978-3-030-27001-8

This Springer imprint is published by the registered company Springer Nature Switzerland AG
The registered company address is: Gewerbestrasse 11, 6330 Cham, Switzerland

Acknowledgements

From its conception, this book sought to account for the diverse scientific work on the study of environment–society relationship in the Yucatán Peninsula in the recent years. The editors of this volume are sincerely grateful to all authors and set of co-authors for believing in us and joining this idea. We acknowledge the gentle, professional, and patient support that Ms. Graciela Valentin provided us with checking the style of the references, headings, subheadings of the text, tables and figures, and the making of indexes and glossaries of all the chapters of the book.

Few interdisciplinary areas of the knowledge, such as human ecology, could have achieved to unify theoretical and philosophical works from different disciplines including biological anthropology, bioarcheology, epidemiology, and toxicology. Several of the chapters contained in this volume highlight the adverse living conditions experienced by contemporary Maya groups from the Peninsula. We are heartily grateful to each of the people who participated in the studies that led to the chapters presented in this book. We hope that the works in this book are the voices of historically oppressed people that are currently experiencing dramatic changes in their labor and consumption patterns.

Contents

Contributors

Flor Arcega-Cabrera Unidad de Química Sisal, Facultad de Química, Universidad Nacional Autónoma de México, Sisal, Yucatán, México

Sunny Asaf Richland County Health Department, Sidney, MT, USA

María Luisa Ávila-Escalante Licenciatura en Nutrición, Facultad de Medicina, Universidad Autónoma de Yucatán, Mérida, Yucatán, México

Hugo Azcorra Departamento de Ecología Humana, Cinvestav-Mérida, Mérida, Yucatán, México

Centro de Investigaciones Silvio Zavala, Universidad Modelo, Mérida, Yucatán, México

Barry Bogin School of Sport, Exercise and Health Sciences, University of Loughborough, Loughborough, UK

Member, UCSD/Salk Center for Academic Research and Training in Anthropogeny (CARTA), La Jolla, California, USA

Daniel E. Brown Department of Anthropology, University of Hawai'I at Hilo, Hilo, HI, USA

Diana Cahuich-Campos Sociedad y Cultura, El Colegio de la Frontera Sur, Unidad Campeche, Campeche, México

Luis Alfonso Ramírez-Carrillo Centro de Investigaciones Regionales Dr. Hideyo Noguchi, Universidad Autónoma de Yucatán, Mérida, Yucatán, México

María Teresa Castillo-Burguete Departamento de Ecología Humana, Cinvestav-Mérida, Mérida, Yucatán, México

Julio Roberto Chi-Keb Laboratorio de Bioarqueología e Histomorfología, Facultad de Ciencias Antropológicas, Universidad Autónoma de Yucatán, Mérida, Yucatán, México

Claudia Guadalupe Chi-Méndez Departamento de Ecología Humana, Cinvestav-Mérida, Mérida, Yucatán, México

Departamento de Ciencias de la Salud y Ecología Humana, Universidad de Guadalajara, Centro Universitario de la Costa Sur, Autlán de Navarro, Jalisco, México

Andrea Cucina Facultad de Ciencias Antropológicas, Universidad Autónoma de Yucatán, Mérida, Yucatán, México

Federico Dickinson Departamento de Ecología Humana, Cinvestav-Mérida, Mérida, Yucatán, México

Patricia Olga Hernández Espinoza Centro INAH Sonora, Secc. Antropología, Instituto Nacional de Antropologia e Historia, Hermosillo, Sonora, México

José Luis Febles-Patrón Departamento de Ecología Humana, Cinvestav-Mérida, Mérida, Yucatán, México

Adriana González-Martínez Departamento de Ecología Humana, Cinvestav-Mérida, Mérida, Yucatán, México

Facultad de Ciencias Biológicas, Universidad Autónoma de Nuevo León, Monterrey, México

Alan H. Goodman School of Natural Sciences, Hampshire College, Amherst, MA, USA

Francisco D. Gurri Department of Sustainability Science, Environmental Anthropology and Gender Lab, El Colegio de la Frontera Sur (ECOSUR), Campeche, México

Almira L. Hoogesteijn Departamento de Ecología Humana, Cinvestav-Mérida, Mérida, Yucatán, México

Oswaldo Huchim-Lara Unidad Experimental Marista, Universidad Marista de Mérida, Mérida, Yucatán, México

Laura Huicochea-Gómez Sociedad y Cultura, El Colegio de la Frontera Sur, Unidad Campeche, Campeche, México

Carlos N. Ibarra-Cerdeña Departamento de Ecología Humana, Cinvestav-Mérida, Mérida, Yucatán, México

Karen L. Kramer Department of Anthropology, University of Utah, Salt Lake City, UT, USA

Hugo Laviada-Molina Escuela de Ciencias de la Salud, Universidad Marista de Mérida, Mérida, Yucatán, México

Thomas Leatherman Department of Anthropology, University of Massachusetts, Amherst, Amherst, MA, USA

Sally López-Osorno Facultad de Medicina, Universidad Autónoma de Yucatán, Mérida, Yucatán, México

Nina Méndez-Domínguez Escuela de Medicina, Universidad Marista de Mérida, Mérida, Yucatán, México

Lourdes Márquez Morfín Posgrado en Antropología Física, Instituto Nacional de Antropología e Historia/Escuela Nacional de Antropología e Historia, Ciudad de México, México

Allan Ortega-Muñoz Department of Physical Anthropology, Centro INAH Quintana Roo, Instituto Nacional de Antropología e Historia, Chetumal, Quintana Roo, México

Erik Otárola-Castillo Department of Anthropology, Purdue University, West Lafayette, IN, USA

Janine M. Ramsey Centro Regional de Investigación en Salud Pública, Instituto Nacional de Salud Pública, Tapachula, Chiapas, México

Samantha Sanchez Coordinación de Nutrición, Universidad Vizcaya de las Américas Campus Mérida, Mérida, Yucatán, México

Lynnette Leidy Sievert Department of Anthropology, University of Massachusetts Amherst, Amherst, MA, USA

J. Tobias Stillman Save the Children Foundation, Washington, DC, USA

Vera Tiesler Laboratorio de Bioarqueología e Histomorfología, Facultad de Ciencias Antropológicas, Universidad Autónoma de Yucatán, Mérida, Yucatán, México

Alba R. Valdez-Tah Department of Anthropology, University of California, Irvine, Irvine, CA, USA

Maria Inês Varela-Silva School of Sport, Exercise and Health Sciences, University of Loughborough, Loughborough, UK

Amanda Veile Department of Anthropology, Purdue University, West Lafayette, IN, USA

Chapter 1
Introduction

Hugo Azcorra and Federico Dickinson

1.1 On Human Ecology

This book offers a human ecology perspective, and we are aware that there is no consensus on what is human ecology, and that among the several proposals that had been put forward you could find those of Park (1936) and McKenzie (1925), from sociology, of Lewin (1944), in psychology, of Barrows (1983), in geography, of Steward (1955), in cultural anthropology, and of Adams (1935), in biology. The variety of fields from which we have theoretical proposal on human ecology ranges from domestic economy (Gyeszly 1988), to human biology (Ulijaszek 2013; Weiner 1977), and philosophy (Christensen 2014) including critical approaches (York and Mancus 2009).

Despite this range of theoretical positions, as editors of this book we asked contributing authors to frame their chapters in the understanding of human ecology as a multidisciplinary research field, that studies the interactions between *ecosystems*, sociocultural systems, and human biology (Dickinson 2004; Stinson et al. 2012). These interactions, a case of *complex systems* as understood by García (2006), literally produce changes in the three subsystems and give place, in certain periods and places, to specific well-being levels (Fig. 1.1).

At the top of the triangular pyramid of Fig. 1.1 is well-being, a concept that "… has multiple constituents, including basic material for a good life, freedom and choice, health, good social relations, and security (…). The constituents of well-being, as experienced and perceived by people, are situation-dependent, reflecting

H. Azcorra (✉)
Departamento de Ecología Humana, Cinvestav-Mérida, Mérida, Yucatán, México

Centro de Investigaciones Silvio Zavala, Universidad Modelo, Mérida, Yucatán, Mexico

F. Dickinson
Departamento de Ecología Humana, Cinvestav-Mérida, Mérida, Yucatán, México

H. Azcorra, F. Dickinson (eds.), *Culture, Environment and Health in the Yucatan Peninsula*, https://doi.org/10.1007/978-3-030-27001-8_1

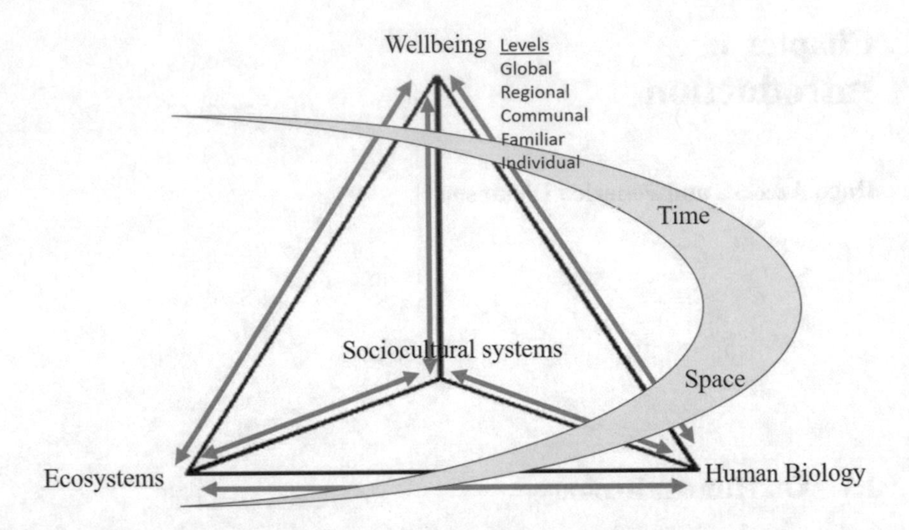

Fig. 1.1 Our concept of human ecology. Source of the yellow figure: White SC (2008) But what is well-being? A framework for analysis in social and development policy and practice: Conference on Regeneration and Wellbeing: Research into Practice. University of Bradford, United Kingdom, Fig. 4, p. 10

local geography, culture, and ecological circumstances." (Alcamo et al. 2003, Box 1, p. 2). Among the properties of each of the three subsystems represented in Fig. 1.1, there is a certain status at a given moment, status that could be understood as its "health."

For human populations, we follow the World Health Organization when saying that health "…is a state of complete physical, mental and social well-being and not merely the absence of disease or infirmity" (World Health Organization 1995), taking into account the critical review of Alcántara Moreno (2008); for ecosystems we follow the proposal of Lu et al. according to which the health of an ecosystem is its "…status and potential (…) to maintain its organizational structure, its vigor of function and resilience under stress, and to continuously provide quality ecosystem services for present and future generations in perpetuity" (Lu et al. 2015, p. 3). In terms of health of a sociocultural system, we resort the concept of human development proposed by the United Nations Development Program. According to the UNDP human development includes the possibility of enjoying, on equal opportunities, a prolonged, healthy, and creative life (UNDP 1995). Unlike other conceptions of development that emphasize results—i.e., to reach a certain level of income or consumption, or have a set of goods and services—, the UNDP approach conceives the well-being of people as a process in which persons are able to maintain and expand the options to do what they value and want to do.

In a hypothetic case in which both the human population and the ecosystems it occupies enjoy "health," but the sociocultural system is "sick," suffering inequity, violence, polarization of wealth… the well-being of the whole system would be,

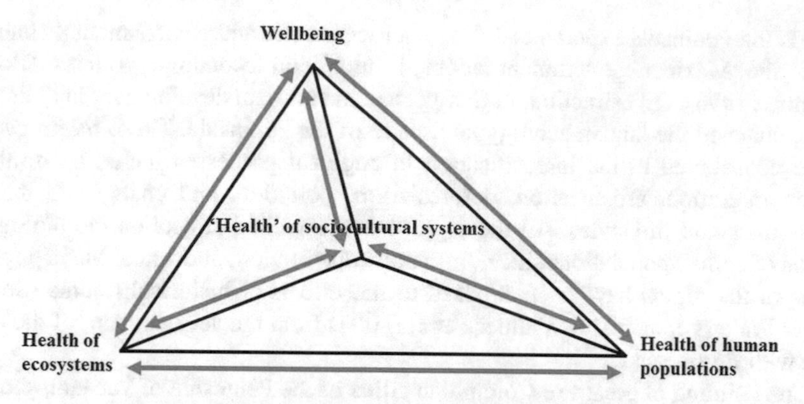

Fig. 1.2 A hypothetic sum of subsystems in particular time and circumstances

from a human ecological point of view, poor (Fig. 1.2). Similarly, if the ecosystems or the human population or both are "sick," the well-being of the system as a whole will be poor.

Authors in this book present cases of human biology and health that are described as results of complex interactions of the human ecology in the Yucatan Peninsula from the Late Classic period (600–800 CE) to present. Each of these cases may be interpreted as an historical product of relations in given places and times.

1.2 On the Region of Interest

The Peninsula of Yucatan, encompassing the current Mexican States of Campeche, Quintana Roo, and Yucatan, has been occupied by human populations for at least the last 13,000–12,000 years (Chatters et al. 2014; Tiesler 2016), and it was the scenery in which the *Maya peoples* developed relevant astronomic, mathematical, architectonical, and agricultural achievements long before the sixteenth century, when Europeans contacted them.

Even there are not precise calculus of the size of the population of the Peninsula of Yucatan at the time of the first contact with Europeans, it is well known that along the colonial times, native population suffered a demographic shrinkage due to droughts and famines (Hoggarth et al. 2017), and epidemics of unknown diseases in America before the arrival of Spaniards and African slaves (Farriss 2012). During European domination, native population lived under a system of oppression which brought them to endure very hard living conditions (Farriss 2012). In the second half of the nineteenth century, the current Yucatan state was one of the richest in Mexico, thanks to the sisal monoculture and the exploitation of peasants and industrial sisal workers, but after the decline of the sisal agroindustry after the II World War, the Yucatecan economy faced a standstill (Wells 2017). By the late

1970s the Peninsula experienced deep socioeconomic and environmental changes after the Mexican government adopted neoliberal economic policies (Baños Ramírez 1996). Oil extraction, in Campeche, and tourism development in Quintana Roo, changed the human ecology landscape of the Peninsula. These recent events have contributed to the intensification of regional processes including rural-to-urban migration, urbanization, deforestation, pollution, and changes in market economies and life styles. All these processes have had effects on the biological status of human populations since, from our point of view, aboriginal Maya populations of the region have been exposed to the effects of historical trauma (Sotero 2006; Walters et al. 2011; Whitbeck et al. 2004) from the very moment of the contact with Europeans.

The building of great Pre-Columbian cities of the Peninsula of Yucatan, such as Calakmul, Chichen Itzá, Uxmal, and Edzná, to mention just the most well known, by Maya people, that lacked metallic tools and draft animals, required a lot of human labor for long time periods, and involved not just stonecutters, bricklayers, and foremen but also architects and astronomers to design the cities and to align buildings to certain constellations, and drawers and painters to ornament the buildings, and women to cook for the workers. And all this human effort was sustained by a surplus that had to be produced by laborers. We think that this effort had to have an impact on the biological status of workers, mainly laborers.

1.3 On Human Corporality and Mirrors of Society

We understand human *corporeity* as the result of historical courses that are multi-determined, and in which take part biological, ecological, and social processes (Dickinson 1983). In a more philosophical field, corporeity is a totality constituted by two natures: organic and social. Human corporeity simultaneously appears as a social product, as social appropriated nature—literally taken possession of nature, and as objectified society (Dickinson 1983, p. 63–64).

In human bodies are expressed social relations, and we can read these relations in the human bodies, looking for the presence or absence of certain characteristics of biological tissues or systems, and by measuring certain phenotypical traits at the population level.

Since at least the 1830s (Tanner 1987) there is evidence of the effects on children growth of labor and, in general of the living conditions of European workers, particularly at England (Engels 1975). The union trade actions, and official rules promoted by laboral inspectors and legislators, step by step took women and children from factories giving place a better children growth. As we know, children and adolescent growth is highly sensible to environmental conditions, and this is the base for the statement that "children growth is a mirror of society" (Tanner 1987). In other words, as several chapters of this book will show, it is possible to know a lot of a society through the study of how their children and adolescents grow.

1.4 The Book Content

This book is organized in three parts and a previous chapter about the history, society, and environment of the Yucatan Peninsula. In this first chapter, Ramírez presents a historical review of the relationship between society and nature in the region. The author reviews the most important stages of the society including the climax of development of the Maya civilization, the colonial society during sixteenth to nineteenth centuries, and the independent period until 1970. Ramírez also discusses the social and environmental implications of complex processes such as the privatization of land, tourism development, and urbanization. Finally, the author evaluates the contemporary model of economic development and its implications on poverty levels, Maya identity, and the change in consumption patterns. This chapter situates the reader on the geographical and historical context of the peninsular population.

The first part of the book includes eight chapters that analyze how recent and long-term changes in political and sociocultural systems in the region shape the physical growth, nutrition, body composition, fertility, somatic symptoms, and consumption patterns in the population. Chapters 3 and 6 analyze the impact of changes in the economic and labor dynamics on the dietetic pattern among the Maya using two different but complementary theoretical approaches. Bogin and coauthors discuss the impact of globalization as a force that tend to transform local food patterns towards a modern pattern characterized by the presence of highly processed food. Leatherman and Goodman address the phenomenon from the perspective of the political ecology; the authors analyze how tourism development in rural areas transform the social relations and disrupt agricultural systems with increasingly delocalized and commoditized food systems.

Chronic undernutrition (mainly stunting) is an unresolved problem in Yucatan affecting more severely the Maya population and several chapters of this book address this issue. In Chap. 4 Varela-Silva et al. analyze the implications of the arbitrary cutoff points that classify people into categories of "stunted" or "non-stunted," discussing how these may include or exclude individuals from one of the categories, depending on the standard/references' databases used and lists the inherent health implications. In Chap. 5, Azcorra and collaborators argue that current biological status of Maya people, particularly the coexistence of stunting and overweight at population level, is the result of the historical adverse living conditions experienced by this ethnic group. The authors resort to the Historical Trauma Theory to guide a review of the events that occurred during sixteenth, seventeenth, and eighteenth centuries and the almost two centuries after Mexican Independence (1821) that have shaped the phenotype of contemporary Maya people. On the other hand, working in agricultural communities from Campeche, Gurri analyzes the relationship between linear enamel hypoplasias (LEH), a proxy for early developmental stress, and fat accumulation in adolescents. Females adolescents, but not males, with LEHs had higher BMIz scores, and these differences seems to be explained by sex-dependent differences in energy-demanding activities.

Women's health in the context of ecological variation and socioeconomic change is addressed in three chapters of this book. In Chap. 7, Ortega and Gurri studied the impact of commercial agriculture and circular migration on reproductive histories of Maya women from Maize and Citrus regions of Yucatan. Their results indicate that after development of commercial agriculture, women from Citrus Region reduced their fertility while salaried work combined with subsistence agriculture in the Maize Region made large families desirable to incorporate labor opportunities into the household's traditional survival strategy. Sievert and colleagues, in Chap. 10, analyze the complex relationships between physical activity, biological and sociocultural factors, and somatic symptoms in women from urban and rural contexts in the State of Campeche. Their results show that even when more hours of physical activity are associated with more somatic symptoms, socioeconomic conditions are more often related to symptoms reported by women. Finally, in Chap. 9, Veile and collaborators analyze the history of water use in a Maya community in northern Campeche and how current forms of water access and storage relate with breastfeeding practices and child morbidity outcomes.

The second part of the book includes three chapters that address the relationship between physical environment, social status, and human biology in Maya pre-Colonial populations. In Chap. 12, Marquez and Hernandez analyze some trends in demographic characteristics and indicators of biological disruptions in relation to population density, exposition to infectious agents, and characteristics of diet in bone remains from several sites of the Yucatan Peninsula at the time of the Terminal Maya Classic. In the same line of analysis, Cucina analyzes caries and linear enamel hypoplasia (LEH) in two Classic Period coastal sites, in the Yucatan peninsula's northern coastline, in comparison with inland sites. In both chapters, the absence of clear patterns drives researchers to make specific-populations interpretations. In terms of Cucina *results (...) highlight the concept that well being is the result of the entangled and intertwined interaction between the environmental, biological and cultural factors that shaped each and every human community.* Moving to the analysis of modern skeletal series, Tiesler and colleagues compare, in Chap. 13, two cemetery series along with their civil records from the city of Merida, Yucatan. The earlier series materializes the living conditions of locals prior to 1925 and the recent materializing the conditions experienced during the twenty-first century. The authors highlight the shifts in lifestyle, life expectancy, and health problems in Merida, Yucatan.

Finally, the third part of the book is heterogeneous in terms of the topics included, but all these analyze the result of interaction between natural and sociocultural environment and human health. In Chap. 14, Laviada-Molina and colleagues address the impact of human behavior on profiles of type 2 diabetes mellitus in Maya people and decompression sickness in fishermen divers, and the diseases transmitted by *Aedes* mosquitoes, including dengue, chikungunya, and zika viruses' infections. The interest of authors of addressing pathologies with different etiologies lies on how communicable and noncommunicable diseases *of epidemiologic relevance in Yucatan can be understood with precision when approached from the human ecology perspective*. In Chap. 15, López and collaborators explore the influence of seafood consumption on total mercury in hair of adult women and men from the city of Merida. The results show that seafood consumption predicts positively concentrations of mercury, taken into account the influence of several covariables such as age, height, and body composition, and that near to 50% of participants had mercury levels above the safe dose, which seems to be a problem that is not taken into account by local, regional, and national authorities.

The editors and authors hope that this book will be of interest to students, practitioners, and researchers from different fields and help them to the understanding of biocultural processes that have repercussions on the health and well-being of populations. Although the chapters contained in this book represent case studies on the Yucatan Peninsula, we think that the theoretical approaches and interpretations made by authors can be useful for other regions of the world.

References

Adams CC (1935) The relation of general ecology to human ecology. Ecology 16:316–335

Alcamo J, Ash NJ, Butler CD et al (2003) Ecosystems and human well-being: a framework for assessment. Island Press, Washington, DC

Alcántara Moreno G (2008) La definición de la salud de la Organización Mundial de la Salud y la interdisciplinarcidad. Sapiens. Revista Universitaria de Investigación 9:93–107

Baños Ramírez O (1996) Neoliberalismo, reorganización y subsistencia rural. El caso de la zona henequenera de Yucatán: 1980–1992. Universidad Autonoma de Yucatán, Mérida

Barrows HH (1983) Geography as human ecology. In: Young G (ed) Origins of human ecology. Hutchinson Ross, Stroudsburg, PA, pp 49–62

Chatters JC, Kennett DJ, Asmerom Y et al (2014) Late Pleistocene human skeleton and mtDNA link paleoamericans and modern Native Americans. Science 344:750–754

Christensen CB (2014) Human ecology as philosophy. Hum Ecol Rev 20:31–49

Dickinson F (1983) Una discusión teórica en Antropología Física. Elaboración de los lineamientos de una Epigenética Histórica. Dissertation, Escuela Nacional de Antropología e Historia

Dickinson F (2004) Ecología humana en México. Una frontera con (casi) todo por hacer. Avance y Perspectiva 23:5–11

Engels F (1975) La situación de la clase obrera en Inglaterra. Ediciones de Cultura Popular, México, DF

Farriss N (2012) La sociedad maya bajo el dominio colonial. Artes de México y del Mundo – Conaculta, México, DF

García R (2006) Sistemas complejos. Conceptos, método y fundamentación epistemológica de la investigación interdisciplinaria. Gedisa, Barcelona

Gyeszly SD (1988) Human ecology and home economics journals: a selected and annotated bibliography. Ser Rev 14:45–54

Hoggarth JA, Restall M, Wood JW et al (2017) Drought and its demographic effects in the Maya lowlands. Curr Anthropol 58:82–113

Lewin K (1944) Constructs in psychology and phsychological ecology. In: Lewin K, Meyers CE, Kalhorn ML, French JRP (eds) Authority and frustration. University of Iowa Press, Iowa, pp 17–23

Lu Y, Wang R, Zhang Y et al (2015) Ecosystem health towards sustainability. Ecosyst Health Sustainability 1:1–15

McKenzie RD (1925) The scope of human ecology. In: American Sociological society (ed) 20th Annual Meeting, 1925, Papers and Proceedings. American Sociological Society, Washington, pp 141–154

Park RE (1936) Human ecology. Am J Sociol 42:1–15
PNUD (1995) Informe sobre Desarrollo Humano 1995. La revolución hacia la igualdad en la condición de los sexos. Harla, México, DF
Sotero MM (2006) A conceptual model of historical trauma: implications for public health practice and research. J Health Dispar Res Pract 1:93–108
Steward JH (1955) Theory of culture change. University of Illinois Press, Urbana, IL
Stinson S, Bogin B, O'Rourke D et al (2012) Human biology: an evolutionary and biocultural perspective. In: Stinson S, Bogin B, O'Rourke BD (eds) Human biology. An evolutionary and biocultural perspective. Wiley-Blackwell, Hoboken, pp 3–22
Tanner JM (1987) Growth as a mirror of the condition of society: secular trends and class distinctions. Acta Paediatr Jpn 29:96–103
Tiesler V (2016) New findings and preliminary results of a submerged Pleistoce skeleton from the underwater site of Hoy Negro, Tulum, Quintana Roo, Mexico. Mexicon XXXVIII. p 5–6
Ulijaszek S (2013) Biological and biocultural anthropology. In: Banwell C, Ulijaszek SJ, Dixon J (eds) When culture impacts health. Global lessons for effective research. Academic Press, London, pp 23–31
Walters KL, Mohammed SA, Evans-Campbell T et al (2011) Bodies don't just tell stories, they tell histories: embodiment of historical trauma among American Indians and Alaska Natives. Du Bois Rev 8:179–189
Weiner JS (1977) Human ecology. In: Harrison GA, Weiner JS, Tanner JM, Barnicot NA (eds) Human biology. Oxford University Press, Oxford, pp 387–483
Wells A (2017) Los informes sobre su desaparción no son exagerados: vida y época del henequén yucateco. In: Marichal C, Topik S, Frank Z (eds) De la plata a la cocaína. Cinco siglos de historia económica de América Latina, 1500–2000. El Colegio de México y Fondo de Cultura Económica, México, DF, pp 426–454
Whitbeck LB, Adams GW, Hoyt DR et al (2004) Conceptualizing and measuring historical trauma among American Indian people. Am J Community Psychol 33:119–130
World Health Organization (1995) Constitution of the World Health Organization
York R, Mancus P (2009) Critical human ecology: historical materialism and natural laws. Sociol Theory 27:122–149

Chapter 2
The Thin Broken Line. History, Society, and the Environment on the Yucatan Peninsula

Luis Alfonso Ramírez-Carrillo

The line that separates is also that which joins
But if the line that joins man with the land of the Maya is thin
That which joins man to man is yet more fragile…

Ramírez L. A.

Abstract This chapter begins by describing the formation of the Yucatan Peninsula. It continues with the arrival of human beings, the growth of Mayan civilization to its peak about 1000 CE, and its decline until the arrival of the Spanish in the sixteenth century. A summary of the effects of three centuries of colonial rule is provided followed by the first one hundred and fifty years after independence from Spain (1821 to 1970). Emphasis is placed on the relations between private/communal property and the commodification of the forests, land, and water of the Maya. In the modern period (1970–2020) the environmental impacts of five processes are described: privatization of property; proliferation of extensive cattle ranching; urbanization; occupation of the coasts; and demographic changes. It concludes with a rapid assessment of the contemporary regional development model that includes analyses of poverty, ethnic identity of the Maya, oil exploitation, tourism, urban sprawl, and changes in consumption patterns.

2.1 The Peninsula's Physical Scenario

Beginning this discourse requires a brief presentation of the region's physical and geographical space. What is the Yucatan Peninsula? It is the most northerly portion of Mesoamerica, and consists of low wide peninsula that immerses itself in the sea, dividing the Gulf of Mexico from the Caribbean Sea. Approximately 600 km long north to south, it covers about 145,000 km^2. Its central and northern portions

L. A. Ramírez-Carrillo (✉)
Centro de Investigaciones Regionales Dr. Hideyo Noguchi, Universidad Autónoma de Yucatán, Mérida, Yucatán, México

H. Azcorra, F. Dickinson (eds.), *Culture, Environment and Health in the Yucatan Peninsula*, https://doi.org/10.1007/978-3-030-27001-8_2

encompass three states in Mexico: Yucatan (39,524 km^2); Campeche (57,507 km^2); and Quintana Roo (34,205 km^2). In the south it includes the country of Belize and the Petén region of Guatemala.

The Mexican portion of the Yucatan Peninsula consists of 131,236 km^2, less than 2% of national territory, and it is one of the country's five physiographic zones. Average annual rainfall on the Peninsula increases from northwest to southeast, which influences the vegetation. The more arid vegetation in the northwest transitions into dry forest, then into semi-evergreen forest, and finally into rainforest in the south. The weather is warm, humid to subhumid. Summer sees rain and hurricanes sweep the Peninsula while in winter cold "northerlies" bring intermittent precipitation. In other words, summers on the Yucatan Peninsula are rainy and winters mostly dry, meaning the seasonal drought extends from winter through spring. Despite occasional rains, mid-March to mid-May is the driest period. Temperatures are highest with some isolated refreshing showers, but the driest month is April during which the heat feels the most intense (Batllori and Canto 2018). The saying runs that "April is the cruelest month" because it brings the least amount of rain, everything is dry and appears dead, and the whole peninsula reverberates with heat that can easily exceed 40 °C for various days.

The Peninsula is just south of the Tropic of Cancer and thus receives long-term vertical radiation all year generating generally uniformly hot temperatures with few changes. The heat is accentuated by the flat topography scarcely above sea level. In addition, the Peninsula is bathed by warm marine waters on all three coasts, which raises the temperature of winds that blow in from the sea (Orellana et al. 1999).

Except for the Hondo River, the entire Peninsula lacks surface water sources such as rivers, lakes, and lagoons. This is partially due to its flat topography, without mountains, and low elevation above sea level. Both are the result of its late geological formation; it only recently emerged from the ocean to join the continent. It is not without water, however, since it is dotted with cenotes (a Spanish adaptation of *tz'ono'ot* in the Peninsular Maya language), underground cavities between 2 and 20 meters deep that provide year-round access to the water table. These sinkholes are one of the most notable features of the geography and are central to understanding why human society developed in the region. Although much of the Peninsula lacks deep soils, and those that do exist are unfit for plowing except in very small areas, it has one of the most important aquifers in southeast Mexico.

Some cenotes are open-air and were formed naturally by collapse of the stone vaults that covered them. However, most are enclosed underground chambers that humans can access by small passageways and caverns. These open up into large, high vaults, with enough space to breathe, that cover pools of different sizes. The silence and darkness in these vaults are tomblike. Openings can also be made in the roof of the vault by breaking through the limestone surface rock. Cenotes are permanent water sources fed by underground veins of rainwater draining from the southern highlands and seeking an exit to the sea. These underground rivers maintain a constant flow as they move towards the Gulf of Mexico or the Caribbean Sea. Without cenotes the rise of the Mayan civilization in the lowlands, especially the Yucatan Peninsula, would be incomprehensible.

An estimated 3000 cenotes on the Peninsula can be accessed in different ways, although no accurate census has been done and there may be more. Considering this, it is hard to imagine that any substantial ancient Maya city on the Peninsula would not have had access to water via one or various cenotes. Exceptions to this would be the coasts, where freshwater springs exist, and in the wetlands of Quintana Roo (Gómez and Ruiz 2010; Lugo et al. 1999). Over the centuries a huge underground network of cenotes has been documented, many of which interconnect across large portions of the Peninsula.

The strongest hypothesis explaining the origin of cenotes is the impact of an immense meteorite about 65 million years ago precisely in the north of what is now the Yucatan Peninsula. At the time it was still submerged under the ocean and formed part of the continental shelf. The impact created what is now known as the Chicxulub Crater and triggered the mass extinction of dinosaurs in the Tertiary Cretaceous Period. As the Peninsula emerged from the ocean over millions of years the edges of this impact crater, consisting of minimally consolidated limestone, eroded and filled in the crater. Essentially invisible on the surface, regional geology still holds signs of the crater which forms a large semicircle centered on todays' ports of Progreso and Chicxulub, and the outer edge of which is marked by the presence of cenotes (Poveda and Espejo 2007; Frankel 1999; Bottke et al. 2007; Schulte et al. 2010). Known today as the ring of cenotes, it is the strongest confirmation of the hypothesis, based on extensive research, of the link between meteorite, crater, and cenotes.

2.2 Sedentism: The First Line Joining Society and Nature

The natural physical world exists in and of itself. But this is inconsequential for human society without the existence and occupation of human beings who endow it with meaning through social relations. The world that was the Yucatan Peninsula when humans first arrived and began to settle it was propitious for their establishment due to the ready availability of fresh water, a warm climate, diverse vegetation and food resources on the coasts. It also had another very important resource: an extensive chain of caves for shelter that were associated with the cenotes.

Human occupation on the Yucatan Peninsula occurred at the beginnings of human sedentism in Mexico. Limestone caverns usually associated with cenotes were one of the earliest human refuges on the Peninsula. Indeed, the oldest human remains found to date on the Peninsula date back between 12,000 and 13,000 years. These are the bones of an adolescent woman (15–17 years of age) found in the underwater chamber of a cenote known as the "Black Hole" in Tulum, Quintana Roo. Known today as Naia, she was malnourished and had already given birth before she died of a fall. This makes her one of the oldest humans in Mexico and the Americas and demonstrates the antiquity of humans on the Peninsula (Chatters et al. 2014; Watson 2017). It was during this period, about 10,000 years ago, that humans in the region became increasingly sedentary as they developed and refined agriculture and animal husbandry. Over several centuries the small societies on the Peninsula slowly developed a diverse complex of crops, but centered on the combination of corn, beans, and squash. Domestication of maize and beans occurred between 7000 and 10,000 years ago. Planting of one or more crops a year allowed early humans on the Peninsula to enjoy a food surplus sufficient to leave the caves and establish growing and increasingly numerous sedentary populations. From this moment on a more complex civilization began to unfold.

2.3 Mayan Society and Culture: Its Transformations

Sedentism and agriculture fueled the emergence of increasingly large and powerful urban centers and the development of complex societies. Leadership initially based on physical and intellectual skills transitioned over time to a ruling elite of priests and warriors in which knowledge and religion, increasingly complex and polytheistic, were vital to controlling the masses. Constant warfare and religion, both large-scale collective undertakings, were the two most important factors in maintaining unity in society. Mesoamerican societies were generally theocracies in which the governing class were both priests and warrior leaders. Manifestations of a more complex culture on the Peninsula can be found in the vestiges of the first towns and cities.

As in other parts of Mesoamerica, 3000–4000 years after maize domestication on the Yucatan Peninsula villages and then small cities with some stone structures began to develop. The urbanization process coincided with demographic growth and development of increasingly sophisticated cultural traits that became more prominent in different regions of Mexico. Known as the Preclassic, this period extended, broadly speaking, from 2500 to 2000 BCE. Some cultures matured before others and subsequently influenced them; the oldest include the Olmec and the Maya.

2.4 Milpa and Forest Management: The Second Joining Line

Maya culture developed in what is today the south and southeast of Mexico, in the states of Chiapas, Tabasco, Campeche, Yucatan, and Quintana Roo. It was also present in what are today the countries of Guatemala, Belize, Honduras, and El Salvador. This region has supported human occupation for at least 13,000 years, perhaps more, and sedentary agricultural populations for some 7000 years. Around 2500 BCE populations began settling in small villages with very well-defined cultural characteristics that allow reference to the Preclassic Mayan period (2500–250 BCE). That said, recent archaeological research suggests that civilizing principles in the region are older and the Maya Preclassic may actually begin nearer 2700 BCE. Archaeological data also supports linguistic research indicating that the Maya is one of the oldest cultures in the Americas.

During the twenty-odd centuries of the Preclassic, urbanism and early monumental architecture, hieroglyphic writing, mathematics, astronomy, and fine arts developed. It was during the Classic Maya period (250 BCE to 950 CE) that these cultural facets flourished, consolidating into one of the most complex and sophisticated civilizations in Mexico and the Americas. Cities grew to contain from 10,000 to 20,000 inhabitants, with large cities reaching populations of 50,000 to 100,000. Major centers like Palenque in Chiapas attained populations as high as 90,000. Chichen Itza in Yucatan and Calakmul in Campeche were large cities on the Peninsula which held as many as 50,000 people at their peaks. The Maya achieved great sophistication in bas-relief sculpture in stone, construction of monumental pyramids employing the architectural principle of the false arch, a vigesimal mathematical system that included the concept of zero, an astronomy capable of calculating calendars with precise dates for eclipses and planetary trajectories, and a complex hieroglyphic writing system not yet fully deciphered. Indeed, their mathematical and astronomical knowledge was far superior to contemporary Western civilizations.

On the Yucatan Peninsula, Maya society was organized into city-states that controlled a surrounding territory and were relatively autonomous (Robles 2014; Ringle 2014). Their cultural development occurred amid continual warfare and shifting alliances. At the end of the Classic period this complex cultural evolution began to falter. Between 950 and 1000 CE, a combination of social rebellions, excessive intercity warfare and a series of severe droughts descended on the Maya world. The entire planet experienced a period of intermittent droughts for more than a century, and these strongly afflicted the Peninsula. This broke the delicate equilibrium that had existed up to that point in exploiting the forest, and devastated productivity of the milpa agricultural system, a seasonal approach based on slash-and-burn techniques (Gill 2008; Andrews 2014). Famines extended across the region, destabilizing the strict and violent social system. Monumental construction, the placing of stelae, and inscription of hieroglyphics all began to diminish in importance. As the ruling theocracy declined much of the scientific and cultural knowledge accumulated over centuries was partially or completely lost. This period of decline is known as the Postclassic. On the Yucatan Peninsula this period extended from about 1000 CE to arrival of the first Europeans among the Maya in 1511, when a Spanish ship wrecked on the east coast of the Peninsula.

2.5 Conquest and Colony on the Yucatan Peninsula: 1521–1810

The conquest of central Mexico was consolidated in 1521, but many other peoples remained independent throughout what was to become Mexico. These included the Chichimeca in the north, the Purépecha in what is today Michoacán and several dozen other indigenous cultures, including the Maya. Once it had conquered central Mexico, the Spanish crown began a series of long campaigns to control these peoples; in cases like the Yucatan Peninsula it took over a century to gain control (Chamberlain 1974; Chuchiak IV 2014). In fact, the Maya were not completely conquered until 1697, with the fall of the last Itzae Maya at Tayasal lake in the Petén region of Guatemala (Jones Grant 1988). The Chichimecas, Tarahumara, Yaquis, and Apaches in the far north remained in relative independence and isolation in their deserts and mountains until well into the twentieth century. A similar process occurred among the Maya of the Yucatan Peninsula, who retreated into the rainforests of what is today Quintana Roo, and the Lacandon Mayas, who melted into the forests of Chiapas. In this way many of Mexico's native nations conserved their culture and social and political organization for centuries.

The Spanish conquest therefore created a multi-ethnic, multicultural, and multilingual society; for example, the Mestizo society that developed on the Yucatan Peninsula over the following centuries. Perhaps one of the most lethal weapons the Spanish unwittingly wielded to conquer the Americas was biological. As much as 90% of the indigenous population may have perished during the first century of conquest; the extent of the demographic impact of disease is still debated (Rosenblat 1967; Sanders 1976; Whitmore 1992; McCaa 1995). This period known as the Colony encompasses the three centuries between the fall of Tenochtitlan and the independence of Mexico, begun in 1810 and consolidated in 1821. These 300 years saw numerous political, cultural, and economic transformations in Spain, New Spain and what was the General Captaincy of Yucatan (García-Bernal 2015).

The promise of riches which brought the Spaniards to Yucatan, and which brought the destruction of complex Mayan society and its elaborate culture, were never realized. During the three centuries of the Colony the Yucatecan economy never really flourished, and the Peninsula as a whole was considered one of the poorest regions of New Spain. This generalized poverty remained chronic throughout the Colony, and the Spaniards and Creoles living on the Peninsula never ceased to complain to the Crown about their lack of money. For example, in 1796, a few years before the colonial period ended, Yucatan imported almost a million pesos in merchandise but exported little more than half a million; this represents a deficit of almost 50% in its balance of payments (Betancourt 1970).

The Peninsula's economic poverty was due to the absence of precious minerals such as gold and silver, the practical impossibility of developing high intensity irrigation agriculture, and to lack of commercially valuable manufactured products that could be exported on a large scale. In response those who sought to enrich themselves looked to Yucatan's true wealth: its native Maya population. Mayan labor, be it directly in productive activities, personal domestic service, or as tribute, was how the colonizing population reaped benefits from the region; it also functioned as a mechanism of political control. Exploitation of the indigenous workforce was intimately linked to land access and possession of physical space. These took various forms over time: the *encomienda*, the hacienda, the *obvenciones*, the *cofradías*, and the *repartimientos*.

However, the greatest wealth extracted by the colonizers was direct exploitation of Maya labor through what were called "personal services." Under this concept the entire indigenous population was obliged to provide a certain number of days per week, month, or year of free labor to the *encomenderos* (who controlled the *encomiendas*), the Catholic church and its priests, and the government. The value of this work, in addition to the products delivered initially as tribute and then as taxes to the government, constituted the foundation of the wealth accumulation process through which the Yucatecan elites and the Spanish government derived benefits (Bracamonte 2007; Bracamonte et al. 2011). This continuous extraction of surpluses and even subsistence resources for generations was also the origin of the cycle of historical reproduction of poverty among the Maya of the Yucatan Peninsula; it is one that continues into the present.

2.6 Independence in Yucatan: The Nineteenth Century

On 15 September 1821, in the city of Merida, capital of the Captaincy, the last colonial governor of Yucatan, Juan María Echeverri, presided over a special assembly in which the city council, provincial deputies, military and ecclesiastical authorities and representatives of different political groups declared Yucatan's independence from Spain and their adherence to the Plan of Iguala, thus joining Mexico. The first half of the nineteenth century in Yucatan was marked by the struggle between federalism and centralism. This was active throughout Mexico as an ideological conflict and in the political project of nation building. The divergence generally coincided with the quarrel between the liberals and conservatives. Federalists were inclined towards a democratic republic in which the sovereignty of the provinces and the states within them were respected. Centralists sought for a political regime in which the highest authority would centralize the most important decisions regarding the nation and its future.

In Yucatan this apparently bipolar conflict played out among several political groups and involved coups and long wars. These two overarching labels also functioned to hide the clash between groups with opposing economic and political interests. Yucatan's money and power elites were disputing control of the region's government and wealth. Campeche, where mariners and exporters; fought against the commercial and agro-livestock capital of Merida by securing economic privileges that could only be obtained through management of state institutions. This civil war spread across the Peninsula and dominated its first decades of independence. Federalism tended to include those who sought to separate Yucatan from Mexico while centralism encompassed those who preferred to remain within the nation.

2.7 The Caste War of 1847

From 1821 to 1847 Yucatan, then encompassing the current Mexican states of Campeche, Quinta Roo, and Yucatan, lived through over 25 years of political and military instability as it fluctuated between separating from or remaining in Mexico. The two political factions had formed armies consisting mostly of Mayan peasants from each region, who over the decades were armed and taught the arts of war. However, the Maya remained a subjugated and exploited population and were the portion of the population that suffered most from the conflict. In 1847 the indigenous population rose up in what became known as the Caste War that in few months spread across the territory of Yucatan, and its magnitude was such that it reconciliated the separate factions of the ruling class, who reached an agreement in August 1848 to fully reincorporate Yucatan into Mexico.

Multifold were the causes of the Caste War. On a material level the advancement of private property into the communal lands of indigenous peoples continued unabated. As the Maya lost land they also lost the ecological equilibrium established over millennia between subsistence agricultural and the forest. Dwindling land availability negatively affected the seasonal milpa cycles by reducing the fallow time allowed for forest recovery. This is indispensable in a slash-and-burn system, and if too short will reduce productivity. At the same time the Mayan peasant communities had to pay higher taxes and tributes to the government and church during the transition from the colonial regime to independent government. In other words, the Mayan peasants had less corn and less money (González-Navarro 1979; Patch 1990; Rugeley 1996; Dumond 2005; Taracena 2013).

The Caste War was phenomenally violent, so much so that it transformed human society on the Peninsula. Indeed, it was one of the most violent indigenous rebellions in Mexico, and in all Latin America. Its demographic impact is an excellent example. The Peninsula was home to just over half a million people in 1846 but the population dropped by almost half in 10 years be it from mortality, migration, or dispersion into the forest; in other words, the population attainable for census was between 280,000 and 320,000 people (Regil and Peón 1852; Rejón 1862; Baqueiro 1990).

For half a century Mayan rebels remained independent in the forest of what is today the Mexican State of Quintana Roo, with a complex, organized, and hierarchical society. It was not until 1902, when General Ignacio Bravo's soldiers took the Maya political and spiritual capital of Chan Santa Cruz, that the Maya were completely stripped of political autonomy. Most responded by relocating to small villages where they could maintain their social and cultural independence, and their complex and rich symbolic and linguistic life. A collateral consequence of these years of civil war and then the Caste War was the political division of the Yucatan Peninsula when the state of Campeche was established on 29 April 1863.

2.8 Henequen and the Transformation of the Landscape and Ecology Between 1850 and 1910

For centuries the Maya had used an agave endemic to the Yucatan Peninsula, known as henequen, or *ki* in the Mayan language. This species, *Agave fourcroydes*, is the product of manipulation for thousands of years by the Maya of its genetic ancestor *Agave angustifolia*. Henequen string and rope were common on the Peninsula from long before Spanish conquest, and were produced with rudimentary techniques. At the Colonial times its use spread beyond the region due to its utility as ship rigging and rope. It appears as an export starting in the eighteenth century, with most shipped out the Port of Campeche and, beginning in 1811, from the Port of Sisal in Yucatan. Trade was steady since most of the sailing ships that stopped at the Peninsula's ports took advantage of the opportunity to replenish their moorings. Production was stimulated by proximity of the important commercial port of Havana, the durability and low price of henequen rope, and its wide acceptance among ship captains. Its quality was only surpassed by Russian hemp and was on a par with abaca from the Philippines, both of which were more expensive and less available in the New World.

The agro-industrial revolution and mechanization of agriculture in the United States further raised demand for henequen fiber, consequently intensifying production, since machines like the cotton gin used henequen rope. By 1870 large-scale planting of henequen was already underway on numerous farms and haciendas near the city of Merida, previously dedicated to corn and cattle production. At this time a race was begun in Yucatan to mechanize defibering of the henequen leaf. Drastic expansion of the market made it impossible to supply using traditional Mayan techniques and rudimentary hand tools. The invention of the first shredding machines, perfected very soon thereafter, was the cornerstone that allowed the building of a cross peninsula henequen industry. Mechanization of the defibering process turned henequen plantations into real agro-industrial production centers linked to the expanding global capitalist market during the nineteenth century, especially that of the United States.

Towards the turn of the twentieth century, henequen production occupied approximately 320,000 hectares distributed among 1000 haciendas of different sizes, of which 850 had defibering machinery, a packing machine, and extensive Decauville narrow-gauge railroad networks (Joseph 1982; Joseph and Wells 1986). These properties were concentrated in the hands of 300 to 400 landowning families (Cámara 1977), although between 20 and 30 families controlled half of all production. More importantly, these same families were also the owners of or partners in the exporters and banks, held the financial capital, and had commercial contacts with the U.S. market (Ramírez-Carrillo 1994). In 1910 the most important of these businesses was Molina Montes, owned by Olegario Molina and family. Molina was a Porfirista governor of Yucatan until 1906 and then served as Minister of National Development until 1910 (Pérez 2006).

2.9 The 1910 Revolution and the Situation on the Peninsula to 1940

The revolution that swept Mexico affected the Peninsula and produced shifts in political power in Campeche, Yucatan, and the territory of Quintana Roo. After the First World War demand for Yucatecan henequen fiber contracted, and prices consequently declined. It remained good business for another decade, but the Great Depression that began in 1929 shrank U.S. henequen markets. Yucatecan agribusiness sank into a deep crisis that lasted throughout the 1930s, and from which it never really recovered. Bankruptcy descended upon hundreds of henequen businessmen, but misery and even famine began to appear in rural areas across the Peninsula, especially among Mayan peasants. When Lázaro Cárdenas became president of Mexico in 1934 the entire henequen industry and the Yucatan Peninsula's economy were depressed.

Cárdenas began implementing a national agrarian reform in May 1935 by creating the National Bank of Agricultural Credit of Yucatan. He organized 48 local *ejido* agricultural credit societies, with 15,364 Mayan farmers as partners, an endowment of almost 30,000 hectares in henequen and another 450,000 hectares of uncultivated land (Lara and Lara 1942). President Cardenas and most of his cabinet came to Yucatan to carry out the final act in August 1937. During fierce public debate, an additional 100,000 hectares of henequen fields were granted to 185 new *ejido* societies. A total of 247 credit societies and 272 *ejido* groups were created. The henequen agro-industry had come under state control, which organized the Mayan peasants into collective *ejidos* to cultivate the resource.

The reform directly affected the hacienda owners who had survived a decade of depression: approximately 500 owners of 583 haciendas. Cardenas' policies indisputably established the state as the principal economic and political agent in Yucatan and became a dynamic force for social change. The state was at the forefront of all productive processes in rural areas and its development policies determined land and natural resources use on the Yucatan Peninsula from 1940 to 1990 (Baños 1989; Sauri 2016). In fact, during this half century and across the Peninsula, businessmen and private capital were only secondary political actors compared to the state. The government maintained an especially powerful clientelistic relationship for controlling the henequen farmers, using them and other *ejido* groups as a political force in its internal conflicts (Brannon and Baklanoff 1987).

2.10 Marginalization of the Peninsula

The prolonged henequen crisis and insecure land tenure drove private capital out of rural areas for 50 years. The delicate ecology of the Peninsula's forests, and poor quality soils of Yucatan already depleted by a century of intense henequen cultivation, hindered cultivation of other large-scale market crops. Sugar cane was re-planted at La Joya plantation in Campeche, and cultivation of many other crops and livestock were attempted but without any noticeable agricultural diversification. Private capital began to invest in extensive cattle ranching outside the henequen area, in the east and south of Yucatan and in large forest areas in Campeche and Quintana Roo. This quickly replaced forests with grazing pastures.

Creating pastures for livestock decreased the land available to Mayan communities. This in turn compromised the seasonal slash-and-burn cycle of the milpa system by necessarily reducing fallow periods for the forest. Corn production declined causing a negative impact on food availability among the rural Mayan population. From 1930 to 1990 the henequen crisis increased poverty and hunger among the Maya participating in this industry, particularly in the Peninsula's northwest. Already chronic malnutrition increased. Simultaneously, the growth of extensive livestock production began to provoke hunger and malnutrition among Mayan rural communities in the Peninsula's east and south that were dependent on corn for subsistence.

The Second World War provided some respite for the henequen industry. Demand for and prices of the fiber and its main manufactured product, rope, began increasing in 1940. The boom experienced by the henequen rope industry for several years in the 1940s led to some improvement in living conditions throughout the Peninsula. The war also raised demand for and the price of another regional raw material, the resin of the chicozapote tree, the main ingredient in chewing gum. It had been exploited since the beginning of the century, but the industry quickly expanded to meet demand, leading to rapid growth in gum collection camps in the forests of Quintana Roo and Campeche. Demand for rope and chewing gum declined after the war ended, and both industries experienced increasing problems throughout the 1950s and almost disappeared.

Rope production had become unprofitable but survived for another 20 years with the creation of Cordeleros de México in 1962. This was initially a joint venture in that it was started with private capital from the surviving rope manufacturers and state funds. In 1964 it was transformed into Cordemex, when the government took complete control of the company. Cordemex kept the entire henequen production cycle afloat by operating in the red; in other words, it promoted a cascade of subsidies to all levels of the industry and became the largest single industry on the entire Peninsula. Its liquidation in 1992 put an end to direct state intervention in the henequen agro-industry. This marked the end in Yucatan of a century of henequen exploitation as the primary agricultural and industrial activity. It is now only grown on small areas.

2.11 "Stabilizing Development" in Yucatan: 1940–1970

The Peninsula's marginal development occurred within a new economic policy model called "stabilizing development." Growth rates in Mexico reached 6% in some years, but development was uneven. Southeast Mexico benefited only minimally from both the restructuring of regional economies that begun in the center, north, and west of the country in the 1940s and from the so-called "Mexican Miracle" stretching from after the Second World War to 1970, when the industrialization through import substitution policy helped bring immense growth to over half of Mexico. The Peninsula as a whole remained a relatively isolated, depressed region with high emigration rates and minimal urbanization.

Entrepreneurs, capital and qualified workers noticeably left for other regions, especially Mexico City. This was a process that had begun in the late 1920s. Hope had been emigrating for almost half a century; in other words, the youngest, most talented and dynamic members struck off on the road to Mexico City, other states or abroad, no matter social class or ethnic group: Mestizo Yucatecans, Lebanese immigrants and their descendants; the poor, middle class and wealthy; and the Maya from the countryside, coasts and cities of Yucatan, Campeche and Quintana Roo (Ramírez-Carrillo 2012, 2018).

For half a century the Peninsula offered few possibilities to new generations, not just in terms of social mobility but of surviving. The future was dark for those who did not already have a trade, property, or a job, and most of those who did were determined to keep them. Many remained relatively motionless in the social class into which they were born, and even calm and resigned to dying in it. Most of the Peninsula's population became poorer between 1930 and 1970, those who were dissatisfied sought to escape and a very few managed to climb the social ladder.

In this period local social mobility was linked to one's ability to participate in one way or another in government spending on henequen processing, livestock, bureaucracy, or trade. The most prosperous businesses were sheltered by their proximity to power through the dominant political party in each state. Initially that meant the Socialist Party of the Southeast in Yucatan or the Agrarian Socialist Party of Campeche, and from the 1950s the Institutional Revolutionary Party (PRI) in Yucatan, Campeche, and the territory of Quintana Roo. In this impoverished region, business success and even just attaining a job required political and/or kinship ties.

2.12 Development on the Peninsula from 1970 to 1982: The "Shared Development" Parenthesis

Winds of change began to blow in 1970 as the end of economic and social stability marked an initial misstep in the fifty-year-long federal government-propelled economic growth model. During the administrations of Luis Echeverría and José López Portillo, from 1970 to 1982, a "shared development" approach was promoted, and supported by public debt. This brought profound changes to southeast Mexico and the Yucatan Peninsula. The two most important were the building of offshore platforms for large-scale oil extraction in Campeche Sound, and the beginning of construction of a tourist destination on the beaches of the Mexican Caribbean in the form of the port and city of Cancun, Quintana Roo. Both ventures were completely controlled by the federal government, through the state oil company (Pemex) and the Bank of Mexico. In contrast, the federal government failed to find a vocation or destiny for Yucatan during these 12 years. There was a clear need to end the agony of henequen and diversify the economy, but the government was unable to achieve it and the state remained mired in henequen production and subsidies (Ramírez-Carrillo 2000, 2004).

Merida, the capital of Yucatan, benefited from the economic growth of other portions of the Peninsula since it remained the urban center with the best infrastructure in the region. It had the best medical, hospital and financial services, a full range of educational options and a well-organized business environment, as well as being central to the Peninsula's communications network. All these reasons allowed the three states of the Peninsula to break out of the decades of stagnation they had been condemned to. Migration began to flow more within the Peninsula, initiating new but incipient processes of urbanization and demographic restructuring.

Indirectly and unintentionally, federal influence began to shape a new social and economic space not limited to the processes experienced within each state but encompassing those of the entire region. The foundations were laid for creation of a Peninsula-wide system which gradually began to take shape as a territory and a market that had to operate in a unified way at various levels. With this a new regional social order arose.

Regional demographics responded to these transformations, exhibiting significant modifications of the previous dynamic. During these two federal administrations the intensity of out-migration from the region gradually diminished. Urban centers began to grow and attract people not just from the immediate surrounding rural hinterland but from other parts of the country. It was during this time that Ciudad del Carmen and Cancun took shape as new regional urban centers, in addition to the three state capitals on the Peninsula: San Francisco de Campeche, Merida, and Chetumal. As Lopez Portillo's presidency ended, in 1982, the Peninsula had new prospects. Among the many social, economic, and political transformations occurring in this period (1970–1994), three were central and laid the foundations for what seems to be a new model of regional development and an integrative territorial process.

2.13 Three Major Transformations from 1970 to the Signing of NAFTA

Such major changes occurred on the Peninsula were: first, the state became increasingly prominent not only as a political axis but as the main generator of ideas, investments and development projects aimed at changing the region. Second, development was built on external debt and not just private investment from regional elites or direct injection of fiscal capital from the federal government. Third, regional ecosystems and natural resources began to manifest worrying changes.

Prior to 1970 most investment in the region had been either private capital or federal funds. For example, the major regional economic projects at the end of the Porfiriato and during the first decades of the twentieth century had been achieved through the active initiative of and strong investment from private entrepreneurs. Although they frequently took advantage of state money and credit, they also participated as direct actors with their own initiatives and projects and risked their own resources to borrow from national and international private banks. In large part they were behind the important productive events that triggered regional development during the early twentieth century, such as the henequen plantations of Yucatan, the gum and lumber companies of Quintana Roo, the shrimp fleet and sugar mills of Campeche, and the plantations and extensive cattle ranches throughout the Peninsula.

Beginning in 1970 the forces guiding regional development changed, as did the financial strategy. There were two principal investment axes: offshore oil in Campeche Sound aimed at exploiting the Cantarell superfield; and creation of a Caribbean tourist destination at Cancun, Quintana Roo. Under the belief that oil prices would remain high for many years, based on the bonanza experienced by the countries belonging to OPEC and the 1974–1976 oil crisis, Mexico acquired an immense amount of debt from international banks. Mexico's 1982 economic crash ended any illusions of oil propping up the economy, but by then Campeche Sound was dotted with oil platforms in full production and Pemex was the region's main political and economic actor.

Promoting tourism by creating Cancun was a different case. Initially the project was protected by the conservative policies of the Bank of Mexico, and investments were sought that were not so onerous for the state. However, once control switched to a government trust called Fonatur, further development was supported through high external debt. This kept it from being a profitable short-term investment but did keep the project under federal control into the mid-1990s.

Economic revival of the three peninsular states and the process of building a more integrated territorial space among them thus remained in federal hands, that is, under the power of political forces external to the region and financial powers outside Mexico. In a centralized state like Mexico, neither condition is novel, but the leading and directing role of the state as the principal agent of development between 1970 and 1994 and the strength with which it acted were remarkable. This quarter-century established economic policy on development throughout the region that continues to the present, despite numerous changes.

The 12 years from 1970 and 1982 also saw a third major transformation that would radically affect the region's present and future. Deterioration of the environment and natural resources occurred on a scale and at a rate unseen in the previous decades and perhaps centuries. Henequen plantations, extensive cattle ranching, and increasingly intense milpa (slash-and-burn) agriculture had destroyed forest across the entire peninsula. In addition, the complete destruction of blackwood trees, extraction of resin from chicozapote trees, and logging of precious woods extensively modified the Peninsula's high forests. For years, geography and ecology studies had been reporting the deterioration of natural resources throughout the region (Beltrán 1959; Revel-Mouroz 1980).

However, nothing seen before compared to the post-1970 deforestation implemented as part of the National Program of Manual Clearing, carried out by federal contractors. This accompanied a massive expansion of extensive ranching on *ejido* and private lands that imitated the Chontalpa Plan in the neighboring state of Tabasco, and UN development programs in all three peninsular states. These are just some examples of the "development" programs that helped to destroy hundreds of thousands of hectares of forest across the Peninsula in just 25 years. Caribbean beaches and coastline began showing signs of the boom in hotel construction, oil extraction and transport was polluting the seas, mangrove and swamps, and regional fisheries were becoming overexploited. Rapid expansion of urban areas in the five main regional cities before 1990 also contributed to the ecological imbalance by erasing nature reserves and the green areas surrounding them. Equilibrium in regional ecosystems was affected quickly and deeply. This broke the thin line joining ecological equilibrium with human exploitation of the Peninsula's natural resources, and compromised their capacity for resilience. Indeed, it has been affected at such a deep level that future development in the region must be implemented considering this natural mortgage.

Development policies have changed in Mexico and the region since the North American Free Trade Agreement (NAFTA) came into effect in January 1994. Nonetheless, the three transformations mentioned above have established the Peninsula's trajectory since 1970, before NAFTA was signed, and laid the foundations for subsequent growth processes, which continue to follow these paths. Despite these minimal changes in development trajectories new models have brought some important changes. New political and economic actors have emerged, and money comes from and enters different sets of pockets.

Funding came less from the state acquiring foreign debt and more from direct investment of foreign capital. Government-controlled companies disappeared and ceded their place to national and international corporate and private monopolies. The federal government receded as a protagonist as state and local governments took on a greater role (highlighting the ineptitude of municipal governments). However, the central points of the overall development vision for southeast Mexico, if treated as a unique and integrated space, remain essentially unchanged from those established during the 1970s and 1980s: oil, tourism, foreign investment and tertiary specialization of urban centers concentrating on trade and services.

2.14 Development on the Peninsula and Its Demographics: The Transition to 2020

In the second decade of the twenty-first century, population movements and changes in demographic behavior now clearly exhibit transformations in various spheres of social and economic life caused by the regional development model begun in 1970. But over these forty plus years the population has reacted differently in the three states of the Peninsula than in the rest of Mexico, especially in its main urban areas.

Although the entire region participated and populations moved along similar routes beginning in the 1970s, this demographic revival has had different intensities. It has been visible since 1980, but it was the 1990 census that most accurately revealed the shared changes and differences, especially between urban and rural areas.

The Peninsula's total population has steadily increased over the last four decades. In the three states on the Peninsula, the combined population grew from 1,710,271 in 1980 to 2,391,402 inhabitants in 1990. By 2000, the figure had reached 3,223,862 inhabitants, in 2010 it was 4,103,596 and in 2015 it was 4,498,931 (INEGI 1980, 1990, 2000, 2010a, b, 2015). Soon (2020) it will probably have reached 5,025,985 (CONAPO 2018). This is well over a one hundred percent growth rate for thirty-five years and was higher than the national average in the same period: about 77%, that is, almost 67 million in 1980 to 119 million inhabitants in 2015 (INEGI 2015). Greater economic dynamism versus the rest of the country cannot be credited with this growth. Rather it was the region's emergence from the lag in and stagnation of production, investment and infrastructure in which the whole region had been trapped since the Second World War. Modernization slowed emigration and helped the population to double. It has continued to grow in response to the new development conditions in effect since 1980.

Demographic growth in the region was multifactorial and not due only to new employment opportunities; rather, it also responded to greater connectivity and ease of communication, an increase in life expectancy and a slow steady decline in mortality and morbidity rates in the native population. Although living conditions on the Peninsula remained, and continue to this day, to be inferior to the national average, improvement was perceptible, particularly when compared to previous regional figures. Never a significant proportion of the national population, the Peninsula's combined population represented a little more than 2.5% of the national total in 1980 but reached more than 3.6% of the national total in 2010. This is a significant increase at the regional level since the intercensal growth rates between 1980 and 2015 were higher on the Yucatan Peninsula than in Mexico.

2.15 The Maya and Poverty in the Twenty-First Century

One cannot speak of poverty in southeast Mexico without immediately highlighting that it is particularly acute among native inhabitants, and especially the Maya (poverty here refers to multidimensional poverty and vulnerability as defined in Mexico by the CONEVAL (2013, 2016). Understanding the social and cultural costs of their remaining in poverty requires a description of their location, number, and characteristics. Poverty in southeast Mexico is historic and for centuries any mention of the poor was synonymous with the Maya. Poverty diversified throughout the twentieth century and continues to do so in the twenty-first century. The poor no longer constitute the population considered indigenous, be it through language or self-description, but now also include the descendants of indigenous people who have ceased to consider themselves such because they have entered the milieu of mestizos, acculturated, and/or lost their culture and language.

These newest members of the poor, who are no longer considered indigenous and do not see themselves as such, have been joined by immigrants from all over Mexico. Of diverse backgrounds, they have created a new regional population from an admixture of cultural, biological, and ideological sources. As such they can no longer be considered Maya or any other indigenous ethnicity, but nonetheless have added their poverty to that of the southeast's native inhabitants.

Hence, in the twenty-first century, poverty in the southeast is no longer unique to or mostly suffered by the Maya or the indigenous population. That said, the opposite assertion generally remains true: being Maya, indigenous or an original inhabitant does identify you as and/or keep you close to poverty. Because the regional native population is relatively large, the various forms of poverty in the southeast have a conspicuous ethnic component, affecting a population that has been marginalized for centuries and is particularly vulnerable. Therefore the fight against poverty and the reduction of vulnerability in the region needs to incorporate a series of cultural, linguistic, legal, and human rights components that are specific to this original population and that also reduce marginalization (Ramírez-Carrillo 2015a).

Marginalization in the three states of the southeast has remained very high over the last few decades, making very little progress and or remaining static compared to national trends. Among the thirty-two states in Mexico, the level of marginalization in Campeche located it in tenth place, followed by Yucatan in eleventh place. In eighteenth place, Quintana Roo has been slightly above the national average. Both Campeche and Yucatan had high degrees of marginalization in 2000 and in 2010, but Quintana Roo has maintained stable with the same average degree of marginalization in 2000 and 2010. At the municipal level marginalization was distributed similarly in 2010. However, these generally high marginalization levels conceal different situations in the three states. For example, in the first three most serious categories (very high, high, and average), Yucatan (total population = 1,955,577 in 2010) had 79,239 in very high marginalization and 129,245 in high marginalization, all in municipalities with a majority indigenous population. Those living in average marginalization numbered 723,558, those in low marginalization were 192,803 and only among the 830,732 in Merida was it very low. Although the Mayan-speaking population in Yucatan was distributed throughout all the state's municipalities, there was a very high correlation between a larger number of speakers of an indigenous language and greater marginalization and poverty, except in Merida (CONAPO 2014; CONEVAL 2016).

Yucatan had the largest Maya population of the three states on the Peninsula in 2015. Of its 106 municipalities, ten were in the very high marginalization category, 23 were high, 68 were average, four were low, and only one, Merida, was very low (CONAPO 2014). The state's distribution of Maya-speakers generally corresponded to that of municipality-level marginalization, apart from Merida, where the correlation between language and marginalization was not high. Little had changed during the previous decade since in 2000 there were 44 municipalities in which more than 70% of the population spoke Maya, of which 20 had a high degree of marginalization and eight a very high level, the most glaringly miserable being the municipalities of Cantamayec and Tahdziú. In another 44 municipalities between 30% and 69% of the inhabitants spoke Maya, of which ten had a high marginalization level and the remaining 34 had an average level. In the remaining eighteen municipalities less than 29.0% of the total population spoke Maya and they had better conditions: thirteen had an average marginalization level; three had a low level; and two had a very low level, one of these being Merida (Bracamonte et al. 2011; Bracamonte 2007). Of course, not all Maya speakers in Yucatan were poor. A certain percentage has benefited from social mobility processes over the last twenty years, but most have clearly remained in poverty at levels more acute than in other states.

Campeche is a different situation. In 2010, just 12.3% of its population spoke a native language, or 91,094 people. Of these, 12,837 lived in the city of Campeche, an additional 2945 in Ciudad del Carmen, and the remaining 75,312 were scattered across the rest of the state. Marginalization in its eleven municipalities ranged from high (2 municipalities), to average (7), to low (Ciudad del Carmen) and very low (the city of Campeche). Excluding the few indigenous speakers in Carmen and Campeche, most indigenous language speakers lived in the two highly marginalized municipalities that encompassed 68,076 inhabitants, and the balance was dispersed among the seven municipalities with average marginalization. Compared to Yucatan and Quintana Roo, the poor in Campeche accounted for a greater percentage of the total population than the number of indigenous people living there. However, the poorest people in Campeche remained the Maya.

In Quintana Roo, a different dynamic surrounds the link between ethnic identity based on indigenous language and poverty and marginalization. High immigration from other parts of southeast Mexico produced a relative increase in its native population up to 2000, and an absolute increase up to 2010. In other words, over the last half century the indigenous population of Quintana Roo has increased based on self-description and on the speaking of an aboriginal language. Speakers of indigenous languages, mainly Yucatec Maya, accounted for 16.7% of the state's total population. Using round figures, that is equal to about 196,000 people, of which a little more than half (101,000) lived among the 900,021 people concentrated in the three main urban centers: 61,190 Mayans lived in Cancun (6000 more than in 2000); 18,000 in Playa del Carmen (2000 more); and 22,000 in Chetumal (3000 less than a decade earlier). Neither the state, which has remained in position number eighteen in terms of marginalization, nor its three main urban municipalities are in situations of marginalization. However, in these three urban municipalities, all considered of very low marginality, Maya speakers largely belong to the poor. The remaining 95,000 Maya speakers were concentrated in the state's three rural municipalities, which are of average marginality. Here they represented the majority of the 136,538 combined population. A smaller number of Maya speakers lived in the remaining three municipalities, all of low marginality, which had a combined population of 314,560.

Two conditions stand out in Quintana Roo compared to Yucatan and Campeche. First is that in territorial terms marginalization and poverty are the lowest on the Peninsula. This reflects the fact that the large number of Maya speakers who immigrated to work in Cancun and Playa del Carmen were able to escape rural poverty. Second, the indigenous population (i.e., speakers of Maya or another indigenous language) which remained in rural areas of the state were poorer and more marginalized than the non-indigenous population living in the same areas. A parallel dynamic occurs in Cancun, Playa del Carmen, and especially Chetumal, where the population in multidimensional poverty has been growing, although they are no longer considered to be marginal.

If the criteria for defining someone as Maya goes beyond speaking the Maya language to encompass those who self-describe as Maya (i.e., those who state in a census that they are Maya), then their absolute numbers may near and even surpass that of the population in multidimensional poverty; however, each state has a different situation. In 2012, for example, 48.9% of Yucatan's population was poor (i.e., 996,900 people), 30.2% spoke an indigenous language (i.e., 537,516), most of whom lived in poor and marginal municipalities. But when self-described indigenous people are included the number of Mayas climbs to 62.7% of the total population. This would lead us to conclude that most of Yucatan's poor remained indigenous (largely Yucatec Maya), but that 15% of the indigenous population was no longer poor. This suggests that this process of social mobility is a sign of the transformation of ethnic identity and its simultaneous revaluation as a positive marker of social ascent; it is expressed as higher education levels, better access to the labor market, and less discrimination towards Mayans. Although not a new process on the Yucatan Peninsula, in the past it was tenuous and reserved for a small Maya elite. Change is beginning to appear in this situation with the growth of a middle class in Yucatan through incorporation of Mayas who continue to identify themselves as such; these are Mayas who have not abandoned their ethnic identity as they have participated in social mobility.

In 2012, Quintana Roo's total indigenous population (based on language) was 190,060 people, or 16.7% of the total population. The total number of poor people was 563,300, or 38.8% of the total population. Using the self-description criterion makes 33.8% of the state indigenous, a proportion greater than half the state's poor. The change is even more pronounced in Campeche where, based on language, only 12.3% of its population (91,094 people) were indigenous, that is, just over a quarter of the state's 387,900 poor (44.7% of total population). However, when self-description is used, 32% of the population is considered indigenous, equal to about three-quarters of the poor population. This correlation between poverty and acceptance of a Mayan identity is a mere hypothesis since the self-description criterion alone cannot be assumed to be synonymous with poverty. And, unlike the linguistic criterion, it is not easy to link it to populations and municipalities exhibiting marginalization.

That leaves language as the hardest core of the link between ethnic identity and poverty. Speaking Maya and being poor would therefore be synonymous in the southeast. A more diffuse criterion, a claim with a "soft" core, would be that those who are considered or self-defined as Maya are poor. Supporting this statement would require verification in great detail due to the myriad contradictory situations it encompasses; for example, these would include social mobility into the middle class of many who self-describe as Maya, revaluation of the Mayan identity as a result of higher education level, or acceptance of marginality by young people through membership in urban gangs. Therefore, the universe of those who self-describe as Maya does include a contingent of the poor, but this group also highlights the need to consider the phenomena of the historical mobility of an ethnic minority into other social classes, maintenance of culture and positive reconstruction of social identities in urban and global contexts (Ramírez-Carrillo 2015b).

2.16 Social Transformation of the Environment Over the Centuries

The history of the Yucatan Peninsula shows how over thousands of years human society wove a tightly interlaced relationship with its environment. Every advance of human society was a point which, upon being attained, created a line of long-term survival that joined society to nature. Within this union the Maya culture took full advantage of the natural resources offered by the Peninsula's natural world. But this thin line that joined them also separated them. Because they manipulated the environment to their ends, the presence of the Maya was intrusive; perhaps their greatest impact was the use of fire to clear forest as part of their agricultural technology. Yet, even at the peak of Maya civilization, the ecological and environmental damage they wreaked was never enough to destroy the balance between society and nature, which would break the line joining the two sides. The droughts that contributed to the collapse of the Classic Maya were apparently natural rather than the result of social manipulation.

Three centuries of Spanish domination produced vast changes and introduced new technologies, but even so human impacts on the natural environment did not reach an intensity that would cause large-scale ruptures in the ecological equilibrium. In contrast, the Peninsula's modern history, extending from the nineteenth century to the present, if treated as a long-term process, has been aggressive to the point of beginning the break in the thin line. In the almost two centuries from 1821 to the present there have been five major processes that have marked the modern relationship between society and nature (Ramírez-Carrillo 2010). Each of these processes has had its own pace and rhythm but has also been subject to a limited number of events.

2.16.1 *Private Property*

The first process that transformed the relationships between society and nature on the Peninsula has been development of private property. This was implemented through establishment of legal appropriation frameworks that, like any normative system, were defined by the political powers that be. Legal changes in land ownership have been particularly useful in this sense since manipulating forms of land tenure is the most efficient social instrument for transforming the environment and society. Periods of large-scale biodiversity loss and change have been linked to alterations in land ownership. Modernity has involved a constant struggle to legitimize unequal access to and exploitation of resources through different forms of ownership that have ensured that only a limited number of individuals in a society control their rights of possession and use. On the Yucatan Peninsula changes in the forms of ownership began with independence from Spain and continue in the current situation of globalization. These fall into four general trends.

Establishment of a liberal property regime. Beginning in 1821 an accumulation of legal dispositions modified the means of appropriation inherited from Spanish rule, and the usufructs granted Mayan communities. These took the form of various federalist and centralist laws and projects approved between 1827 and 1851 (Aznar y Pérez 1849–1851). The privatization inherent in many of these was a strategy that continued to be encouraged in the laws created during the Reform and the Second Empire (1853 and 1867).

Henequen hacienda expansion. The boom in henequen cultivation from 1870 to 1916 drove the creation of an integrated widespread transformation of legal-territorial organization on the Peninsula. The second change in property that impacted the environment and biodiversity, it constituted a body of law that allowed privatization of the entire northern portion of the Peninsula, effectively stripping two thirds of the indigenous population of their lands (Orosa 1960).

Creation of ejido ownership. Post-revolution implementation of land reform began as early as 1916 on the Peninsula with the communal land ownership instrument of the *ejido*. Creation of *ejidos* intended to allow milpa subsistence agriculture accelerated with the redistribution of half a million hectares during the Yucatan state administration of Felipe Carrillo Puerto between 1922 and 1923, and continued with the agrarian reform of president Lazaro Cardenas between 1938 and 1940 (Lara and Lara 1941). The final wave of expansion of the *ejido* form of ownership washed over Yucatan with the creation of new *ejido* population centers and collective *ejidos* during the Echeverría regime (1970 and 1976).

Agrarian neoliberalism. Constitutional reforms during the Salinas de Gortari administration started a new agrarian reform process and created modifications in land ownership between 1993 and 2006 (Baños 2003; Bolio 2016). This has been the most profound transformation in the Peninsula's land and resource ownership landscape so far in the twenty-first century, and has triggered accelerated growth of a neo-plantation dynamic in rural and urban settings. Any future efforts aimed at preserving the environment and biodiversity must assume that the space in which society and nature interact in Yucatan is now largely privately owned.

2.16.2 Market Economy

The second process has been a deepening of the market society. This has been especially true in rural environments with the introduction of plants, animals, and new genetic material that offer commercial value to markets. Six major events can be identified in this process between independence and thc present.

Planting and commercial use of cotton. This occurred during the entire colonial period, but efforts were made to expand it in the early nineteenth century.

Sugarcane. A boom in sugarcane cultivation occurred between 1821 and 1847 when this crop and its accompanying technological complex greatly impacted the Peninsula's forests and indigenous communities.

Henequen plantations. Begun in 1853, henequen cultivation expanded rapidly until covering 320,000 hectares in 1916. However, exploitation of the surrounding forest to support henequen production and processing impacted a surface of about two million hectares, almost half the state of Yucatan. These impacts continued until 1990 under the *ejido* system. The ecological and social deterioration caused by this plantation economy was and remains devastating. Indeed, the overall impact of the henequen cultivation system on the forests, biodiversity, and Maya rural communities can be fairly compared to that of a plague.

Agricultural modernization. The 1970s saw increasing application of modern technology, irrigation agriculture, and agrochemicals, especially fertilizers, herbicides, and pesticides. New varieties of genetically manipulated seeds have also come increasingly into use since the turn of the twenty-first century.

Extensive cattle ranching. This system has at least three negative effects. Establishing pastures requires massive clearing of the forest and the introduction of new grass varieties that affect existing species. Clearing of the forest also shrinks the area available to Mayan communities to rotate land, resulting in fallow periods too short to allow the forest to adequately recover. Pastures require fencing which has increased social conflict between landowners and farming communities, as well as compromising biological diversity (Ramírez-Carrillo 1994, 2000; Patch 1990; Bracamonte 2007).

Declining milpa production. The milpa system is no longer viable in vast areas since traditional techniques cannot be used, fallow periods are too short, and multicropping is not used. As a result, the current milpa system has become its antithesis: a predatory agricultural system that returns neither nutrients nor biodiversity to the forest, which is even more acute if fire is used.

2.16.3 Demographic Growth

Human populations affect and modify the environment. Data on population—whether it is urban or rural, growing or shrinking—is fundamental to understanding changes in physical and social spaces. However, the environment also exists as a human creation, as a real and symbolic appropriation of the medium and the cultural uses that society gives it. As human beings relate to their physical environment, they recreate it. Independence from Spain sparked sociopolitical and economic phenomena on the Yucatan Peninsula that drove population growth up to 1847. The Caste War almost halved the population (from half a million to 280,000 people) with the added catastrophe of abandonment of rural areas. As the labor demands of the henequen boom increased, the countryside was slowly repopulated.

Human populations also moved to different locations. The Peninsula's northwest, around the city of Merida, became more densely populated while much its south and east were left nearly uninhabited. This was a long-term demographic change and extremely important. Total population of the Peninsula did not fully recover until eighty years later, in 1930, but in all three states was concentrated in urban areas, causing drastic demographic expansion in cities. Urbanization was a true revolution, not just in demographics but in the kind of relationships between society and the environment.

2.16.4 Urbanization

By 1970 slightly more than half the Peninsula's population was "urban," although many of these urban centers were little more than small towns. However, by 2015 the population was concentrated in the municipalities of the six main cities: Mérida (892,363); Cancun (743,626); Campeche (283,025); Playa del Carmen (209,634); Ciudad del Carmen (248,303); and Chetumal (224,080). Almost sixty percent (i.e., 2,601,031) of the Peninsula's total population of 4,498,931 now lived in urban areas. Urbanization continues in the twenty-first century, particularly in the Merida and Cancun-Riviera Maya urban areas. Environmental impacts, resource use, social inequality, and poverty all need to be understood within this changing demographic dynamic, its consumption patterns, and economic activities.

2.16.5 Globalization

Beginning in 1990 Mexico has participated in world markets by implementing a new model of growth, and culturally inserting itself into a global society. How this has affected the relationship between society and the environment on the Yucatan Peninsula can be highlighted in five major changes. First, new trade rules have affected existing productive and natural processes by allowing a growing and varied flow of new agrochemicals, with long-term direct and collateral effects. Second, traditional production patterns have been modified by the introduction of new genetic material, particularly new varieties and crops with demand in international markets. Third is the decline of natural resources through extraction of local varieties for industrial, pharmaceutical, and commercial use in other countries.

The fourth is tourism, one of the most promising economic options for the Yucatan Peninsula. It brings new environmental and social impacts with unprecedented risks. Unlike urbanization, the greatest effect of which is to concentrate energy and deeply but locally affect the environment at specific points, tourism extends as much as possible into the whole territory, bringing environmental and social impacts to vast areas. Increased waste, pollution, extraction of natural resources, new legal and illegal forms of consumption, and abandonment of local productive traditions are just a few of the perverse effects of tourism development.

The fifth is linked to the impacts of oil exploitation, which has been intensely active in the waters of the Yucatan Peninsula since 1970. For almost half a century the oil rigs in the Campeche Sound have polluted the air by burning millions of cubic meters of natural gas, polluted the sea with small and constant spills that leave crude oil residues on the seafloor, and polluted the beaches and mangroves with spreading oil, burnt diesel fuel, and industrial waste. The "petroleumification" of the economy and social relations in innumerable communities on the Peninsula has been intense and had largely negative effects, especially in the municipality of Carmen. International demand for crude oil has been variable over the past fifty years, as has the price per barrel of Mexican crude oil, but it is quite probable that demand for oil will continue for at least several decades more. Consequently, the Yucatan Peninsula will continue to be increasingly affected by the oil industry for at least another half-century before new energies replace petroleum and its derivatives.

2.17 Conclusion

The Yucatan Peninsula has changed profoundly from the time NAFTA came into effect in 1994 and the present. In this time there have been clear improvements in the telecommunications infrastructure, life expectancy is greater and extreme poverty has declined, although the pace of development on the Peninsula is palpably behind that of the country's central and northern regions. Foreign franchises, shopping centers, and publicity for foreign goods inundate the streets of the main cities. A large part of the population, including the urban and rural poor, are now more connected to the rest of the world and participate in a globalized society via communication media and social networks. They have been incorporated and internationalized, although most remain only spectators, virtual neighbors, or instant witnesses to the lives and consumption of others. A quarter of a century after Mexico entered the world of free trade, Yucatan's population is still struggling with poverty and vulnerability. Employment remains scarce and income levels and purchasing power remain stagnant at levels similar to those of the 1980s. Higher consumption levels can be explained by the advent of new lifestyles and an increase in social inequality; under this new scenario a small minority of the population is responsible for most of the consumption.

If the inhabitants of the entire Yucatan Peninsula could be reduced to a metaphorical one hundred people at the time of this writing, only twenty would not be poor or vulnerable. Thirty-two would be vulnerable in that they do not have access to adequate health, education, housing, transportation, and/or social security, and remain at risk of falling into misery with the slightest misfortune. They could easily slide back into poverty if, for example, a family member becomes seriously ill, their children reach school age, the father dies or ages without a pension, the head of household lose her/his job and is not compensated, etc. The remaining forty-eight people would be suffering the same indignities as the vulnerable but with the added burden of poverty (CONEVAL 2016; Ramírez-Carrillo 2015b).

Regional development has reduced hunger and the percentage or relative number of the poor, but has increased social inequality, creating a small group of high-income households while most remain part of the vulnerable and poor (CONEVAL 2016; INEGI 2013; ENIGH 2010b). As has occurred throughout Latin America, the middle class on the Yucatan Peninsula has grown only slowly and does not yet account for a quarter of the total population. Social well-being has not remained static but has progressed slowly, and is above all unfair, since social inequality has actually increased as the absolute number of poors has grown along with the overall population.

Poverty persists on the Peninsula and the slow improvement in living standards since the 1930s has driven emigration to other parts of Mexico and, to a lesser extent, the United States. The shared development model implemented beginning in 1970 caused internal migrations, partially redirect within the southeast in search of employment, education and investment opportunities and/or social mobility. These new economic opportunities strongly pushed intra-peninsular population redistribution towards the six urban areas. Although this has improved living standards, the decline in poverty has been very slow. Between 1990 and 2010, years for which relatively precise data is available for the poverty line methodology, poverty decreased by 8% of the total regional population, although half remained poor. Employing the multidimensional poverty methodology, relative poverty between 1990 and 2015 would be lower, at 46–48%, but the social deficiencies of the 32 to 34% vulnerable population would stand out more clearly.

Intra-peninsular migrations has been oriented towards urban centers, the economic enclaves and state-driven development centers promoted by the government since 1970. This has continued after NAFTA began changing Mexico's economic growth model. The result in the second decade of the twenty-first Century has been a population redistribution that is modifying ethnic origins and identities, especially of the Maya who constitute most migrants. It has also amalgamated people and cultural practices in a unique social crucible that is creating a new region which is increasingly integrated and interdependent. The Yucatan Peninsula is a large region, but in the twenty-first century should still be considered as a single social and spatial unit with a common socioeconomic identity. This does not necessarily make it a more egalitarian region since for some social groups pre-existing inequalities have been restructured and even intensified between rural and urban, and, within each state, between the major urban centers and the rest of the territory. Poverty and social inequality are particularly acute for the Peninsula's Maya population.

References

Andrews EW (2014) El colapso maya. In: Quezada S, Robles F, Andrews AP (eds) Historia General de Yucatán. La Civilización Maya Yucateca, tomo 1. Universidad Autónoma de Yucatán, Mérida, p 277

Aznar y Pérez A (1849–1851) Colección de leyes, decretos, y órdenes o acuerdos de tendencia general del Poder Legislativo del Estado Libre y Soberano de Yucatán de 1832 a 1850. 3 tomos. Imprenta de Rafael Pedrera, Mérida

Baños O (1989) Ejidos sin campesinos. Universidad Autónoma de Yucatán, Mérida

Baños O (2003) Modernidad, imaginario e identidad rurales. El Colegio de México, México

Baqueiro S (1990) Ensayo histórico sobre las revoluciones de Yucatán desde el año de 1840 hasta 1864, vol 5. Universidad Autónoma de Yucatán, Mérida

Batllori E, Canto N (2018) Espacio geográfico y físico. In: Florescano E, Esma J, Quezada S (eds) Atlas histórico y geográfico de Yucatán. México, Conaculta/Gobierno del Estado de Yucatán

Beltrán E (1959) Los recursos naturales del sureste y su aprovechamiento. Instituto Mexicano de Recursos Naturales Renovables, México

Betancourt A (1970) Historia de Yucatán. Gobierno del Estado de Yucatán, Mérida

Bolio J (2016) En unas cuantas manos. Urbanización neoliberal en la periferia metropolitana de Mérida, Yucatán 2000–2014. Universidad Autónoma de Yucatán, Mérida

Bottke W, Vokrouhlicky R, Nesvorny D (2007) An asteroid breakup 160 myr ago as the probable source of K/T impactor. Nature 449:23–25

Bracamonte P (2007) Una deuda histórica. Ensayo sobre la pobreza secular entre los mayas de Yucatán. Centro de Investigaciones y Estudios Superiores en Antropología Social/Miguel Ángel Porrúa, México

Bracamonte P, Lizama J, Solís G (2011) Un mundo que desaparece. Estudio sobre la región maya peninsular. Centro de Investigaciones y Estudios Superiores en Antropología Social/CDI, México

Brannon J, Baklanoff E (1987) Agrarian reform and public enterprise in Mexico. Alabama University Press, Tuscaloosa, AL

Cámara G (1977) Historia de la industria henequenera hasta 1919. In: Echánove C (ed) Enciclopedia yucatanense, tomo 3. Gobierno del Estado de Yucatán, Mérida, p 657

Chamberlain RS (1974) Conquista y Colonización de Yucatán 1517–1550. Porrúa, México

Chatters JC, Kennet DJ, Asmerom Y et al (2014) Late Pleistocene human skeleton and mtDNA link paleoamericans and modern native Americans. Science 344(6185):750–754

Chuchiak JF IV (2014) La conquista de Yucatán. In: Quezada S, Castillo J, Ortiz I (eds) Historia General de Yucatán. Yucatán en el orden colonial, tomo 2. Universidad Autónoma de Yucatán, Mérida, pp 29–59

CONAPO (2014) Índices de marginación 2000–2010. Consejo Nacional de Población, México

CONAPO (2018) Dinámica demográfica 1990–2010 y proyecciones de población 2010–2030. Consejo Nacional de Población/Secretaría de Gobernación, México

CONEVAL (2013) Pobreza en México en las entidades federativas 2008–2010. Consejo Nacional de Evaluación de la Política Social, México

CONEVAL (2016) Resultados de pobreza en México 2008–2016 a nivel nacional y por entidades federativas. Consejo Nacional de Evaluación de la Política Social. https://www.coneval.org.mx/Medicion/Paginas/Pobreza_2008-2016.aspx. Accessed 9 Sep 2018

Dumond D (2005) El machete y la cruz. La sublevación de campesinos en Yucatán. UNAM/Plumsock Mesoamerican Studies, México

Frankel CH (1999) The end of the dinosaurs: Chicxulub crater and mass extinction. Cambridge University Press, Cambridge

García-Bernal MC (2015) Población y encomienda en Yucatán bajo los Austrias. Consejo Superior de Investigaciones Científicas, Sevilla

Gill R (2008) Las grandes sequías mayas. Agua, vida y muerte. Fondo de Cultura Económica, México

Gómez G, Ruiz J (2010) Cenotes y grutas de Yucatán. Compañía Editorial de la Península, Mérida
González-Navarro M (1979) Raza y tierra. La guerra de castas y el henequén. El Colegio de México, México
INEGI (1980) X Censo General de Población y Vivienda. Instituto Nacional de Geografía y Estadística, México, DF
INEGI (1990) XI Censo General de Población y Vivienda. Instituto Nacional de Geografía y Estadística, Aguascalientes, México
INEGI (2000) XII Censo General de Población y Vivienda. Instituto Nacional de Georafía y Estadística, Aguascalientes, México
INEGI (2010a) XIII Censo General de Población y Vivienda. Instituto Nacional de Georafía y Estadística, Aguascalientes, México
INEGI (2010b) Encuesta Nacional de Ingreso y Gasto de los Hogares (ENIGH). Instituto Nacional de Estadística y Geografía, Aguascalientes, México
INEGI (2013) Encuesta Nacional de Ocupación y Empleo (ENOE). Instituto Nacional de Geografía y Estadística, Aguascalientes, México
INEGI (2015) Conteo de población y vivienda 2015. Instituto Nacional de Geografía y Estadística, Aguascalientes, México
Jones Grant D (1988) The conquest of the last Maya kingdom. Stanford University Press, Stanford
Joseph G (1982) Revolution from without. Cambridge University Press, Cambridge
Joseph G, Wells A (1986) Control corporativo de una economía de monocultivo. In: Joseph G, Wells A (eds) Yucatán y la International Harvester. Maldonado Editores, Mérida
Lara d, Lara H (eds) (1941) El ejido henequenero de Yucatán. Su historia desde el 1o de febrero de 1938 hasta el 30 de noviembre de 1940, vol 1. Editorial Cultura, Gobierno de Yucatán, México
Lara d, Lara H (eds) (1942) El ejido henequenero de Yucatán. Su historia desde el 1o de febrero de 1938 hasta el 30 de noviembre de 1940, vol 2. Editorial Cultura, Gobierno de Yucatán, México
Lugo J et al (1999) Geomorfología de la península de Yucatán. In: Chico P (ed) Atlas de procesos territoriales de Yucatán. Universidad Autónoma de Yucatán, cap. 3.1, Mérida, pp 155–162
McCaa R (1995) ¿Fue el siglo XVI una catástrofe demográfica para México? Una respuesta basada en la demografía histórica no cuantitativa. Cuad Hist 15:123–126. http://users.pop.umn.edu/~rmccaa/nocuant/nocuant.htm. Accessed 6 Sep 2017
Orellana R, Balam M, Bañuelos I et al (1999) Evaluación climática. In: Chico P (ed) Atlas de procesos territoriales de Yucatán, México. Universidad Autónoma de Yucatán, Capítulo 3.2, Mérida, pp 163–194
Orosa J (1960) Legislación Henequenera de Yucatán. 5 tomos. Universidad de Yucatán, Mérida
Patch R (1990) Descolonización, el problema agrario y los orígenes de la guerra de castas 1812–1847. In: Baños O (ed) Sociedad, estructura agraria y estado en Yucatán. Universidad Autónoma de Yucatán, Mérida, pp 45–96
Pérez M (2006) El continuismo yucateco: La reelección de Olegario Molina en 1905. Hist Graf 27:47–73
Poveda A, Espejo F (2007) El cráter de Chicxulub y la extinción de los dinosaurios. Gobierno del Estado de Yucatán, Mérida
Ramírez-Carrillo L (1994) Secretos de familia. Libaneses y élites empresariales en Yucatán. Conaculta, México
Ramírez-Carrillo L (2000) Historia regional de Yucatán. Limusa/SEP, México
Ramírez-Carrillo L (2004) Las redes del poder. Miguel Ángel Porrúa, México
Ramírez-Carrillo L (2010) Las relaciones peligrosas: Sociedad, naturaleza y construcción de la modernidad. In: Durán R, Méndez M (eds) Biodiversidad y desarrollo humano en Yucatán, México. CICY/PNUD/CONABIO/SEDUMA, Mérida, pp 29–34
Ramírez-Carrillo L (2012) Empresarios y regiones en México. Miguel Ángel Porrúa, México
Ramírez-Carrillo L (2015a) Pobres pero globales. Desarrollo y desigualdad social en el sureste de México. Miguel Ángel Porrúa, México
Ramírez-Carrillo L (2015b) Nuevos nómadas. Desarrollo regional, migración interna y empleo en el sureste de México. Miguel Ángel Porrúa, México

Ramírez-Carrillo L (2018) Migraciones y población en Yucatán. In: Sánchez F, Martín E (eds) Enciclopedia yucatanense. Actualización, tomo IV, en prensa. Sedeculta/Secretaría de Cultura/ Gobierno del Estado de Yucatán, Mérida
Regil J, Peón A (1852) Estadística de Yucatán. Bol Soc Mex Geogr Estad 1(3):237–340
Rejón A (1862) Memoria del estado que guarda la administración pública de Yucatán. Secretaría General de Gobierno del Estado de Yucatán, Mérida
Revel-Mouroz J (1980) Aprovechamiento y colonización del trópico húmedo mexicano. Fondo de Cultura Económica, México
Ringle WM (2014) La influencia externa en el norte de Yucatán. In: Quezada S, Robles F, Andrews AP (eds) Historia General de Yucatán. La Civilización Maya Yucateca, tomo 1. Universidad Autónoma de Yucatán, Mérida, pp 247–275
Robles F (2014) Génesis de la civilización maya yucateca. In: Quezada S, Robles F, Andrews AP (eds) Historia General de Yucatán. La Civilización Maya Yucateca, tomo 1. Universidad Autónoma de Yucatán, Mérida, pp 63–91
Rosenblat Á (1967) La población de América en 1492: Viejos y nuevos cálculos. El Colegio de México, México
Rugeley T (1996) Yucatán's Maya peasantry and the origins of the caste war. University of Texas Press, Austin
Sanders WT (1976) The population of the central Mexican symbiotic region. The basin of México, and the Teotihuacán valley in the sixteenth century. In: Denevan WM (ed) The native population of the Americas in 1492. University of Wisconsin Press, Madison, pp 85–150
Sauri D (2016) Elites y desigualdad regional. Los casos de Yucatán y Nuevo León. Dissertation. Centro de Investigaciones y Estudios Superiores en Antropología Social, Mérida
Schulte P, Alegret L, Arenillas I et al (2010) The Chicxulub asteroid impact and mass extinction at the cretaceous-paleogene boundary. Science 327(5970):1214–1218
Taracena A (2013) De héroes olvidados. Santiago Imán, los huites y los antecedentes de la guerra de castas. UNAM, México
Watson T (2017) Ancient bones reveals girl's tough life in early Americas. Nature 544(7648):15–16
Whitmore TM (1992) Disease and dead in early colonial México. Simulating Amerindian depopulation (Dellplain Latin American Studies). Westview Press, Boulder

Part I
Living Conditions and Human Biology

Chapter 3
Globalization and Children's Diets: The Case of Yucatan, Mexico

Barry Bogin, Hugo Azcorra, María Luisa Ávila-Escalante, María Teresa Castillo-Burguete, Maria Inês Varela-Silva, and Federico Dickinson

Abstract Globalization is an economic force to bring about a closer integration of national economies. Globalization also has effects on human biology. Food globalization brings about nutritional transitions, the most common being a shift from a locally grown diet with minimally refined foods, to the modern diet of highly processed foods, high in saturated fat, animal products and sugar, and low in fiber. Food globalization also changes the social, economic, and political ecology and increases poverty for some. This chapter examines the influences of food globalization using the Maya children of Yucatan as a case study. Yucatecan Maya children often live in poverty and suffer the dual-burden of stunting (low height-for-age) and overweight/

B. Bogin (✉)
School of Sport, Exercise and Health Sciences, University of Loughborough, Loughborough, UK

Member, UCSD/Salk Center for Academic Research and Training in Anthropogeny (CARTA), La Jolla, CA, USA
e-mail: b.a.bogin@lboro.ac.uk

H. Azcorra
Departamento de Ecología Humana, Cinvestav-Mérida, Mérida, Yucatán, México

Centro de Investigaciones Silvio Zavala, Universidad Modelo, Mérida, Yucatán, México

M. L. Ávila-Escalante
Licenciatura en Nutrición, Facultad de Medicina, Universidad Autónoma de Yucatán, Mérida, Yucatán, México
e-mail: marialuisa.avila@correo.uady.mx

M. T. Castillo-Burguete · F. Dickinson
Departamento de Ecología Humana, Cinvestav-Mérida, Mérida, Yucatán, México
e-mail: maria.castillo@cinvestav.mx; federico.dickinson@cinvestav.mx

M. I. Varela-Silva
School of Sport, Exercise and Health Sciences, University of Loughborough, Loughborough, UK
e-mail: m.i.o.varela-silva@lboro.ac.uk

H. Azcorra, F. Dickinson (eds.), *Culture, Environment and Health in the Yucatan Peninsula*, https://doi.org/10.1007/978-3-030-27001-8_3

obesity. This may be due to eating processed foods with insufficient essential nutrients for normal metabolism and growth, to an energy imbalance related to sedentary behavior, to social and emotional stress of poverty that inhibit height growth, or some combination of these. The case of the Yucatan is not isolated, and we must come to terms with food globalization if we are to translate research into better child health and well-being.

3.1 Introduction

On 14 May 2018, Chinese president Xi Jinping spoke at an international summit about China's $US 900 billion "Belt and Road initiative," which he hailed as, "…a means of building a modern-day version of the ancient Silk Road and a new 'golden age' of globalization." (The Guardian, https://www.theguardian.com/world/2017/may/14/china-xi-silk-road-vision-belt-and-road-claims-empire-building). The "Belt and Road" initiative is an immensely ambitious transportation and infrastructure development campaign through which China wants to boost trade and stimulate economic growth across Asia and to countries around the globe. Xi compared the initiative to the Silk Road, a terrestrial trade route originating with the Han Dynasty (~200 BCE) that connected Europe and the Middle East to China and India. Xi said that, "The glory of the ancient Silk Road shows that geographical dispersion is not insurmountable." The new "Belt and Road initiative" is, perhaps, the single largest vision and expenditure on a global trade network. The $900bn question is: Will this "Belt and Road" initiative end with a totally globalized world?

Globalization has many definitions (Al-Rodhan 2006). An anonymous author of a Wikipedia page writes that, "Globalization is the process of international integration arising from the interchange of world views, products, ideas" (http://en.wikipedia.org/wiki/Globalization). This international integration is the total of "…all those processes by which the peoples of the world are incorporated into a single world society" (Albrow and King 1990, p. 8).

This chapter is limited to food globalization, the process that integrates the products of multinational food producing corporations with the biological, social, and ideological processes of dietary change. It was once thought that a "single world society" of food, with a similar diet, would be unachievable. The hundreds of human cultures, with their multiplicity of religions, social classes, and ethnicities, would demand that people eat culturally specific foods. Moreover, the diversity of physical environments for food production would mean that biological species available to eat and the final choices of edible items would reflect local bio-social ecologies. In addition, transport costs would make the price of imported foods beyond the means of most of the world's people, who are of low income.

Today, however, there is abundant evidence that people are increasingly coming to eat a homogenized diet that is designed, produced, and distributed by a few international, globalized companies. China's "Belt and Road" initiative is designed to allow Chinese products, social preferences, and political ideology to sweep across the world, especially over the lowest income nations of Asia and Africa (see The Guardian reference cited above). Food globalization is a part of the Chinese initiative and joins with efforts of other governments and multinational corporations to integrate the products of agro-industry with processes of dietary change. These efforts and initiatives are not new—the ancient Romans "globalized" foods within their empire. The current integration of globalized food and diet may not produce a "single world society" of food, but as diets transition to ever-greater similarity based on an ever-diminished variety of living species there will be influences on the biology, social organization, and ideology of people.

Archaeological and historical sources do, in fact, support the local nature of food availability, food production methods, and consumption choices (Bogin 1998; Dufour et al. 2013). A 'modern' shift toward global foods may have begun about 300 years ago. In his book *Sweetness and Power*, Mintz (1985) contends that in the eighteenth century European Colonial powers practiced a trade of goods and slaves which marked the beginning of food globalization. The mass transport of the food crops of tea, sugar, tobacco and, to a lesser extent, chocolate and coffee, made these commodities the first mass produced, imported foods, of global consumption. The production was done, primarily, in the tropical zone colonies and the consumption was principally achieved, at first, in the metropoles of Europe and later in America, Asia, and Africa.

Today these same food commodities are part of a globalized trade. The products of tea, sugar, tobacco, chocolate, and coffee are found in virtually all regions of the planet that have human inhabitants. Their consumption crosses all ethnic, social, economic, and political boundaries. There are remarkably few religious prohibitions against their consumption (e.g., Mormons, Seventh-Day Adventists, and Rastafarians avoid coffee, tea, and tobacco). These food products are grown, manufactured, packaged, transported, and sold through networks that are largely controlled by multinational corporations. Consumers purchase these products in containers bearing corporate logos, which aid and promote the instant public identification of the company.

These corporate logos have symbolic meanings to the consumers of the products. Corporations use their economic, social, and political power to shape and enhance meaning to include notions of quality, purity, social good, social status, health, ethnic or religious identity, and more. The globalized soft drink *Coca-Cola®* is one well-researched example of the meaning imbued in a product and its corporate logo. We focus on Coca-Cola (also referred to as Coke) because the Mexican people are currently the greatest per capita consumers of Coca-Cola of any nationality (see data below). It is claimed that the Mexican State of Yucatan may have the highest Coke consumption of all Mexico (http://www.yucatanliving.com/culture/mexico-sweet-mexico.htm). The Maya people of Yucatan, Mexico have been one of the targets of the Coca-Cola Corporation's sales campaigns for at least the past 20 years (Leatherman and Goodman 2005; Verza 2013). The globalization of Maya diets has not been caused by Coca-Cola alone, but this globally most popular of all carbonated, sugar-sweetened soft drinks is part of the story of the globalization of children's diets.

3.2 The "Coca-Colonization" of the Yucatan

The term "coca-colonization" may have first been used in 1949 by the newspaper *L'Humanité* (Kuisel 1991) and then again by the magazine *Time* in 1950 to describe the post-World War II sales offensive of the Coca-Cola Company in Europe (http://content.time.com/time/magazine/article/0,9171,812138-1,00.html). Since then the term has come to mean both the globalization of this particular beverage and also the process of commodity globalization in general.

Leatherman and Goodman (2005) titled their analysis of diets in four Yucatec Mayan communities "Coca-colonization of diets in the Yucatan." By this title they mean "…the pervading presence of *Coca-Colas* ®, *Pepsis*®, and an assortment of chips, cookies, candies, and other high-sugar, high-fat snack foods, collectively called 'comida chatarra' (junk foods)" (p. 883). Mexicans have one of the highest per capita consumption of soft drinks of all nations, estimated in 2011 to be 163 l per capita/year. By comparison the per capita consumption of soft drinks in the United States reaches only 118 l (https://www.bbc.co.uk/news/magazine-35461270).

Not surprisingly, Mexico is one of the world leaders for diseases associated with poor diet (Sánchez-Romero et al. 2016). According to a United Nations report, 7 out of every 10 Mexican adults are overweight or obese and diabetes is Mexico's number 1 cause of death, taking some 70,000 lives a year. According to the 2012 Mexican National Survey of Health and Nutrition (INSP 2013), 34.4% of Mexican children ages 5–11 years are obese. This is a higher obesity rate than any other country. The comparable figure in the United States is 16.9% (http://www.therecord.com/living-story/2618264-mexico-facing-a-diabetes-disaster-as-obesity-levels-soar/).

The Mexican government recognized the tyranny of liquid calories from soft drinks. In October 2013 the Mexican government enacted an eight percent "sugar tax" on soft drinks. The tax is lowering the per capita consumption of sugar-laden drinks and impacts the lower income people the most as they spend proportionately more of total income on food than do higher income people (https://www.theguardian.com/society/2017/feb/22/mexico-sugar-tax-lower-consumption-second-year-running). This economic effect may be justified by data showing that poor parents tend to buy more sugary soft drinks than wealthier parents (Jimenez-Cruz et al. 2010; Han and Powell 2014). Poor families in Mexico, the United States, Europe, and elsewhere may feed their children soft drinks and other high sugar snacks because these are cheap way to satisfy and reward children.

Globalized corporations provide the large and complex infrastructure to support the pervasive presence of snacks and soft drinks. Coca-Cola, Pepsi, and other companies supply kiosks to sell their product. The companies supply refrigerators, tables, and chairs with their logos to small stores, cafes, and even schools. The companies also pay for these outlets to be painted with their corporate colors and logos. The investment by the beverage giants seems to be working (Fig. 3.1). Sugary soft drinks are replacing traditional beverages in the Yucatan Peninsula. Even in the most remote rural areas advertisements invite people to try high sugar, high caloric

Fig. 3.1 The "coca-colonization" of Yucatan. Children playing in an abandoned kiosk. The Spanish phrase "toma lo bueno" translates literally to English as "take the good." The phrase "toma lo bueno" is found in Spanish language passages from the Judeo-Christian Bibles and the Islamic Koran (https://www.linguee.com/spanish-english/translation/toma+lo+bueno.html), which imbues religious symbolism and a sacred authority into the consumption of the advertised product. Photographer: Inês Varela Silva. Location: Celestun, Yucatan, Mexico. Date: 2007

drinks. These drinks are only slightly more expensive than bottled water, which is often produced by the same companies that manufacture the soft drinks.

Leatherman and Goodman (2005) found that in the Yucatan village of Yalcoba, with about 1500 inhabitants, there are at least 40 "sales points" (small stores called *tiendas*) for Coca-Cola, Pepsi-Cola, and other sugary soft drinks. In their survey of 75 school-aged children in Yalcoba, Leatherman and Goodman found that reported daily intakes of junk food included about 360 ml (~12 oz.) of soft drinks (mostly Coke or Pepsi), 1.5 packages of chips, cookies, or other snack foods, and about two small candies. The maximum daily intake of junk food reported was about 1.8 l of soda, seven packages of snack foods, and six candies (ibid).

3.3 The Maya of Yucatan

The Yucatan Peninsula is historically a center of the Maya culture. There are an estimated seven–eight million Maya living in Guatemala, the Yucatan Peninsula of southern Mexico, Belize, El Salvador, and western Honduras (Lovell 2010). This makes the Maya the largest Native American ethnic group. Common features of rural lifestyle, economic activities, kinship and marriage systems, religion, philosophy, and a brutal history of repression since the Conquest of America binds all Maya together into a shared cultural identity (Glittenberg 1994; Bartolomé 1998; Guzmán-Median 2005; see Chap. 1). There are, however, 30 or so Maya languages, each associated with a specific Maya group such as the Yucatec Maya of southern Mexico, and many variations in cultural behavior.

The living Maya are the biological and cultural descendants of the people inhabiting the same culture area prior to European contact in the year 1500 CE. Archaeology of the region indicates that hunter-gatherers and small-scale farmers existed in the region for thousands of years. It is not certain which of these groups became the Maya. By about 250 CE a Maya cultural identity was well established and the people were organized in several state-level societies, ruled by priest-kings and an elite class of political-religious leaders. Each Maya state group-maintained armies and a workforce of peasants that produced food using a mosaic system of "… agricultural fields, raised wetland fields, kitchen gardens, terraced hills, and …managed forests" (McNeil et al. 2010). Maya plant and animal husbandry cultivated a diversity of species to provide food; medicinal plants; and wild animals for protein food and honey, firewood, and building materials.

This is how the Spanish Conquistadors encountered the Maya in the Yucatan Peninsula in the year 1500. New infectious diseases introduced by the Conquistadors, such as smallpox, bubonic plague, and measles, spread rapidly among the Maya, and without any biological or social resistance whole villages of Maya were decimated. Between 1519 and 1632 at least eight epidemics spread across Yucatan and Guatemala (Lovell 2010). It is estimated that 90 percent of the Maya population died between 1500 and 1625, totaling about 1.8 million people in Guatemala (Lovell and Lutz 1996).

In the past 200 years or so, Maya culture has been characterized by subsistence and market-oriented agriculture, small-scale animal holdings (pigs, chickens, turkeys) augmented by craft specialization. An example of Maya women, and a girl, wearing traditional woven and embroidered clothing, and making tortillas, a traditional food, is provided in Fig. 3.2.

Other characteristics of traditional Maya culture are social behavior relating to household economy, endogamy (marriage within the community), collective religious practice, use of the Maya calendar, and communication in a Maya language. Some aspects of modern-day traditional Maya culture predate the Conquest; others are postcolonial syncretic blends between various Maya and Spanish social-religious practices. New practices are derived from globalization, such as use of mobile phones, the internet, and drinking Pepsi-Cola and Coca-Cola.

3.4 The Maya of Merida

Since the European Conquest, the Maya of Yucatan have experienced, and continue to experience, adverse socioeconomic conditions, including marginalization and poverty (Siniarska and Wolanski 1999; Bracamonte and Lizama 2003; Ramirez 2006). During much of the nineteenth and twentieth centuries the Maya lived in

Fig. 3.2 Women and girls make tortillas on the *ko ben*, a sheet of metal placed on three stones above a wood fire. Dried corn kernels are ground to a flour, mixed with calcium carbonate, and then boiled to a paste-like consistency. The resulting mixture is dough called *masa* (corn dough) (pictured on the table). Small balls of *masa* are hand-patted into disks that are cooked. The tortilla is the most commonly eaten food item of the Maya and of all Mexicans. Photographer: Miguel Cetina. Location: San José Oriente, Yucatan, Mexico. Date: 2011

rural areas of the Yucatan and worked in the henequen agro-industry. Henequen (*Agave fourcroydes*) production was the primary industry of the region. Henequen fibers were used to make rope and twine, paper, cloth, mattresses, wall coverings, carpets, and other products. Maya workers lived on the henequen plantations, often under terrible conditions. During fieldwork in 1978 an elderly Maya man told one of us (FD) that the henequen workers called the plantation era "the age of slavery." There was local resistance and violent protest by the Maya in the nineteenth century, but it was the Mexican Revolution of the early twentieth century and the global economic depression of 1929 that destroyed much of the plantation system. Competition from less expensive henequen produced in Brazil and Africa and from other lower cost natural fibers, and then synthetic fiber such as polypropylene, caused the Yucatan henequen industry to collapse in the 1980s. Since the collapse, migrants from rural Maya villages searching for jobs and new opportunities have flooded into the city of Merida (Lizama 2012) or toward tourist resorts such as Cancun. According to the Mexican Census, the population of Merida rose from 241,964 inhabitants, in 1970, to 830,732 inhabitants in 2010 (Azcorra et al. 2013), and likely more than a million today.

Many of the Maya of Merida live in the southern neighborhoods of the city. This is a low socioeconomic status area and geographically segregated from the central and northern regions by the International Airport, a Federal prison, and a military base. Surveys in the year 2000 found that at least 19% of the residents of this southern region were Maya-speaking people, the highest percentage in the city, though most were fluent in Spanish as well. Other pockets of Maya language speakers were found in some areas of the East and North of Merida (Azcorra et al. 2013).

3.5 The Nutritional Dual-Burden

In 2005, several of the present authors (BB, IVS, MTCB, FD) met in Merida and began to work together to better understand the biocultural living conditions of Maya families in Merida. This new work was supported by our previous studies with Maya living in rural Yucatan and Guatemala, and Maya migrants to the United States (Wolanski et al. 1993; Bogin et al. 2002; Castillo-Burguete et al. 2008). Our new research focused on the nutritional dual-burden, which may be broadly defined as the coexistence of undernutrition (mainly stunting) and overnutrition (overweight and obesity) in the same population/group, the same household/family, or the same person (Varela-Silva et al. 2012). In the field of public health, stunting is defined as a height-for-age and sex <2 standard deviation scores (SDS, or z-scores) of a growth reference median value. Overweight may be defined as weight-for-age and sex, or fatness-for-age and sex >2 SDS of a growth reference. Obesity may be defined as >3 SDS or a growth reference.

The Maya are one of the shortest stature, non-Pygmy, populations in the world. Pygmies are short statured due to genetic variations associated with the lack of specific hormones, their carrier proteins, or cell-binding agents, but there are no such known genetic reasons for Maya short stature. The average height of contemporary, rural-living Maya men and women in Mexico and Guatemala is 160 cm and 148 cm, respectively. This sex difference in stature is expected, as on a worldwide basis women tend to average 12 cm less than men from the same population (Bogin and Keep 1999; Bogin et al. 2017). But, the average height of both sexes is very low; a global analysis of 200 samples reported current mean heights of 171.28 cm for men and 159.49 cm for women (NCD-RisC 2016). Maya women have the shortest average height in this database and Maya men are the fourth shortest after East Timor, Yemen, and Lao PDR (Laos)—nations that like Guatemala have suffered histories of violence and poverty.

There is some evidence that Maya were taller in the past. Skeletal remains at the Maya archaeological site of Tikal indicate that the height of men averaged 166 cm at the start of the Classic Period in 250 AD, a time of relative prosperity for the Maya. Maya men of high social status, indicated by burials within ornate tombs or pyramids, averaged 170 cm in height, and at least one man was 177 cm tall. These high-status Maya were about the same height as "tall" African populations, such as the Tutsi of Rwanda and Burundi, measured in the early twentieth century (170 cm), and taller than many Europeans at the end of the nineteenth century (Bogin 2013). By the end of Classic Period, at about 900 CE, Maya societies were suffering from centuries of warfare and environmental change. Maya kingdoms in lowland Guatemala were disbanded, but centers such as Uxmal in the Yucatan peninsula remained vibrant until about 1200 CE. During this period the average stature of Maya men decreased to 163 cm and declined again to its present value after the European Conquest of 1500 CE (Bogin and Keep 1999).

Maya adults and children measured in the early to mid-twentieth century were usually thin, with very low body fat. The combination of short stature and thinness were indications of an inadequate diet. Surveys in rural Guatemala Maya villages in the 1950s–1960s found that total energy intake averaged only 87% of requirement, with less than 15% of energy derived from fat or protein. There were also deficiencies of riboflavin (vitamin B_2), niacin (vitamin B_3), and vitamin C (Bermudez et al. 2008; Bogin 2013). These Maya diets would result in both short stature and thinness, especially as the inadequate diets were combined with heavy workloads and recurrent respiratory and gastrointestinal infections. In addition, periodic epidemics of measles took a heavy toll in terms of reduced body growth (Bogin 2013).

By the early twenty-first century the relationship of weight to height for Maya children and adults changed. Based on a cross-sectional survey conducted in Merida in 2006, we found that Maya mothers remained short, with 69.4% of the 206 women measured below 150 cm, an indicator of stunting in adult women. Stunting was detected in 21% of their children, aged 4.0–6.9 years old, and overweight in 33%, with 2.4% both stunted and overweight. Mother's height below 150 cm and the child's birth weight below 3000 g both were statistically significant predictors of stunting in the child (Varela-Silva et al. 2009). We interpreted this to indicate that there is an intergenerational transmission of poor growth in this Maya sample.

In 2010 we collected additional data from 58 Maya mother-child pairs in Merida. The 31 boys and 27 girls were 6.8–9.9 years old. Based on references for height and weight produced by the Centers for Disease Control, United States, 31% of the children were stunted, 12.1% were overweight, and 15% were obese. Using these same references, 81% of the mothers were stunted (height < 150 cm), 91.4% were overweight, with 39.7% of the overweight women classified as obese (Varela-Silva et al. 2012).

In principle it should be virtually impossible for a 7–9-year-old child to be simultaneously stunted and overweight because an excess intake of total food energy should provide other needed nutrients, such as amino acids, lipids, vitamins, and minerals. Accordingly, it should be highly unlikely to find so much stunting and obesity, even in different children, within the Maya of Merida. Our findings are not isolated, as similar reports come from several other low socioeconomic status groups in both developing and developed nations (Varela-Silva et al. 2012). Several hypotheses have been proposed to explain the nutritional dual-burden (Frisancho 2003; Varela-Silva et al. 2007; Said-Mohamed et al. 2012; Wilson et al. 2013). In the following analysis we examine the most basic of these hypotheses—that the diet of the Maya mothers and children participating in our studies is excessive in total energy, but deficient in nutrients essential for skeletal growth.

We employed a food frequency questionnaire (FFQ) to collect data about the diet of our sample of 58 mother-child dyads. The age of the mothers was between 22 and

49 years (mean = 34.30 ± 6.28 years). The 58 children (boys = 31) were between 6.8 and 9.9 years old (mean = 8.42 ± 0.79).

FFQs may be used to identify patterns of consumption of specific foods at the individual and community level. The FFQ methodology may be less accurate than other methods, such as diet diaries, 24-h recall, and weighed food intakes, especially for determining individual diets. In terms of costs and benefits, FFQs are useful to identify the main food components of the diet of groups of people, and when applied to representative samples a FFQ may provide a general idea about the dietary characteristics of the population (Dary and Imhoff-Kunsch 2012). The kind and number of foods or products included in the questionnaire depend on the purpose of the study.

Our FFQ included 78 foods, grouped in seven categories: (1) cereals, breads, and rice; (2) beans, meats, and fish; (3) milk and eggs; (4) animal and plant fats (e.g., cooking oil and lard); (5) sugar, honey, and sugar-sweetened soft drinks; (6) local and nonlocal fruits; and (7) vegetables. We also determined if the food eaten was prepared at home or was commercially processed and packaged.

We visited the families at their homes and applied the FFQ to the mothers to obtain information about their own food consumption during the previous week and that of their son/daughter who participated in the study. We followed a standardized methodology to apply the FFQ (Madrigal and Martínez 1996) and we are confident that virtually all the food consumed at home by the mother and the child were recorded. We likely missed some foods consumed out of the home, such as soft drinks and snack foods consumed by children. Not recoding these items will underestimate the weekly frequencies and more conservatively estimate total food intake.

Our methods of analysis are described elsewhere along with detailed tables of the food frequencies (Azcorra et al. 2013). All analyses were done separately for the mothers as a group and then for children as a group. There were no significant differences between the boys and girls in height, weight, or frequencies of foods consumed. For each group, the frequency of consumption of each of the 78 foods was categorized in four levels: (1) no consumption (or less than weekly consumption), (2) low consumption, at 1–2 days/week, (3) medium consumption, at 3–4 days/week, or (4) high consumption, at 5–7 days/week.

The findings from our food frequency questionnaire are that, in general, the Maya mothers and their children eat:

1. A high consumption of maize *tortillas* and locally produced wheat baguettes (called *pan francés*), a medium consumption of rice, pasta, and sugar-sweetened refined wheat bread (called *pan dulce*), and low consumption of refined wheat white bread (called *pan blanco*) or high fiber carbohydrates, such as oats.
2. Beans were consumed in low to high amounts, pork and chicken were consumed in medium to low amounts, and beef and fish in low to no amounts.
3. Milk consumption was reported to be low for the mothers and high to medium for the children, with 84.5% of children and 60.3% of mothers consuming only whole milk.

4. Vegetable oils and mayonnaise had high consumption, while lard and margarine had low consumption.
5. More than 70% of mothers and 60% of children had high to medium consumption of sugar (sucrose) and sugar-sweetened soft drinks. Honey consumption was low to absent.
6. Except for the consumption of bananas and oranges by children, both mothers and children reported very low frequencies of consumption of fruits. The lowest frequencies of consumption were for nonlocal fruits (apple, pear, grapes, and strawberry) which are more expensive than local fruits (banana, watermelon, papaya, and orange).
7. Vegetable consumption tended to be low to absent. The exceptions were for cucumbers, tomato, and onion which had a medium consumption by both mothers and children. Even so, these vegetables tend to be used in relatively low quantities, mostly to add flavor to other foods, in the culinary culture of Yucatan.

We analyzed the consumption of soft drinks and packaged "junk foods" (*comida chatarra*) in greater detail. The frequencies of soft drink consumption are shown in Table 3.1. Mothers have, in general, a medium to high intake, with these two categories accounting for 70.7% of frequency. The children have a more variable intake, but 53.5% have a medium to high consumption of sugar-sweetened soft drinks. The children's data are based on the mothers' reports and do not include soft drinks consumed out of the home and out of sight of the mother.

In Table 3.2 are shown the data for consumption frequency of "junk foods." These are packaged snack foods such as biscuits (*galletas*), fried chips or crisps of corn, potato, or wheat (*frituras de maíz, papa y trigo*), candies and sweets (*dulces*), such as caramels and chocolates. Most of these processed, packaged products are purchased from the *tiendas* (small shops) located in neighborhoods where our participants reside. Mothers report, generally, no or low consumption of these junk foods. One mother reported that she and her child only ate junk foods on special occasions, such as a birthday. Children, in contrast, are reported to have a low to high consumption of junk foods. About half the children have a medium or high consumption (50.1%). Again, this is based on mothers' reports and does not include biscuits, chips, or sweets eaten outside the home.

We compared and interpreted the results from our FFQ against the Mexican National Health and Nutrition Survey (*Encuesta Nacional de Salud y Nutrición*) of

Table 3.1 Consumption of soft drinks by mothers ($n = 58$) and children ($n = 58$)

	Mothers		Children	
Consumption	*F*	%	*F*	%
No consumption	2	3.4	5	8.6
Low	15	25.9	22	37.9
Medium	16	27.6	8	13.8
High	25	43.1	23	39.7
Total	58	100	58	100

F frequency; Low: 1–2 days per week; medium: 3–4 days per week; high: 5–7 days per week

Table 3.2 Consumption of snack foods by mothers ($n = 58$) and children ($n = 58$)

Consumption	Mothers		Children	
	F	%	*F*	%
No consumption	15	25.9	6	10.3
Low	30	51.6	22	37.9
Medium	7	12.2	10	17.3
High	5	8.6	19	32.8
Special occasions	1	1.7	1	1.7
Total	58	100	58	100

F frequency; Low: 1–2 days per week; medium: 3–4 days per week; high: 5–7 days per week

2006 (INSP 2007). The Mexican National Health and Nutrition Survey was a carefully constructed, representative investigation of the Mexican population. In Yucatan, the survey included 6985 persons living in 1553 households. A detailed FFQ, including portion sizes of foods eaten, was included. The FFQ included 101 foods and participants were asked for each food item how many times it was eaten in past 7 days, the number of days per-week the item was eaten, the number of times per day, the portion size (later converted to a weight) eaten, and number of portions eaten each day (Rodríguez-Ramírez et al. 2009). This allowed for the estimation of intakes of specific nutrients.

The findings for 5–11 years old children of the lower socioeconomic status (SES) were that total energy intake was only 81% of requirement and that there were inadequate intakes of vitamin A, folates, heme iron, zinc, and calcium (INSP 2007). These estimates are for the whole of the Yucatan region, both rural areas as well as the city of Merida, and for children from all low SES families, not only Maya families. Another FFQ study based on further analysis of the Mexican National Health and Nutrition Survey data of 2006 plus a sample of 126 children from impoverished Yucatan families found nearly identical results (Cuanalo et al. 2014).

3.6 The Globalized Diet of the Maya

The diet of the Maya mothers and children of our sample seems typical of diets of low socioeconomic status (SES) people of the Yucatan. All these people have experienced the effects of diet globalization. Compared with rural diets of Maya villagers surveys in the 1950s (Bonfil 2006; Bermudez et al. 2008; Bogin 2013) our Merida Maya sample has undergone a nutritional transition, with a shift from locally grown foods high in fiber and low in fats, to the global diet of highly processed foods, high in saturated fat, animal products and sugar, and low in fiber. In the traditional rural villages of the middle twentieth century, Maya health was not very good as evidenced by stunting and thinness. In twenty-first century Merida, with a globalized diet, health problems remain as evidenced by short stature and obesity.

Our hypothesis for this nutritional dual-burden is not fully supported. There is no evidence of excessive energy intake, but there is evidence of vitamin and mineral deficiencies.

3.7 Physical Activity and Inactivity of Maya Children

An alternative hypothesis is that the participants have a total energy imbalance due to low levels of physical activity, so that despite low food energy intakes they expend even less energy and there is a net gain of body weight.

We objectively measured the physical activity of 33 of the 58 children in the FFQ sample (17 boys and 16 girls) over a one week period using an instrument called the Actiheart (Wilson et al. 2011, 2013). We found they were highly active, overall, spending an average of 120 min per day in moderate-to-vigorous physical activity (MVPA), which exceeds international recommendations for their age and sex. Stunted children, however, spent significantly less time in MVPA, and five stunted girls did not achieve the minimum level for MVPA of 60 min per day. Further analysis found that children with short stature, measured as a continuous variable, and lower fat-free mass (meaning, the skeleton, muscle, and organs of the body, but not the fat) have lower total energy expenditure compared to taller children with higher fat-free mass. The children had a mean energy expenditure of 1986.5 (SD = 332.6) kilocalories per day. This estimate was based on Actiheart reference equations, which were developed based on white British children, so may not be accurate for Mexican-Maya children.

We estimated the energy intake adequacy of the Maya children in our sample using their mean group values for age, height, weight, and level of physical activity. These values are 8.25 years old, 122 cm, 27 kg, and "active" (defined as sedentary activities plus 1–2 h per day of additional activities reflecting their MVPA). Based on these values, and using the United States references for energy requirements, the Maya children should consume an average of 1920 kcal per day. This would equal a shortfall of 66.5 kcal per day, based on our estimate of mean energy expenditure of 1986.5 kcal/day.

If these Maya children were deficient in total energy intake they should be underweight or, at best, normal weight. However, more than 25% are overweight or obese and only 5% are underweight (Varela-Silva et al. 2012). It is possible that we either overestimated the energy expenditure of the children in our sample, or the children were consuming more kilocalories than reported in the FFQ.

But, even if there were an energy imbalance it cannot explain the short stature and overweight/obesity of our Maya children and their mothers. If the children were energy deficient then they should be thin, not fat. The explanation may be that their diet is deficient in essential nutrients, but not in fact deficient in total energy. As shown in Fig. 3.3, essential nutrient deficiencies, episodic energy imbalances due to illness and food insecurity, and a globalized diet—all of which characterize the Maya of Yucatan—may lead to shorter stature and decreased muscle mass. The

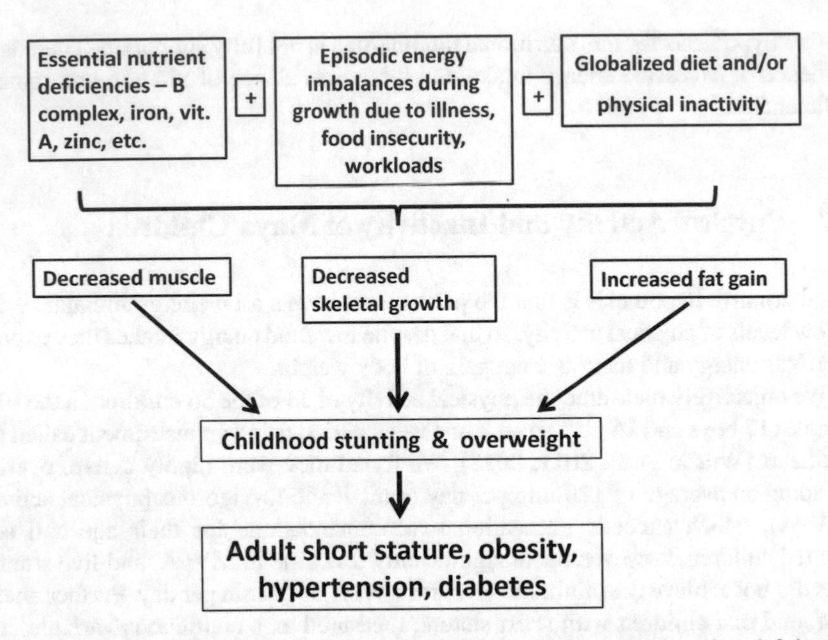

Fig. 3.3 Possible association in low-income populations between essential nutrient deficiencies, energy imbalances, globalized diet, physical inactivity, childhood stunting/short stature and overweight, and the adult phenotype of short stature, obesity, hypertension and diabetes (modified from Sawaya et al. 2009)

fat-free mass, of which muscle mass is the largest fraction, is a major component of total energy expenditure (TEE). TEE may be expressed simply as the sum of basal metabolism plus physical activity. Short stature and decreased muscle mass in Maya children will reduce total energy expenditure and the requirements for total energy intake (Wilson et al. 2012). An 8-year-old girl, for example, who is 124 cm tall requires about 150 kcal less per day than a girl of the same age who is 131 cm tall. If the short girl consumes the "extra" 150 kcal per day, and assuming that an extra 3500 kcal not used for any metabolic activity will be stored as 0.45 kg of body fat, then she may gain about 0.57 kg of body fat in 30 days. If this is true, then we should not have used the United States references for energy requirements, which are based on taller children, with more muscle mass, and higher energy requirements than for the Maya.

A further complication is that an increased risk for elevated body fat for short statured children may be exacerbated by high amounts of sedentary behavior. While we measured physical activity, we did not measure sedentary behavior. The British Heart Foundation explains that, "Sedentary behaviour is not defined simply as a lack of physical activity. It is a group of behaviours that occur whilst sitting or lying down and that require very low energy expenditure" (http://www.bhfactive.org.uk/userfiles/Documents/whatissedentarybehaviour.pdf). Reading a book while seated is an example of a sedentary behavior. Exercising while seated, such as using a rowing machine, is physical activity and not sedentary behavior. Maya children may run, play, and do other activities to meet the 60 min per day requirement for MVPA, but our ethnographic observations, and those of many other fieldworkers since the time of Sol Tax's studies of highland Maya weavers (Tax 1953), find that Maya children and adults practice more sedentary behavior than better nourished Mexican and European children and adults. Tax reported that Maya girls learning to weave sit almost motionless for hours at a time while watching an older women weave. During our fieldwork in Guatemala, Mexico, and the United States we have observed Maya children sit, almost motionless, while we took measurements on their classmates. Non-Maya children in these same countries are more often moving and playing while waiting to be measured (Bogin and Varela-Silva, field notes). The synergistic effects of smaller stature, lower fat-free mass, sedentary behavior, and a globalized diet may work to increase body fat, resulting in the nutritional dual-burden that plagues the Maya communities of Merida and elsewhere (e.g., Smith et al. 2002).

3.8 Globalized Food Consumption Behavior

The globalized diet of the Maya is especially problematic, as this diet is high in fat, relative to carbohydrates, and high in sugars, especially from processed foods such as soft drinks and packaged junk foods. Even the tortillas which are ubiquitous in the diet are a source of concern. Traditionally made tortillas are not a problem, whether made at home (Fig. 3.2) or at a factory and sold in shops. Tortillas provide energy, protein, fiber, folic acid (folate), calcium, iron (as well as some niacin, thiamine, pantothenic acid, magnesium, potassium, phosphorus, zinc, and copper), and are not high in fat. As shown in Table 3.3, tortillas and factory-produced white bread (the Pan Bimbo brand) are nutritionally similar. The white bread is fortified with folic acid, calcium, and iron, but also has much more salt. Our FFQ found high consumption of the tortillas and low consumption of the white bread, which is nearly three times as expensive than tortillas per 100 g. The FFQ found a high consumption of locally produced *pan francés* (wheat baguettes) and *pan dulces* (sugar-sweetened refined wheat breads and cakes). These heavily consumed foods are high in salt and total energy, especially from fats, but low in most other nutrients (Fig. 3.3).

Fats are more than twice as energy dense as carbohydrates, but often are nutrient poor. Sugars provide an attractive taste, but few essential nutrients. Based on our ethnographic observations and interviews with the mothers and, occasionally other family members, we understand that the patterns of food consumption reported represent a strategy to adapt to the low purchasing power of the families. Many of the products and foods frequently consumed are sold at low prices in the small stores (*tiendas*) of the neighborhood. The stores have a limited variety of products for sale and those products available satisfy immediate hunger and energy requirements of the family and satisfy desire for good tasting food and drink. During fieldwork in Yucatan we were told that when people eat bread (*pan francés* or *pan dulce*) and

Table 3.3 Nutrient composition of tortillas made from yellow corn masa, prepared with calcium carbonate (*tortilla de maíz amarillo con cal*), processed white bread (Pan Bimbo®), store-bought sweetened bakery goods (Pan dulce) without and with fat added

Food	Tortillas de maíz Amarillo con cal[a]	Pan bimbo® (white)[b]	Pan dulce without fat[b]	Pan dulce with fat "Orejas"[b]
Amount	1 piece	1 slice	1 piece	1 piece
Weight (g)	30	27	64	68
Energy (Kcal)	62	71	248 †	324 †
Protein (g)	1.7	2.2	7 ◊	7 ◊
Lipid (g)	0.4	0.8	8 †	18 †
Carb. (g)	14	14	45 †	45 †
Fiber (g)	1.3 ◊	0.5	0.4 †	0.4 †
Folic acid (mg)	Trace	28 ◊	0 †	0 †
Calcium (mg)	47◊	101◊	0 †	0 †
Iron (mg)	0.8◊	1.1 ◊	0 †	0 †
Sodium (mg)	0	149.9†	236 †	236 †

Data Sources: [a]INCAP. Tabla de Composición de Alimentos de Centroamérica. 2012, 2nd. edition, 3rd. reprint
[b]Pérez A, Palacios B, Castro A. Sistema Mexicano de Alimentos Equivalentes. 2008. 3rd. edition
Especially desirable nutrients are highlighted with ◊ and undesirable levels of nutrients are highlighted with †

drink a Coke, their stomach becomes full and the sensation of hunger disappears (FD, field notes). Food economists working with low SES populations have shown that the importance of taste is as important, or a bit more important, than the cost and the nutrient density of foods purchased (Drewnowski and Darmon 2005). "Good taste, high convenience, and the low cost of energy-dense foods, in conjunction with large portions and low satiating power, may be the principal reasons for overeating and weight gain" (ibid, p. 900).

Soft drinks represent one of the most common processed items in the Maya diet. Based on data collected in November 2013 we found that the difference in price between a liter of still water and a liter of soft drink is $3 Mexican pesos (about US$0.25). The still water is cheaper, but people prefer the sweet taste of the soft drinks. The authors visited Maya homes in March and April 2010. In many homes sugar-sweetened soft drinks were visible, either in bottles around the home or being consumed. In one of the homes we saw a stack of 4 or 5 cases of 2-l Coca-Cola bottles. Each case contained 4 bottles. The father of this home explained that he purchased these cases at Wal-Mart, which is a United States based, global department store and supermarket. The father assured us that these Coca-Colas were for family consumption and not for sale.

The consumption of sugar-sweetened beverages begins in infancy. In our fieldwork we have seen one-year old infants given small amounts of Coca-Cola by their mothers. One method used by the mothers is to dip a finger in the Coke and then allow the infant to suck the finger. Another method is to pour Coca-Cola into a baby bottle and allow the infant to drink. One of the present coauthors (MTCB) observed during her fieldwork in Yucatan that mothers may alternate feeding their babies Coca-Cola and sugar-sweetened orange- or apple-flavored soft drinks. Some of the mothers explained that they do this because physicians informed them that Coca-Cola is not good for children, but children want to drink it. So, the mothers "negotiate" by giving sometimes the fruit-flavored soft drinks and sometimes the Coca-Cola.

In every home we visited there are items in use that have the Coca-Cola logo, such as plastic table and chairs. One of the primary schools in the neighborhood had a refrigerator painted with the Coca-Cola. We visited this school on a day when the children were presenting dances and songs based on traditional Mexican folktales. There were several vendors both inside the school grounds and just outside selling Coke and Pepsi beverages and packaged snack foods (*comida chatarra*). In a few homes we were offered refreshment, which is long-standing custom of hospitality in Yucatan and Chiapas, Mexico. In every instance this was a soft drink such as Coke or another commercial brand drink, such as "sparkling apple juice"—a sugar-sweetened, carbonated apple-flavored beverage.

In the Mexican state of Chiapas, the Coca-Cola and Pepsi-Cola companies have worked successfully to convince people to offer the globalized soft drinks instead of a locally produced, mildly alcoholic beverage distilled from fermented sugar cane and corn, called *pox* (pronounced as "posh," Verza 2013; FD and MTCB, field notes). Offering friends and guests a drink of *pox* was an important part of traditional religious and social activities. The corporate strategy to replace *pox* with soft drinks is reinforced by Christian Evangelical groups who campaign against traditional Maya and Catholic religion and the drinking of alcohol (Eber 2000; Kovic 2005; Verza 2013). The Evangelicals are another type of globalization, as these religious groups originate from the United States and other wealthy nations and export their practices and ideologies to the poorer nations. The Evangelicals prefer that Coke, Pepsi, or a similar soft drink be offered at social/religious gatherings. Similar pressure from global corporate and religious entities in Merida may be encouraging the shift from locally made beverages to globalized soft drinks.

3.9 Social, Economic, and Political Hypotheses for the Nutritional Dual-Burden

There are some benefits associated with the nutrition transition to globalized foods. The Mexican National Health and Nutrition Survey of 2012 reports a decline in rates of stunting from 27.9% in 2006 to 15.8% in 2012 for 1–4-year-olds residing in urban Yucatan. This decline is explained, in part, by improvements in the total health care system. These improvements include attention to preventative medical care. There was also investment to improve access to safe water, sanitary toilets, and cement floors in low-income family homes. Some of these public health benefits were provided by the *Oportunidades* (Opportunities) Program. *Oportunidades* was initiated in 1996 as a new political policy and socioeconomic intervention to prevent and reduce poverty by providing cash payments to families in exchange for regular school attendance, health clinic visits, and nutrition support (INSP 2007). The program recipients are mothers, who are the caregiver directly responsible for children and family health decisions. Cash payments are made from the government directly to families to decrease overhead and corruption. Nearly 40% of homes in the state of Yucatan received benefits from *Oportunidades.*

Mexican social, economic, and political policy has biological and behavioral effects. *Oportunidades* reduced stunting and the tax on sugar-sweetened drinks lowers consumption. Even so, the prevalence of stunting remains high. In the state of Chiapas, which has a high percentage of native peoples in its population, there has been no impact of the *Oportunidades* program on rates of stunting (García-Parra et al. 2016). The Mexican National Health and Nutrition Survey of 2012 also reported that since 2006, rates of overweight/obesity in the same 1–4-year-old age group increased from 14.4% to 15.1% (INSP 2013). Obesity in older age groups increased by greater amounts. The overweight-obesity side of the nutritional dual-burden is an increasing public health problem. Even with *Oportunidades,* most Maya families continue to live in poverty.

The social, economic, and political consequences of poverty offer another possible explanation for the persistence of short stature and overweight for the Maya. One Yucatan anthropologist who researched the impact of globalization on the Maya concludes that it is not globalization per se that is the health problem, but rather it is the persistence of poverty in a globalized world (Ramirez 2006). Ramirez Carrillo writes that (2006, p. 8),

Que cuando decimos que los mayas yucatecos se encuentran instalados en la globalización nos referimos a fenómenos específicos pero diferenciados entre sí, como formas de comer y vestir, tipos de subordinación económica y en especial un imaginario de vida que conforma una nueva identidad, instalada en la modernidad, distinta a la anterior ...Decir que un maya yucateco es moderno o pertenece a la modernidad y no a la tradición, es decir nada más que hay un acto de autodefinición histórica...Es por ello que su modernidad no solo es resultado de la pobreza y la desigualdad, sino que la reproduce y perpetua, aunque le haga cambiar sus formas de vida y sus expectativas de consumo, y le haga construir otro imaginario.

[When we say that the Yucatec Maya are positioned in globalization we refer to specific but differentiated phenomena, ways of eating, of dressing, types of economic subordination and especially a concept of life that forms a new identity, installed in modernity, different from the previous one...To say that a Yucatec Maya is modern or belongs to modernity and not to tradition, is to say nothing more than an act of historical self-definition...That is why modernity is not only the result of poverty and inequality, but that it reproduces and perpetuates it, even as it changes ways of life and consumption expectations, and builds another concept of life.]

We agree—modernity and its globalization process perpetuate poverty and inequality. The food products and other goods and services of globalization may be healthful, but only if families can afford to purchase these imports and if the families have the educational and social preparation to use the products correctly. Low-income people lack both the money and the education required.

Scholarly and popular publications of Sen (2002), Wilkinson and Pickett (2009), and Marmot (2015) elegantly and passionately make the case for the pernicious effects of poverty, especially its social and economic inequalities, on health. Based on his analysis of gradients in social disadvantage, Marmot writes that, "The gradient in health in rich countries makes clear that we are discussing social inequalities more than absolute amounts of money" (Marmot 2015, p. 2444). Marmot uses the word "pollutant" to describe the impact of inequality and social disadvantage on human well-being. When applied to the lower-income nations, the pollutions of poverty and inequality are magnified and concentrated on the least advantaged, such as the Maya.

In his view, these pollutants disempower people from the resources needed for their own healthy growth and development and for the health and good growth of their children. The poor suffer from poverty (low income), and like all people in modern nation-states exist along a gradient of access to resources that is determined by social, educational, and occupational status as much or more than income. The social status differences not only influence income and wealth, but also decisions and behaviors related to diet, health care seeking, smoking, alcohol consumption, sexual activities, educational attainment, and other similar variables that are associated with physical growth. Economists such as Sen, public health researchers such as Marmot, and many anthropologists, including the present authors, agree that promoting greater equality is the most effective way to narrow the social gradient and improve the well-being of all members of society.

3.10 Community Effects and Strategic Growth

There is evidence that social, economic, and political status differences may directly influence the neuroendocrine activity that regulates growth in height (Bogin et al. 2015; Hermanussen and Scheffler 2016). Considering this evidence, we offer one more explanation for the persistence of stunting even in the face of overweight, obesity, and interventions such as *Oportunidades* in Maya communities. This is the community effects hypothesis. In Fig. 3.4 we present a schematic diagram of the determinants of adult body height. The classic interpretation in the left-hand side of the figure focuses on the individual and holds that average living conditions constrain height of the individual within a range set by the interplay of genetic potential and environmental circumstances. The new interpretation in the right-hand column posits that there is no fixed individual genetic potential for height. Rather, human height may vary within a range of more than 20 cm. The tallest and shortest non-Pygmy people in the world today, the Dutch (men average 186 cm) and Maya of Guatemala (men average 160), vary by 26 cm. In the early twentieth century Dutch men were little taller than contemporary Maya (NCD-RisC 2016). The Dutch never had a genetic potential for short stature and neither do the Maya. Maya children gain

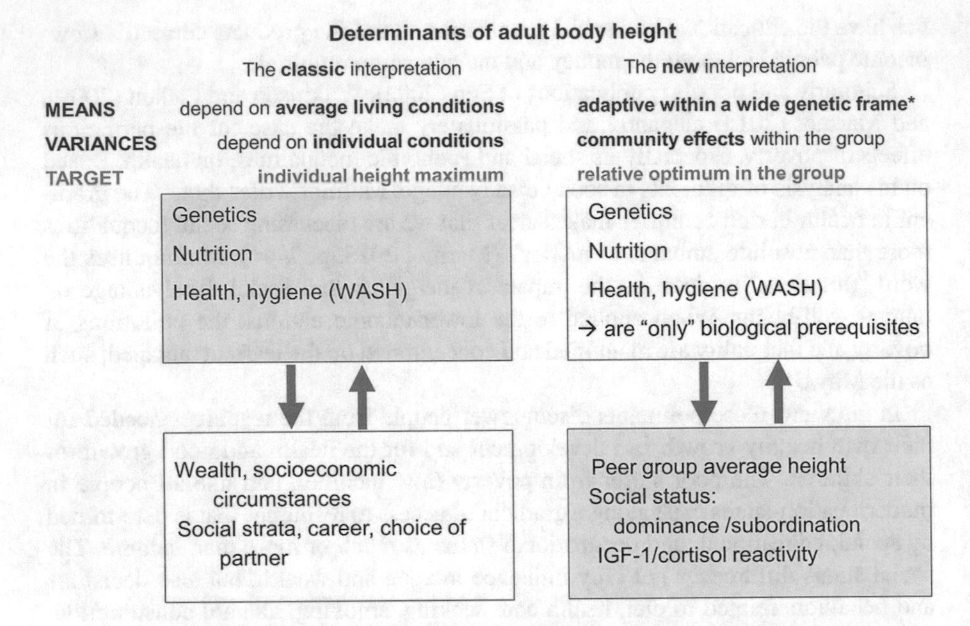

Fig. 3.4 Determinants of adult body height: classic interpretation versus a newer interpretation based on the community effects and strategic growth hypotheses. See text for explanation. *WASH* water, sanitation & health. Source: Hermanussen et al. (2017)

11 cm in height if their families migrate to the United States (Bogin et al. 2002). This height increase occurs within a generation, so cannot have a genetic basis. The new interpretation accepts that there is some genetic basis to the range of height an individual can achieve and that factors such as adequate nutrition, health, and hygiene (WASH = water, sanitation & health) are biological pre-requisites for growth. Within the 20+ cm range of possible height variation, it is social and emotional stimulus from the peer group (friends, family, school-mates) combined with the individual's and peer group's social status that directs height toward a community target. In the case of Maya families of the Yucatan living in poverty, it is possible that the persistent social subordination and racism and the emotional stress of living in a globalized material and food environment stimulate the hypothalamic–pituitary–adrenal axis of the body to produce cortisol and other stress hormones. These endocrine products interact with the hypothalamic–pituitary– insulin-like growth factor-1 (IGF-1) axis. IGF-1 is the primary stimulant for growth in length and high cortisol reactivity suppresses IGF-1 and reduces growth in height.

Social-emotional stress and growth has been well studied in nonhuman species and it has been shown that growth rates of mammals may vary in relation to their social environment. Wild Kalahari meerkats (*Suricata suricatta*) are a cooperative breeding species and only one dominant female and male breed at a time.

Competition for dominance is a direct stimulus to body growth and individuals adjust their growth to the size of their closest competitor (Huchard et al. 2016). Huchard and colleagues conducted experimental feeding studies with the meerkats and found that growth adjustments toward larger body size during social status competition happened before an increase in food intake. The authors discussed similar competitive and strategic growth in other social mammals, including primates and humans.

Community effects in human height growth may be studied using Social Network Theory—how people relate to each other, affect each other, and interact with each other (Christakis and Fowler 2013). Human social networks influence many aspects biology, behavior, and emotion (Meehan et al. 2014; Aral and Nicolaides 2017). Membership within human social communities may set targets for adult height and the tempo of maturation via contention for status, and through an emotional desire and social pressure to conform toward the mean phenotype of the immediate social community (Hermanussen and Scheffler 2016).

Details of the biological mechanisms connecting social competition and community effects with amounts of growth are yet to be determined. The phrases "emotional desire" and "social pressure," used just above, may at first seem unscientific. Copious literature, however, links emotional stressors with endocrine physiology, health status, and future disease risk. Good general reviews are offered by McEwen and colleagues (Lupien et al. 2009; McEwen et al. 2015). There are studies focused on topics relating to community effects on the Maya and other ethnic groups (Goodman et al. 1988; Creel 2001; Del Giudice 2009; Radley et al. 2011; Slavich and Cole 2013; Chomat et al. 2015).

Preliminary data on the hormone IGF-1 suggest its central role in the conversion of emotional and social information into centimeters of height. Socially subordinate male baboons have suppressed levels of IGF-1 (Sapolsky and Spencer 1997) and British university elite athletes who win their sporting events have significantly higher levels of IGF-1 both before and after their sport success (Bogin et al. 2015). IGF-1 closely interacts with the mammalian target of rapamycin (mTOR) signaling pathway. Both the IGF-1 axis and the mTOR pathway sense and integrates a variety of environmental cues, including positive and negative emotional stress, to adjust organismal growth and maintain homeostasis (Saxton and Sabatini 2017; Wang et al. 2017).

A future research project with the Maya will be to assess emotional and hormonal status in relation to growth in both height and fatness.

3.11 Conclusion

The globalized diet of the Maya is not an isolated case. Lower-income and socially marginalized communities around the world are experiencing similar types of nutrition transitions and the associated health consequences. The Maya case is a bit more poignant because following the Conquest and colonization by European powers more than 500 years ago, the Maya suffered undernutrition, with short stature, thinness, and high infant and child mortality. Today, the Maya face a new mixture of nutritional threats from diets that are supplied by multinational corporations. Many local foods remain part of the Maya diet, such as tortillas, but even these local foods are increasingly globalized, manufactured in factories from nonlocal ingredients, and even nontraditional ingredients (Lind and Barham 2004). All societies must come to terms with food globalization. Research is a first step toward a better understanding of the reasons for, and consequences of, food globalization. The next step is to translate research findings into social, economic, and political action to improve child health and well-being. Enabling people to increase local control of their diet is one possible research translation. Almost all the Maya homes we visited have garden space, but very few grow any food. Promoting the development of home vegetable and fruit gardens may be one effective intervention to improve diet and health.

Acknowledgments The research reported here is part of The Maya Project http://mayaproject.org.uk. Created in 2012, The Maya Project is a combination of biocultural research, fine arts, and public engagement. The goal is to accurately represent the living Maya people. The Maya Project achieves this goal through the photography, paintings, music, crafts, and writing of Maya artists, as well as non-Maya artists and researchers. The authors thank the Wenner-Gren Foundation (#IRCG-93) and Santander Universities at Loughborough University for funding this research. Janice Tut-Be, Deira Jiménez.Balam, Hannah Wilson, and Adriana Vázquez-Vázquez provided valuable assistance in fieldwork and ethnographic understanding. The cooperation of the Maya mothers and their children, as well as local primary schools in the Maya neighborhoods, is very much appreciated.

References

Albrow M, King E (1990) Globalization, knowledge and society. Sage, London

Al-Rodhan NRF (2006) Definitions of globalization: a comprehensive overview and a proposed definition. In: Geneva Centre for Security Policy. http://www.css.ethz.ch/en/services/digital-library/publications/publication.html/19462. Accessed 9 Apr 2019

Aral S, Nicolaides C (2017) Exercise contagion in a global social network. Nat Commun 8:14753. https://doi.org/10.1038/ncomms14753

Azcorra H, Wilson H, Bogin B et al (2013) Dietetic characteristics of a sample of Mayan dual burden households in Merida, Yucatan, Mexico. Arch Latinoam Nutr 63(3):209–217

Bartolomé M (1998) La dinámica social de los mayas de Yucatán: Pasado y presente de la situacion colonial. Instituto Nacional Indigenista, Mexico, DF

Bermudez OI, Hernandez L, Mazariegos M et al (2008) Secular trends in food patterns of Guatemalan consumers: new foods for old. Food Nutr Bull 29(4):278–287. http://www.ncbi.nlm.nih.gov/pubmed/19227052. Accessed 27 Dec 2013

Bogin B (1998) From caveman cuisine to fast food: the evolution of human nutrition. Growth Horm IGF Res 8(Suppl. 2):79–86. http://www.ncbi.nlm.nih.gov/pubmed/10990138

Bogin B (2013) The Maya in Disneyland: child growth as a marker of nutritional, economic, and political ecology. In: Dufour DL, Goodman AH, Pelto GH (eds) Nutritional anthropology: biocultural perspectives on food and nutrition, 2nd edn. Oxford University Press, Oxford, pp 231–244

Bogin B, Hermanussen M, Blum WF et al (2015) Sex, sport, IGF-1 and the community effect in height hypothesis. Int J Environ Res Public Health 12(5):4816–4832. https://doi.org/10.3390/ijerph120504816

Bogin B, Keep R (1999) Eight thousand years of economic and political history in Latin America revealed by anthropometry. Ann Hum Biol 26(4):333–351

Bogin B, Scheffler C, Hermanussen M (2017) Global effects of income and income inequality on adult height and sexual dimorphism in height. Am J Hum Biol 29:2. https://doi.org/10.1002/ajhb.22980

Bogin B, Smith P, Orden AB et al (2002) Rapid change in height and body proportions of Maya American children. Am J Hum Biol 14(6):753–761. http://www.ncbi.nlm.nih.gov/pubmed/12400036

Bonfil G (2006) Diagnóstico sobre el hambre en Sudzal, Yucatán: Un ensayo de antropología aplicada. Centro de Investigaciones y Estudios Superiores en Antropología Social, Universidad Autónoma Metropolitana. Universidad Iberoaméricana, Mexico, DF. http://www.ciesas.edu.mx/Publicaciones/Clasicos/libros/sudzal.pdf

Bracamonte P, Lizama J (2003) Marginalidad indígena: Una perspectiva historica de Yucatan. Desacatos 13:83–98

Castillo-Burguete MT, Viga MD, Dickinson F (2008) Changing the culture of dependency to allow for successful outcomes in participatory research: Fourteen years of experience in Yucatan, Mexico. In: Bradbury H, Reason P (eds) The SAGE Handbook of Action Research. SAGE Publications, London

Chomat AM, Solomons NW, Koski KG et al (2015) Quantitative methodologies reveal a diversity of nutrition, infection/illness, and psychosocial stressors during pregnancy and lactation in rural mam-mayan mother-infant dyads from the Western Highlands of Guatemala. Food Nutr Bull 36(4):415–440. https://doi.org/10.1177/0379572115610944

Christakis N, Fowler JH (2013) Social contagion theory: examining dynamic social networks and human behavior. Stat Med 32:556–577. https://doi.org/10.1002/sim.5408

Creel S (2001) Social dominance and stress hormones. Trends Ecol Evol 16(9):491–497. https://doi.org/10.1016/S0169-5347(01)02227-3

Cuanalo HE, Ochoa E, Tuz FR et al (2014) Food intake and nutrition in children 1–4 years of age in Yucatan, Mexico. Ann Hum Biol 41(1):46–52. https://doi.org/10.3109/03014460.2013.824024

Dary O, Imhoff-Kunsch B (2012) Measurement of food consumption to inform food fortification and other nutrition programs: an introduction to methods and their application. Food Nutr Bull 33(Suppl 3):141–145. http://www.ncbi.nlm.nih.gov/pubmed/23193764

Del Giudice M (2009) Sex, attachment, and the development of reproductive strategies. Behav Brain Sci 32(1):1–21., discussion 21–67. https://doi.org/10.1017/S0140525X09000016

Drewnowski A, Darmon N (2005) Food choices and diet costs: an economic analysis. J Nutr 135(4):900–904

Dufour DL, Goodman AH, Pelto GH (2013) Nutritional anthropology, 2nd edn. Oxford University Press, Oxford

Eber C (2000) Women and alcohol in a highland Maya town: water of hope, water of sorrow, 2nd edn. University of Texas Press, Austin

Frisancho AR (2003) Reduced rate of fat oxidation: a metabolic pathway to obesity in the developing nations. Am J Hum Biol 15(4):522–532. https://doi.org/10.1002/ajhb.10191

García-Parra E, Ochoa-Díaz-López E, García-Miranda R et al (2016) Are there changes in the nutritional status of children of Oportunidades families in rural Chiapas, Mexico? A cohort prospective study. J Health Popul Nutr 35(1):1. https://doi.org/10.1186/s41043-015-0038-5

Glittenberg J (1994) To the mountain and back: the mysteries of Guatemalan highland family life. Waveland Press, Long Grove, IL

Goodman AH, Thomas RB, Swedlund AC, Armelagos GL et al (1988) Biocultural perspectives on stress in prehistoric, historical, and contemporary population research. Am J Phys Anthropol 31(S9):169–202. https://doi.org/10.1002/ajpa.1330310509

Guzmán-Median MGV (2005) Una nueva mirada hacia los mayas de Yucatán: identidad, 3th edn. Universidad Autonoma de Yucatan, Mérida, Yucatán

Han E, Powell LM (2014) Consumption patterns of sugar-sweetened beverages in the United States. J Acad Nutr Diet 113(1):43–53. https://doi.org/10.1016/j.jand.2012.09.016

Hermanussen M, Bogin B, Scheffler C (2017) Strategic growth adjustment. Am J Hum Biol 8:e22974. https://doi.org/10.1002/ajhb.22973

Hermanussen M, Scheffler C (2016) Stature signals status: the association of stature, status and perceived dominance – a thought experiment. Anthropol Anz 73(4):265–274. https://doi.org/10.1127/anthranz/2016/0698

Huchard E, English S, Bell MBV et al (2016) Competitive growth in a cooperative mammal. Nature 533:532–534. https://doi.org/10.1038/nature17986

INSP (2007) Encuesta Nacional de Salud y Nutrición 2006. In: Resultados por entidad federativa, Yucatán. Instituto Nacional de Salud Pública, Cuernavaca, México

INSP (2013) Encuesta Nacional de Salud y Nutrición 2012. In: Resultados por entidad federativa, Yucatán. Instituto Nacional de Salud Pública, Cuernavaca, México. http://ensanut.insp.mx/informes/Yucatan-OCT.pdf

Jimenez-Cruz A, Bacardi-Gascon M, Pichardo-Osuna A et al (2010) Infant and toddlers' feeding practices and obesity amongst low-income families in Mexico. Asia Pac J Clin Nutr 19(3):316–323. http://www.ncbi.nlm.nih.gov/pubmed/20805074

Kovic C (2005) Mayan voices for human rights: displaced catholics in highland Chiapas. University of Texas Press, Austin

Kuisel RF (1991) Coca-Cola and the Cold War. The French Face Americanization, 1948–1953. Fr Hist Stud 17(1):96–116. http://web.ics.purdue.edu/~wggray/Teaching/His337/Handouts/Kuisel-Cola.pdf

Leatherman TL, Goodman A (2005) Coca-colonization of diets in the Yucatan. Soc Sci Med 4:833–846. http://www.ncbi.nlm.nih.gov/pubmed/15950095

Lind D, Barham E (2004) The social life of the tortilla: food, cultural politics, and contested commodification. Agric Human Values 21(1):47–60. https://doi.org/10.1023/B:AHUM.0000014018.76118.06

Lizama J (2012) El perfil maya de la blanca Mérida. CIESAS, México, DF

Lovell WG (2010) A beauty that hurts: life and death in Guatemala, 2nd edn. University of Texas Press, Austin, Texas

Lovell GW, Lutz CH (1996) "A dark obverse": Maya survival in Guatemala: 1520–1994. Geogr Rev 86(3):398–407. http://www.jstor.org/stable/215503

Lupien SJ, McEwen BS, Gunnar MR et al (2009) Effects of stress throughout the lifespan on the brain, behaviour and cognition. Nat Rev Neurosci 10(6):434–445. https://doi.org/10.1038/nrn2639

Madrigal H, Martínez H (1996) Manual de encuestas de dieta. Instituto Nacional de Salud Pública, México, DF

Marmot M (2015) The health gap: the challenge of an unequal world. Lancet 386(10011):2442–2444. https://doi.org/10.1016/S0140-6736(15)00150-6

McEwen BS, Gray JD, Nasca C (2015) 60 years of neuroendocrinology: redefining neuroendocrinology: stress, sex and cognitive and emotional regulation. J Endocrinol 226(2):T67–T83. https://doi.org/10.1530/JOE-15-0121

McNeil CL, Burney D, Burney LP (2010) Evidence disputing deforestation as the cause for the collapse of the ancient Maya polity of Copan, Honduras. Proc Natl Acad Sci U S A 107(3):1017–1022. https://doi.org/10.1073/pnas.0904760107

Meehan CL, Helfrecht C, Quinlan RJ (2014) Cooperative breeding and Aka children's nutritional status: is flexibility key? Am J Phys Anthropol 153(4):513–525. https://doi.org/10.1002/ajpa.22415

Mintz SW (1985) Sweetness and power. Penguin Books, New York

NCD-RisC (2016) A century of trends in adult human height. eLife 5:e13410. https://doi.org/10.7554/eLife.13410

Radley JJ, Radley JJ, Kabbaj M et al (2011) Stress risk factors and stress-related pathology: neuroplasticity, epigenetics and endophenotypes. Stress 14(5):481–497. https://doi.org/10.3109/10253890.2011.604751

Ramirez LA (2006) Impacto de la globalización en los mayas yucatecos. Estud Cult Maya 27:73–91

Rodríguez-Ramírez S, Mundo-Rosas V, Jiménez-Aguilar A et al (2009) Methodology for the analysis of dietary data from the Mexican National Health and Nutrition Survey 2006. Salud Publica Mex 51(Suppl 4):S523–S529. http://www.ncbi.nlm.nih.gov/pubmed/20464228

Said-Mohamed R, Bernard JY, Ndzana AC et al (2012) Is overweight in stunted preschool children in Cameroon related to reductions in fat oxidation, resting energy expenditure and physical activity? PLoS One 7(6):e39007. https://doi.org/10.1371/journal.pone.0039007

Sánchez-Romero LM, Penko J, Coxson PG et al (2016) Projected impact of Mexico' sugar-sweetened beverage tax policy on diabetes and cardiovascular disease: a modeling study. PLoS Med 13(11):e1002158. https://doi.org/10.1371/journal.pmed.1002158

Sapolsky RM, Spencer EM (1997) Insulin-like growth factor I is suppressed in socially subordinate male baboons. Am J Physiol 273:R1346–R1351

Sawaya AL, Martins PA, Baccin VJ et al (2009) Malnutrition, long-term health and the effect of nutritional recovery. Nestle Nutr Workshop Ser Pediatr Program 63:95–108. https://doi.org/10.1159/000209975

Saxton RA, Sabatini DM (2017) mTOR signaling in growth, metabolism, and disease. Cell 168(6):960–976. https://doi.org/10.1016/j.cell.2017.02.004

Sen A (2002) Why health equity? Health Econ 11(8):659–666. https://doi.org/10.1002/hec.762

Siniarska A, Wolanski N (1999) Living conditions and growth of Maya Indian and non-Maya boys from Yucatan in 1993 in comparison with other studies. Int J Anthropol 14(4):259–288

Slavich GM, Cole SW (2013) The emerging field of human social genomics. Clinical Psychol Sci 1(3):331–348

Smith PK, Bogin B, Varela-Silva MI et al (2002) Does immigration help or harm children's health? The Mayan case. Soc Sci Q 83(4):994–1002. http://search.ebscohost.com/login.aspx?direct=true&db=ecn&AN=0664513&site=ehost-live

Tax S (1953) Penny capitalism: a Guatemalan Indian economy. Smithsonian Institution Press, Washington DC

Varela-Silva MI, Azcorra H, Dickinson F et al (2009) Influence of maternal stature, pregnancy age, and infant birth weight on growth during childhood in Yucatan, Mexico: a test of the intergenerational effects hypothesis. Am J Hum Biol 21(5):657–663. https://doi.org/10.1002/ajhb.20883

Varela-Silva MI, Dickinson F, Wilson H et al (2012) The nutritional dual-burden in developing countries--how is it assessed and what are the health implications? Coll Anthropol 36(1):39–45

Varela-Silva MI, Frisancho AR, Bogin B et al (2007) Behavioral, environmental, metabolic and intergenerational components of early life undernutrition leading to later obesitia y in developing nations and in minority groups in the U.S.A. Coll Antropol 31(1):39–46. http://www.ncbi.nlm.nih.gov/pubmed/17600917. Accessed 5 Jul 2012

Verza M (2013) The "Coca-Colization" of Mexico, the spark of obesity. Periodismohumano. http://english.periodismohumano.com/2013/03/05/the-coca-colization-of-mexico-the-spark-of-obesity/

Wang L, Zhou K, Fu Z et al (2017) Brain development and akt signaling: the crossroads of signaling pathway and neurodevelopmental diseases. J Mol Neurosci 61(3):379–384. https://doi.org/10.1007/s12031-016-0872-y

Wilkinson RG, Pickett KE (2009) The spirit level: why more equal societies almost always do better. Allen Lane, London

Wilson H, Dickinson F, Griffiths P et al (2011) Logistics of using the Actiheart physical activity monitors in urban Mexico among 7- to 9-year-old children. Am J Hum Biol 23(3):426–428. https://doi.org/10.1002/ajhb.21150

Wilson HJ, Dickinson F, Griffiths PL et al (2013) Maternal short stature does not predict their children's fatness indicators in a nutritional dual-burden sample of urban Mexican Maya. Am J Phys Anthropol 153(4):627–634. https://doi.org/10.1002/ajpa.22463

Wilson HJ, Dickinson F, Hoffman DJ et al (2012) Fat free mass explains the relationship between stunting and energy expenditure in urban Mexican Maya children. Ann Hum Biol 39(5):432–439. https://doi.org/10.3109/03014460.2012.714403

Wolanski N, Dickinson F, Siniarska A (1993) Biological traits and living conditions of Maya Indian and non Maya girls from Merida, Mexico. Int J Anthropol 8(4):233–246

Chapter 4
Growth Stunting and Low Height-for-Age in the Yucatan Peninsula

Maria Inês Varela-Silva, Samantha Sanchez, Barry Bogin, Federico Dickinson, and Hugo Azcorra

Abstract We critically analyze definitions of growth stunting and low height-for-age, focusing on the arbitrary cut-off points that classify people dichotomously as "stunted" or "non-stunted." We discuss how individuals may be included or excluded in one of such categories, depending on the standard or references used, and list inherent health implications. We analyze data on prevalence and predictors of stunting among Maya children and adults in Yucatan, to highlight pros and cons of using dichotomized versus continuous height-for-age data. Stunting among Yucatecan Maya children has decreased since the 1990s (Azcorra et al., Growth status in children and adolescents in Yucatan, Mexico: a human ecology perspective. In: Studies in human ecology, House for Science and Technology, Ha Noi, p. 120–138, 2010) but prevalence of low height-for-age continues to be high, throughout the growing period (Méndez Domínguez, Aspectos del ambiente familiar y escolar relativos a los hábitos de alimentación y actividad física que se asocian al exceso de peso entre escolares de Mérida, Yucatán, 2013; Azcorra-Perez, Intergenerational factors that shape the nutritional status of urban Maya households in Merida, Mexico. A 3-generations study, 2014). Male and female adult Maya average stature has not changed in 100 years and remains extremely low (Steggerda, Maya Indians of Yucatan. Carnegie Institution of Washington, Washington, DC, 1941; McCullough, Am J Phys Anthropol 58:221–225, 1982; Dickinson et al., Estud Antropol Biol.

M. I. Varela-Silva (✉) · B. Bogin
School of Sport, Exercise and Health Sciences, University of Loughborough, Loughborough, UK
e-mail: m.i.o.varela-silva@lboro.ac.uk; b.a.bogin@lboro.ac.uk

S. Sanchez
Coordinación de Nutrición, Universidad Vizcaya de las Américas Campus Mérida, Mérida, Yucatán, México

F. Dickinson
Departamento de Ecología Humana, Cinvestav-Mérida, Mérida, Yucatán, México

H. Azcorra
Departamento de Ecología Humana, Cinvestav-Mérida, Mérida, Yucatán, México

Centro de Investigaciones Silvio Zavala, Universidad Modelo, Mérida, Yucatán, México

H. Azcorra, F. Dickinson (eds.), *Culture, Environment and Health in the Yucatan Peninsula*, https://doi.org/10.1007/978-3-030-27001-8_4

4:123–150, 1989; Varela-Silva et al., Coll Anthropol 36:39–45, 2012). Short stature has been associated with increased risk for cardiovascular diseases and other non-communicable diseases and its effects seem to be potentiated by overweight/obesity (OW/OB). This is especially true among the Maya who also show an elevated accumulation of centralized fat. The association of being short in stature with being OW/OB produces negative health outcomes that are stronger and more pernicious than the ones seen in populations where only one of the conditions happens, in the absence of the other. We highlight some aspects that may help to answer the following questions: i) why is the prevalence of stunting decreasing among Maya children but the prevalence of low height-for-age during growing years remains high, and ii) why has the stature of the adult Maya remained unchangeable in the last century?

4.1 Introduction

Growth stunting is defined as very low height-for-age and it reflects linear growth retardation due to long-term nutritional inadequacies and/or recurrent infections (WHO 2010a).

The term "stunting" was coined by Waterlow (1972). It was proposed with the aim of distinguishing health outcomes caused by changes in weight-for-height during the growing years, from health outcomes caused by growth retardation leading to reduction in final adult stature. This means that the former would be used to classify the severity of undernutrition, and the latter to classify its duration (Perumal et al. 2018).

> A reduction in the rate of linear growth could appropriately be referred to as retardation and a reduction in final stature as stunting—(Waterlow 1972, p. 567).

This is interesting because the word "stunting," in the current literature, is used when assessing children and it is very seldomly applied to classify a very short adult. This contravenes the initial purpose of the term, which was aimed, mostly, to be used at the population level, focusing on individuals whose growth process in length was completely finished.

The most commonly used standards, worldwide, to assess the presence or absence of stunting, in individuals and populations, are the ones proposed by the World Health Organization—WHO (WHO 2006; De Onis et al. 2012). A child is classified as "stunted" if her/his height-for-age falls below the −2 standard deviations (−2SD) of the WHO Child Growth Standards median (WHO 2010a; De Onis et al. 2012).

> *The standards describe normal child growth from birth to 5 years under optimal environmental conditions and can be applied to all children everywhere, regardless of ethnicity, socioeconomic status and type of feeding*—(WHO 2010b).

The WHO growth standards cover children from 0 to 5 years; therefore, when an assessment needs to be conducted in children above 5 years of age, then the 2007 WHO Growth References, covering the ages of 5–19 years, complement the WHO child growth standards (WHO 2013). The WHO references do not include data for individuals above 19 years of age, therefore excluding adult populations. This means that any assumptions regarding the nutritional status of adults and, specifically, data and classifications based on height-for-age will be extrapolated from other databases of reference, such as the ones published by the National Center for Health Statistics (NHANES) or the Center for Disease Control and Prevention (CDC).

From a human biology perspective, the distinction between standards and references is of significance.

> *… the term **standard** implies that the growth curves depicted on the chart reflect desirable growth, while a **reference** chart reflects the growth pattern of the children who were selected as the source sample for the charts, and no judgment or desirability is placed against their growth*—(Frisancho 2008, p. V).

This is an important difference because standards are developed using statistical methodologies for sample selection that allow for data to be normally distributed, while references do no such thing. Therefore, some references include skewed anthropometric data that might impact data analysis and results' interpretation at the individual and the populational levels.

Very low height-for-age affects all body systems and if not curbed early in life will leave permanent and damaging physical, cognitive, developmental, and economic traits that will impact at individual, familial, and national levels (Spurr and Words 1983; Spurr et al. 1983; Spurr 1990; Aguayo et al. 2016).

Height-for-age z-scores (HAZ), based on a standard/reference database, were initially used as a general indicator of a population's living conditions and well-being but, quickly, started being used at the individual level. With this methodological change, healthy children who are short might have been classified as stunted, the same way that some children with substantial growth restrictions that do not fall below the threshold of −2SD will continue unnoticed. To avoid these, Perumal et al. (2018) suggest that analysis using HAZ should check for the slope of the decline in HAZ, through time, and also check for final adult stature. Otherwise, the risk of underestimating the burden of linear growth retardation in children will substantially increase.

4.1.1 Arbitrary Cutoff Points

Higher mortality rates, poor cognition, higher odds for NCDs, less percentage of muscle mass, reduced work capabilities, and reduced earning potential exist along a HAZ continuum without a notable inflection point at –2SD (Spurr and Words 1983; Spurr 1990; Dickinson et al. 1993; Wilson et al. 2012; Datta et al. 2014; Perumal et al. 2018).

The main problem with nutritional status classifications that rely on values of height-for-age (HAZ), weight-for-height (WHZ), and weight-for-age (WAZ) is that the cut-off points that define, respectively, stunting, wasting, and underweight are not biologically defined (Perumal et al. 2018). Furthermore, the databases used to

Table 4.1 Discrepancy in nutritional status classifications, in Maya children and Maya adult women, depending on the references, cutoff points, and indicators used (Adapted from Varela-Silva et al. 2012)

		WHO (*z* scores)	CDC (percentiles)	*p*-value
Children	Stunting	15.5% (HAZ < −2SD)	31.0% (HA < 5th percentile)	0.001
	Underweight	1.7% (WAZ < −2SD)	5.2% (BMI < 5th)	Ns
	Overweight	8.6% (BMIZ > +2SD)	12.1% (BMI > 85th–<95th)	0.001
	Obesity	0% (BMIZ > +2SD)	15.5% (BMI > 95th)	0.001
Mothers	Stunting	55.2% (HAZ < −2SD)	81.0% (HA < 5th percentile)	0.001
	Underweight	0% (WAZ < −2SD)	0% (BMI < 5th)	Ns
	Overweight	BMI ≥ 25.0–<30.0		–
	Obesity	BMI ≥ 30.0		–

define the WHO standards and references, as well as the references used by the Centre for Disease Control (CDC), are different. Additionally, the CDC transforms the raw data of height-for-age into percentiles rather than *z*-scores. In this case, stunting is defined as a value of height-for-age below the fifth percentile, adjusted for sex. In a statistically normally distributed population, the fifth percentile for height-for-age corresponds to a z-score of −1.659SD, not −2SD. This means that individuals who are borderline stunted according to the CDC classification will not be classified as such if the WHO standards or references are used. This has been reported by Varela-Silva et al. (2012) who analyzed data from an urban Maya sample in Yucatan and found that 15% of the children (6.0–9.9 years old) and 25% of the mothers showed height-for-ages that fell between the −2SD and the fifth percentile. This discrepancy was also noted in the other nutritional status indicators (see Table 4.1). Thus, the interpretation of the health status of children and women, if they were classified merely based on a categorical, dichotomic variable, was bound to be heavily influenced by the criteria used to define stunting.

4.2 Causes of Very Low Height-for-Age

4.2.1 Individual Level

Very low height-for-age usually occurs due to a combination of biocultural factors that can be summarized based on when they happen in the life course of the individual:

1. Before conception (intragenerational effects): non-genomic mechanisms are in place among women with poor health outcomes during their growth period. This

results in compromised reproductive systems that will impair these women's pregnancies with a high likelihood of producing low birthweight offspring. If the offspring are female then their reproductive systems will also be compromised, passing down the cycle of negative outcomes to the following generation (Drake and Walker 2004; Drake and Liu 2010).

2. *In utero* (intrauterine growth restriction): inadequate prenatal growth and developmental constrains result in low birthweight, causing a panoply of negative outcomes on the child, which will track into adulthood increasing the odds for noncommunicable diseases later in life. These findings form the basis of the Barker Hypotheses and of the Developmental Origins of Health and Disease—DOHaD hypothesis (Barker 1997; Godfrey et al. 2010).
3. Postnatally (especially during the first 2 years of life): where no prenatal constrains are detected, a full-term pregnancy is reported, and the baby is born with normal birthweight. In this case, when low height-for-age occurs, it is likely due to a combination of the following: (1) postnatal systemic and recurrent infections, (2) high levels of inflammation, (3) inadequate breastfeeding practices, (4) inadequate weaning processes, (5) inadequate nutrition of the mothers during the breastfeeding period (Sanchez 2017).

The Mexican Maya have a high prevalence of normal birthweight but show high prevalence of very low height-for-age during childhood. In a sample of 260 Maya-Mexican children, less than 8% were born with low birthweight. More than half of these low birthweight babies were pre-term, therefore, having an appropriate weight for their gestational age. This clearly shows that low birthweight due to intrauterine growth restrictions appears minimal. The same sample, when assessed at ages 6–8 years, showed 12% of stunting (using the WHO references and the cutoff point of −2SD from the median). More worryingly was, however, that 36.5% were classified as being overweight (when using z-scores for BMI-for-age, 32.3% showed high abdominal fat and 26.5% had high trunk fat (Sanchez 2017)). Varela-Silva et al. (2009) reported, in another Maya sample in the Yucatan, that children within the healthy range of birthweight, but below 3000 g, were over 3 times more likely to be stunted (<fifth percentile HA) suggesting that pernicious effects associated with birthweight do not only happen when the child is below the 2500 g cut-off point for low birthweight. Finally, Azcorra and Mendez (2018) found a positive and significant effect of maternal height on the birthweight of their children, among Maya and Mestizo women in Merida.

4.2.2 *Structural, Political, and Ecological Levels*

There are also structural, political, and ecological causes for very low height-for-age that include, but are not limited to, inadequate sanitation infrastructures, irregular and untrustworthy supply of clean drinking water, unavailability of primary health care facilities, unavailability of education for all, and marked gaps in gender equality and autonomy. At this level, the onus should be placed on the governments and on regional authorities. Until these structural problems are solved, it would be naïve to assume that stunting (at the population level) and low height-for-age (at the individual level) can be efficiently tackled by very well-intentioned, but clearly insufficient, nutritional interventions, or small-scale water, primary health care, and sanitation initiatives. This is not to say that these are not valid and meritorious approaches. If well implemented, they will make a difference in the health and well-being of the groups included but, because they tend to be specific and short-termed public health interventions, their effects may not be visible changes in child's height, although they can have positive effects in other nutritional status indicators. However, per se, they will not change national trends and overall health outcomes at the population level (Aguayo and Menon 2016; Joe et al. 2016; Subramanian et al. 2016; Vir 2016; Perkins et al. 2017; Perumal et al. 2018).

4.3 Low Height-for-Age in Yucatan and Its Associations with Negative Health Outcomes

Table 4.2 summarizes four studies conducted in Yucatan, in which low height-for-age is associated with negative health outcomes. Overall, these studies clearly show a persistent and over two-decade-long evidence of nutritional dual burden, in which height-for-age coexists with OW/OB and/or other NCDs indicators (Dickinson et al. 1993; Azcorra et al. 2010; Azcorra 2014; Wilson et al. 2012; Datta et al. 2014). Rural areas tend to be more affected by high rates of stunting (Azcorra et al. 2010) and being Maya exacerbate all the negative outcomes. Overall, levels of physical activity are above the daily recommended by the WHO, even among the stunted children. This means that physical inactivity does not explain the elevated rates of overweight/obesity. Lower levels of energy expenditure and less fat-free mass (kg) were directly associated with low height-for-age, but not with stunting (Wilson et al. 2012). On the other hand, in one of the studies, stunted individuals had higher percentage of body.

A non-published Ph.D. thesis, by Méndez Domínguez (2013), in which the author analyzed growth and health data of 3243 6–12 years old, Maya and non-Maya, boys and girls, from 16 schools in Merida, showed that an overall average of 50.4% (min = 43%, max = 56.3) of the children were either overweight, obese, or morbidly obese. The same sample also showed a total deficit of stature (low HAZ, stunted and very stunted) of 28.4% with percentages ranging from 17% and 45.7%, depending on the school. The three schools that show percentage of low stature above 40% are all located in the south of Merida, where most of the urban Maya in Yucatan live in very poor conditions. The results also highlighted that having one Maya surname increased the odds of being simultaneously low statured and overweight.

Table 4.2 Low height-for-age associated with negative health outcomes in Yucatan, 1993–2014

Sample	Height	Other health outcomes	Comments
Dickinson et al. (1993)			
N = 216 women Age: >30.0 years Origin: (a) rural-urban internal migrants; (b) nonmigrants in sisal area; (c) nonmigrants in coastal area. Ethnicity: Not-stated Data collection: 1989–1990	Units of measurement: cm Internal migrants: (Mean ± SD) = 145.23 ± 4.97 Sisal: 143.6 ± 4.78 Coastal: 147.70 ± 4.96	Kaup index classification: (weight/height^2) ×100: – Undernourished: 0.9%. – Normal: 12.5%. – Obesity grade I: 35.2%. – Obesity grade II: 49.1%. – Obesity grade III: 2.3%. Hypertension: 17.1%	– Very short average stature coexists with elevated rates of obesity, and hypertension. – The existence of nutritional dual-burden is evident.
Azcorra et al. (2010)			
N = 993 Age: 9.0–17.9 years Location: Merida Ethnicity: Not-stated Data collection: 2005	Units of measurement: HAZ Reference database: WHO references – Very low (>−3SD) = 1.11%. – Low (<−3SD, >−2SD) = 9.67%. – Slightly low (<−2SD, >−1SD) =31.02%. – Normal (±1SD) = 52.47%. – Slightly tall (<+2SD, > + 1SD) = 4.83. – Tall (> + 2SD) = 0.91%.	BMI classification (weight/height^2): – Thin (<−2SD) = 0.50% – Normal (<+2, ≥1SD) = 56.19% – Overweight (> + 1SD, <+2SD) = 26.69% – Obesity (> + 2SD) = 16.62%	– Data shows clear evidence of nutritional dual-burden. – Cites literature stating that Yucatan has the highest mortality rate in Mexico for diabetes, cardiovascular disease, and stroke. – Rural areas more affected by high rates of stunting.

(continued)

Table 4.2 (continued)

Sample	Height	Other health outcomes	Comments
Wilson et al. (2012)			
N = 58 Age: 7.00–9.99 years Location: Merida Ethnicity: Maya Data collection: 2010	Units of measurement: HA Reference database: NHANES III – Stunting (≤5th percentile HA) = 18.9% – Not-stunted: (>5.1th percentile HA) = 81.0%	– Percentage of body fat = 29.81% – Average BMIZ = 0.45 ± 0.90 For a subsample of *n* = 33 – Resting Energy Expenditure (REE) = 172.98 ± 21.50 kj/kg/day – Activity Energy Expenditure (AEE) = 127.38 ± 26.78 kj/kg/day – Total Energy Expenditure (TEE) = 33,373 ± 45.33 – Light physical activity (average minutes per day<3MET) = 1316.94 ± 65.46 – Moderate/vigorous physical activity (average minutes per day≥3MET) = 123.21 ± 65.21	– Clear evidence of nutritional dual-burden. – Stunted children showed lower levels of REE, AEE, and TEE. – Stunted children spent less time in moderate/ vigorous physical activity – Stunting, as a dichotomic variable, does not have an effect on levels of energy expenditure. – Low height-for-age (continuous variable) associates with less fat-free mass and lower TEE and AEE. – Overall, levels of physical activity are above the daily recommended by the WHO (even among the stunted children). This means that inactivity does not explain the elevated rates of overweight/obesity.
Datta et al. (2014)			
N = 321 (n_males = 152, n_females = 154) Age: 15.00–17.00 years Location: Merida Ethnicity: Not stated Data collection: 2008–2009	Units of measurement: cm and HAZ Reference database: WHO Average height(cm): – Males: 169.01 ± 7.44 – Females: 154.75 ± 5.87 Stunting (−2SD HAZ): – Males: 7.69% – Females: 18.20%	Overweight (BMIZ > + 2SD, < + 3SD): – Males: 26.28% – Females: 24.24% Obesity (≥ + 3SD) – Males: 10.26% – Females: 6.06%	– Stunted adolescents show higher % BF – Socioeconomic status positively associated with height, %BF and dry lean mass – Maternal occupation negatively associated with height.

Azcorra et al. 2010. Growth status in children and adolescents in Yucatan, Mexico: A human ecology perspective. In: Studies in Human Ecology, pp. 120–138; Datta, S, Castillo T, Rodriguez L et al. 2014. Body fatness in relation to physical activity and selected socioeconomic parameters of adolescents aged 15–17 years in Merida, Yucatan. Ann Hum Biol 4460:1–9; Dickinson F et al. 1993. Obesity and women's health in two socioeconomic areas of Yucatan, Mexico. Coll Anthropol 17:309–317; Wilson et al. 2012. Fat free mass explains the relationship between stunting and energy expenditure in urban Mexican Maya children. Ann Hum Biol 39:121004093114005

4.4 Summary and Recommendations

A decrease in the prevalence of stunting does not automatically provide a group or population with a clean bill of health.

> … mild to moderate malnutrition [seen by low weight-for-age], is associated with an elevated risk of mortality, an association that has great policy significance considering the overwhelming number and proportion of children who fall into this category. The result is that 45–83% of all malnutrition-related deaths (…) occur to children in the mild-to-moderate category (WA 60–80% of the median). These deaths would not be reduced if policies and programs were directly solely towards treatment of the severely malnourished [i.e. stunting]—(Pelletier et al. 1993, p. 1132).

The Maya in Yucatan do not show mild-to-moderate malnutrition, on the contrary, their rates of overweight and obesity are soaring to historic highs and this is equally problematic and equally deadly. There has been a marked shift to the right of the distribution on the average values for all body-mass-related measurements, through time, among the Maya in Yucatan; the overall prevalence of stunting has decreased but the values of low height-for-age remain high during the growth process and the Maya adult stature is extremely low (Steggerda 1941; McCullough 1982; Dickinson et al. 1989). Adults with very short stature are at very high risk for negative health outcomes because it is during adulthood that the majority of NCD effects will manifest. For example, adult short stature has been shown to be associated with coronary heart disease (Paajanen et al. 2010).

We agree with Perumal et al. (2018) when recommending that the main focus should be placed on raising the average height-for-age of a population, instead of curbing stunting. Furthermore, the use of measures of stunting should be used with parsimony at the individual level even though they remain a very valuable indicator of the severity of malnutrition at the population level. However, to answer more specific research questions, continuous HAZ values should be preferred, both when aiming to find predictors of short height-for-age and when height-for-age is used as a predictor of another health outcome.

References

Aguayo VM, Menon P (2016) Stop stunting: improving child feeding, women's nutrition and household sanitation in South Asia. Matern Child Nutr 12(Suppl. 1):3–11. https://doi.org/10.1111/mcn.12283

Aguayo VM, Nair R, Badgaiyan N et al (2016) Determinants of stunting and poor linear growth in children under 2 years of age in India: an in-depth analysis of Maharashtra's comprehensive nutrition survey. Matern Child Nutr 12(Suppl. 1):121–140. https://doi.org/10.1111/mcn.12259

Azcorra H, Mendez N (2018) The influence of maternal height on offspring's birth weight in Merida, Mexico. Am J Hum Biol 30:e23162. https://doi.org/10.1002/ajhb.23162

Azcorra H, Valentin G, Vazquez-Vazquez A et al (2010) Growth status in children and adolescents in Yucatan, Mexico: a human ecology perspective. In: Studies in human ecology. House for Science and Technology, Ha Noi, pp 120–138

Azcorra-Perez HS (2014) Intergenerational factors that shape the nutritional status of urban Maya households in Merida, Mexico. A 3-generations study. Dissertation, Loughborough University

Barker DJP (1997) Maternal nutrition, fetal nutrition, and disease in later life. Nutrition 13:807–813. https://doi.org/10.1016/S0899-9007(97)00193-7

Datta S, Castillo T, Rodriguez L et al (2014) Body fatness in relation to physical activity and selected socioeconomic parameters of adolescents aged 15–17 years in Merida, Yucatan. Ann Hum Biol 4460:1–9. https://doi.org/10.3109/03014460.2014.897755

De Onis M, Onyango A, Borghi E et al (2012) Worldwide implementation of the WHO Child Growth Standards. Public Health Nutr 15:1–8. https://doi.org/10.1017/S136898001200105X

Dickinson F, Castillo MT, Vales L et al (1993) Obesity and women's health in two socioeconomic areas of Yucatan, Mexico. Coll Anthropol 17:309–317

Dickinson F, Murcia R, Cervera MD et al (1989) Antropometría de una población en crecimiento en la costa de Yucatán. Estud Antropol Biol 4:123–150

Drake AJ, Liu L (2010) Intergenerational transmission of programmed effects: public health consequences. Trends Endocrinol Metab 21:206–213. https://doi.org/10.1016/j.tem.2009.11.006

Drake AJ, Walker BR (2004) The intergenerational effects of fetal programming: non-genomic mechanisms for the inheritance of low birth weight and cardiovascular risk. J Endocrinol 180:1–16

Frisancho AR (2008) Anthropometric standards. An interactive nutritional reference of body size and body composition for children and adults. The University of Michigan Press, Ann Arbor, MI

Godfrey KM, Gluckman PD, Hanson MA (2010) Developmental origins of metabolic disease: life course and intergenerational perspectives. Trends Endocrinol Metab 21:199–205. https://doi.org/10.1016/j.tem.2009.12.008

Joe W, Rajaram R, Subramanian SV (2016) Understanding the null-to-small association between increased macroeconomic growth and reducing child undernutrition in India: role of development expenditures and poverty alleviation. Matern Child Nutr 12(Suppl. 1):196–209. https://doi.org/10.1111/mcn.12256

McCullough JM (1982) Secular trend for stature in adult male Yucatec Maya to 1968. Am J Phys Anthropol 58:221–225

Méndez Domínguez N (2013) Aspectos del ambiente familiar y escolar relativos a los hábitos de alimentación y actividad física que se asocian al exceso de peso entre escolares de Mérida, Yucatán. Dissertation, Universidad Autónoma de Yucatán

Paajanen TA, Oksala NKJ, Kuukasjärvi P et al (2010) Short stature is associated with coronary heart disease: a systematic review of the literature and a meta-analysis. Eur Heart J 31:1802–1809

Pelletier DL, Frongillo EA, Habicht JP (1993) Epidemiologic evidence for a potentiating effect of malnutrition on child mortality. Am J Public Health 83:1130–1133

Perkins JM, Kim R, Krishna A et al (2017) Understanding the association between stunting and child development in low- and middle-income countries: next steps for research and intervention. Soc Sci Med 193:101–109. https://doi.org/10.1016/J.SOCSCIMED.2017.09.039

Perumal N, Bassani DG, Roth DE (2018) Use and misuse of stunting as a measure of child health. J Nutr 148:311–315. https://doi.org/10.1093/jn/nxx064

Sanchez S (2017) Crecimiento en la niñez y su relación con factores maternos y del embarazo, peso al nacer y alimentación en la infancia temprana. Dissertation, Centro de Investigación y de Estudios Avanzados del Instituto Politécnico Nacional – Unidad Mérida

Spurr GB (1990) Physical activity and energy expenditure in undernutrition. Prog Food Nut Sci 14:139–192

Spurr GB, Reina JC, Barac-Nieto M (1983) Marginal malnutrition in school-aged Colombian boys – anthropometry and maturation. Am J Clin Nutr 37:119–132

Spurr GB, Words KEY (1983) Nutritional status and physical work capacity. Yearb Phys Anthropol 35:1–35

Steggerda M (1941) Maya Indians of Yucatan. Carnegie Institution of Washington, Washington, DC

Subramanian SV, Mejía-Guevara I, Krishna A (2016) Rethinking policy perspectives on childhood stunting: time to formulate a structural and multifactorial strategy. Matern Child Nutr 12(Suppl. 1):219–236. https://doi.org/10.1111/mcn.12254

Varela-Silva MI, Azcorra H, Dickinson F et al (2009) Influence of maternal stature, pregnancy age, and infant birth weight on growth during childhood in Yucatan, Mexico: a test of the intergenerational effects hypothesis. Am J Hum Biol 21:657–663. https://doi.org/10.1002/ajhb.20883

Varela-Silva MI, Dickinson F, Wilson H et al (2012) The nutritional dual-burden in developing countries – how is it assessed and what are the health implications? Coll Anthropol 36:39–45

Vir SC (2016) Improving women's nutrition imperative for rapid reduction of childhood stunting in South Asia: coupling of nutrition specific interventions with nutrition sensitive measures essential. Matern Child Nutr 12(Suppl. 1):72–90. https://doi.org/10.1111/mcn.12255

Waterlow JC (1972) Classification and definition of protein-calorie malnutrition. Br Med J 3:566–569. https://doi.org/10.1136/bmj.3.5826.566

WHO (2006) WHO child growth standards: methods and development. World Health Organization, Geneva

WHO (2010a) Interpretation Guide Nutrition Landscape Information System (NLIS) Country Profile Indicators. World Health Organization, Geneva

WHO (2010b) Q. Will the standards be applicable to all children? In: WHO. http://www.who.int/childgrowth/faqs/applicable/en/. Accessed 6 Sep 2018

WHO (2013) WHO | Growth reference data for 5–19 years. In: WHO. http://www.who.int/growthref/en/. Accessed 6 Sep 2018

Wilson HJ, Dickinson F, Hoffman DJ et al (2012) Fat free mass explains the relationship between stunting and energy expenditure in urban Mexican Maya children. Ann Hum Biol 39:432. https://doi.org/10.3109/03014460.2012.714403

Chapter 5
The Urban Maya from Yucatan; Dealing with the Biological Burden of the Past and a Degenerative Present

Hugo Azcorra, Barry Bogin, Maria Inês Varela-Silva, and Federico Dickinson

Abstract Today a considerable proportion of Yucatecan Maya reside in urban areas, with the city of Merida being a site with a long history of settlement. The Maya bring with them a historical legacy of poverty and abusive treatment during the more than 500 years of European and then Mestizo domination. This chapter discusses, from a human ecology perspective, the biological consequences of chronic adverse living conditions experienced by the Maya in the context of a research project in which we studied a sample of grandmothers, mothers, and grandchildren from the city of Merida. Our results highlight three important biological phenomena: (1) the coexistence of undernutrition and overweight, both extremes of malnutrition, at the individual and mother-child dyads levels, (2) the growth and nutrition status of recent generations are shaped by conditions experienced by recent maternal ancestors, and, (3) overall, the phenotype of Maya people studied is the result of historical trauma experienced during the colonial period and decades after Mexican independence.

5.1 Introduction

The common tourist in Yucatan would require several weeks to know and enjoy the diversity of the Maya culture. The same visitor would require just a few days of reading classic academic works to discover that the history of the Maya people from the time of contact with Spaniards can be described in two words: misery and

---

H. Azcorra (✉)
Departamento de Ecología Humana, Cinvestav-Mérida, Mérida, Yucatán, México

Centro de Investigaciones Silvio Zavala, Universidad Modelo, Mérida, Yucatán, México

B. Bogin · M. I. Varela-Silva
School of Sport, Exercise and Health Sciences, University of Loughborough, Loughborough, UK
e-mail: b.a.bogin@lboro.ac.uk; m.i.o.varela-silva@lboro.ac.uk

F. Dickinson
Departamento de Ecología Humana, Cinvestav-Mérida, Mérida, Yucatán, México

H. Azcorra, F. Dickinson (eds.), *Culture, Environment and Health in the Yucatan Peninsula*, https://doi.org/10.1007/978-3-030-27001-8_5

injustice. The contemporary Maya people carry with them the deep traces of several centuries of subjugation and *social inequality*. Perhaps readers with limited knowledge of the Yucatan Peninsula might think that the Maya live in small communities in huts made of perishable materials. This idea is far from reality; nowadays, most of the direct descendants of the first inhabitants of this Mesoamerican region live in urban communities, such as the cities of Merida, Campeche, and Cancun, working in construction, commerce, and tourism services.

The authors of this chapter have dedicated their careers to the study of physical growth in human populations. The *growth* and nutritional status of individuals are good indicators of health and well-being in a population and, in a broader sense, are also good indicators of the quality of socio-environmental conditions prevailing in a specific place. A wide range of studies show that the Maya from Yucatan exhibit high rates of *chronic undernutrition* throughout the growing period (Kelley 1991; Wolanski et al. 1993; Siniarska and Wolanski 1999; Varela-Silva et al. 2009; Varela-Silva et al. 2012) and other studies suggest the absence of a positive secular change in stature since the end of the colonial period (McCollough 1982; Siniarska and Wolanski 1999). The Maya from Yucatan and their long history of *poverty* and inequality provide a useful scenario to study how human biology responds to biosocial stress experienced by several generations.

In this chapter we aim to describe the growth and *nutritional status* of a sample of Maya children, their mothers and maternal grandmothers residing in the cities of Merida and Motul in the Yucatan state. In this chapter we argue that current biological status of studied individuals is the result of the historical adverse living conditions experienced by this *ethnic group*. For this purpose, we resort to the historical trauma theory (Sotero 2006; Whitbeck et al. 2004) to guide our brief review of the events that occurred during the sixteenth, seventeenth, and eighteenth centuries and the decades after Mexican Independence in 1821 that we think impacted upon and have shaped the *phenotype* of contemporary Maya people. Historical trauma theory provides a useful framework to examine the consequences of several traumatic events from a macro-level perspective on the populations exposed to them. It is our conviction that our approach also contributes to understand the causes of the current epidemiological profile observed among the Yucatecan population.

5.2 Traces of Historical Trauma Among the Maya from Yucatan

Most people belonging to minority ethnic groups show poorer health conditions than other demographic groups in the same population (Stephens et al. 2005; Gracey and King 2009). Commonly, these differences are explained by primary causes such as current adverse socioeconomic conditions, inadequate health habits, and poor access to health services. Recent theoretical contributions have proposed that health disparities observed in minority ethnic groups have their origins in the deep traces caused by their historical conditions of poverty, inequality, and discrimination (Kuzawa and Gravlee 2016).

Historical trauma theory (HTT) proposes that populations subjected to long-term, mass trauma show higher rates of disease, even several generations have passed since the primary traumatic event occurred (Sotero 2006). Originally used to explain the persistent hurt among Holocaust survivors and their families and then applied in the study of African American and United States Native-American/Indian populations subjected to chronic discrimination and injustice, HTT has been suggested as a useful approach to analyze the health disparities among minority ethnic groups (Williams et al. 2003; Walters and Simoni 2002).

Based on her review, Sotero (2006) proposes that HTT is supported by four assumptions: (1) mass trauma or damage is deliberately and systematically inflicted upon a population by a dominant group, (2) trauma is characterized not by a single event, but rather by a series of traumatic events during a period, also called secondary traumas, (3) trauma is experienced by almost all the population, and (4) traumatic events tend to derail the population from its course, producing a legacy of physical, psychological, social, and economic disparities over several generations. To illustrate this point, we will provide a brief description of the events that we consider have had a great impact on the biological and health status of the Maya during three centuries of colonial rule and two centuries after the independence. We argue that the phenotype of contemporary Maya descendants has been shaped by the historical sociopolitical conditions experienced over this period.

The contact between the Maya and Europeans at the beginning of the sixteenth century was devastating for the *Mesoamerican* natives. Even though there are no exact figures as to the size of population in Yucatan at the time of colonizers' arrival, historians such as García Bernal (1978) and Cook and Borah (1978) suggest the presence of 230,000–240,000 people. From the arrival of Europeans to the first years of the seventeenth century the native population decreased around 70%. Estimations of population decrease for other Maya regions reached 90% (Lovell and Lutz 1996). The epidemic diseases of the old world, mainly smallpox and measles, for which the Mesoamerican natives had no immunity, were mostly responsible for the brutal decrease in the population. There are relatively clear records that during 1640s the population reached its historical minimum. Population size recovered throughout the eighteenth century and by 1790 had returned to the level pertaining at the time of the Europeans arrival (Farriss 1984).

However, contagious diseases do not fully explain the population losses during the three centuries of the colonial period. Using sociohistorical sources in her classic study *La sociedad maya bajo el dominio colonial* (The Maya Society Under the Colonial Rule), Farriss (1984) describes the occurrence of, at least, five severe famines (1525–1541, 1627–1631, 1650–1653, 1726–1727, 1769–1774) that affected the Maya population. The occurrence of these famines has been explained with reference to a complex combination of droughts, hurricanes, locust plagues, and epidemics. Other famines of lesser intensity were recorded during 1571–1572, 1575–1576, 1604, 1692–1693, 1700, 1730, 1742, 1765–1768, 1795, and 1799.

From the beginning of the Spanish conquest to the mid-eighteenth century, society was run on the basis of a slave-like payment tribute system and forced labor. The Maya were forced to give their harvest surplus to the local governments as a form of tribute payment. Men were also forced to provide "personal services" which included to carry out heavy work such as the construction of churches and roads and providing labor on Spaniards' properties (Bracamonte 2007). From the mid-eighteenth century, the elite social group (government and rich families) changed its system of appropriation of wealth with the establishment of the *hacienda* system. The Maya passed from a forced labor system to a wage labor system; however, the living conditions of Maya people did not improve. Many Maya people were employed in the production of sugar cane, livestock, and, subsequently, the *henequen* agro-industry (Bracamonte 2007).

The presence of epidemics was not limited to sixteenth, seventeenth, and eighteenth centuries. There is evidence of two epidemics of *vibro cholera* coming from Asia and affecting the population of the Yucatan Peninsula in 1833 and 1853 (Peniche Moreno 2016). Although there are no figures for the number of people infected or of deaths due to the disease, estimates suggest that, even considering some variation in the conditions of hygiene and poverty among the native population, in severe untreated cases, cholera killed nearly half of those infected. The disease was present in rural and urban communities. However, it is possible that the affliction was greater in Merida and Campeche given the poor sanitary conditions in the cities during the nineteenth century and because of ease of transmission of infectious diseases in overcrowded areas. Cholera arrived in Yucatan after a year of food shortages caused by a locust plague that affected agricultural production throughout the region. After the loss of crops, there was a drought that worsened the famine among the population. It is clear that cholera affected a population already weakened by hunger.

Famines, as epidemics, in the Yucatan were not limited to the three centuries of the colonial period; there are records of three other famines during the first decade of nineteenth century (1800–1804, 1807, 1809–1810) (Farriss 1984). Also, during the second half of the 19th, the population of the coast experienced an epidemic of cholera (1868), three of smallpox (1870–1875, 1878, 1881) and three of yellow fever (1874–1882, 1888, 1889) (Ancona Ricalde 2017). The domination system and natural disasters impacted profoundly upon the biological status of Maya people for several generations. The Maya had little chance of coping with severe droughts and epidemics due to their poor biological status and their subjugation to the ruling groups.

Between the second half of the nineteenth century and the first half of the twentieth century, the economy of Yucatan state depended on the production of *henequen* fiber (*Agave fourcroydes*), which was exported mostly to the international market. Due to soil characteristics, its production was concentrated in the center and northwest of the Yucatan state. The enormous wealth generated by the production and commercialization of *henequen* remained in the hands of very few families whose profits were increased by government policies that helped to ensure the presence of Maya workers in the large *haciendas* (Wells 1985).

In the context of the *henequen hacienda* system, the Maya experienced miserable living conditions. Some authors, such as Turner (1909), point out that the Maya lived in conditions of slavery. By 1862 there were more than a thousand *haciendas* in Yucatan, the haciendas were not only work centers but also population centers, in some cases with 1000s of inhabitants including the families of Maya laborers (Ransom 2006). Estimations suggest that by the end of nineteenth century, around 100,000–125,000 Maya laborers resided and worked in the *henequen haciendas* (Turner 1909).

The conditions of poverty forced unemployed men in cities such as Merida to enter into debt through loans that could only be paid by physical work on the *haciendas*. In this way landowners were able to force Maya people to relocate to *haciendas* and replace malnourished or sick agricultural workers (Turner 1909). The value of the laborers depended on market conditions and the price of *henequen*. Towards the end of the nineteenth century the price of a Maya peon ranged between 200 and 300 Mexican pesos (USD 10–15), and Maya peons were the property of the hacienda's owner until their debt was covered. A few years later, with the increase of the price of this natural fiber, their value increased to 1500–2000 pesos and during the market crisis in 1930 it fell again to 400 pesos (Katz 1974). The Maya slaves never received a salary, each family received a daily credit of around 25 cents in merchandise that included maize, beans, chili, salt, and blankets. In some other *haciendas* the slaves received daily food rations that were nutritionally insufficient. The daily work in the haciendas used to start very early in the morning and ended when darkness prevented continuing and consisted mainly of cutting and cleaning the henequen leaves, removing its thorns, and cleaning the land. Each slave was assigned a certain number of *henequen* leaves in the day and it was common that women and children provided help to the men to complete their task. The whip and imprisonment in the basement of the main house of the haciendas (some of them now converted to luxury hotels) were the main form of punishment if the slave did not arrive on time, if he did not fulfill the task or did it incorrectly (Turner 1909; Katz 1974).

Not all of the Maya population were settled on the *haciendas*, the rest of population lived in small communities as small-scale farmers, mainly in the south and east of the state. In these contexts, the population was also continuously exposed to droughts, hurricanes, and plagues that maintained the individuals in suboptimal nutritional conditions.

Although the living conditions of contemporary Maya people have visibly improved in recent decades, their current socioeconomic conditions are still disadvantageous. As in other countries, poverty in Mexico is concentrated in minority ethnic groups such as the Maya. In his study *Del pueblo a la urbe. El perfil maya de la blanca Mérida* (From Town to the City. The Maya Profile of the White Merida), Lizama Quijano (2012) describes the adverse conditions experienced by the Maya in terms of income, education, access to health services, household conditions and urban infrastructure. In their recent history, the Maya have contributed significantly to global markets with labor at low cost. However, their work and effort has not been paid fairly (Bracamonte 2007). There have been improvements in their living conditions as the result of government social policies, but these have not been sufficient to break the cycle of poverty.

Moreover, during around three centuries, the Maya were victims of the violence, murder, imprisonment, torture, forced relocations and prohibitions against education and polytheism by the Spaniards as policies of control.

The events that describe the biosocial history of the Maya people in the centuries after the Spanish conquest seem to support the first assumptions of the HTT (Sotero 2006). First, the living conditions were the result of trauma inflicted deliberately and systematically by the colonizers and the political class in power. Although some epidemics came from distant regions, the damage caused to the population and the scarce capacity of response shown by individuals were the result of poor biological conditions imposed by the political system. Second, the trauma was not caused by a single event, in this case the initial contact with the Spaniards. The Maya experienced several famines and a sociopolitical system that enslaved them for decades, which represent chronic traumas. Third, even when we lack precise figures on the number of people who experienced the consequences of the oppressive regime, the sociohistorical evidence allows us to point out that the clear majority of the Maya were victims of the conditions imposed during the colonial period and the subsequent decades, until present day.

5.3 The Intergenerational Influences of Adverse Living Conditions on the Growth and Nutritional Status Among the Yucatec Maya

The data presented in this chapter is taken from an investigation conducted by our research team during 2011–2014 in the cities of Merida and Motul in Yucatan. In this research we aimed to analyze the influence of socioeconomic and intergenerational factors on the growth and nutritional status of Maya families in a sample of children, their mothers and maternal grandmothers.

Merida, the capital city of Yucatan state, was founded in 1542 on the vestiges of a Maya city where, according to historical and archaeological sources, a sizeable and politically important population became established during the Maya post-Classic period (900–1521 AD) (Roys 1957). *T'Ho*, the Maya name of the old city, was inhabited by a small number of families at the time of European arrival. Four hundred and seventy-six years later, the stones that formed the Mayan temples and other buildings can still be seen as they were recut by natives to be used in the construction of Merida's cathedral and houses of conquistadors. Nowadays, Merida is perceived as a modern and safe city; the common visitor will come across a city with numerous shopping malls, cinemas, ice rinks, and avenues with companies offering high-value cars. Merida is a medium-sized city; in 2015 it was inhabited by nearly 900,000 people (INEGI 2016) and its current economic dynamic is based mainly on the supply of goods and services, including health, education, and tourism (Pérez Medina 2010).

The presence of the Maya population in Merida dates from the foundation of this city. Throughout the colonial period the Maya families were relegated to occupy spaces on the periphery of the city. The town center and main avenues of the city could be only used by the Europeans and government employees (Pérez Medina 2000; Fuentes 2005). Even today, the Maya are concentrated in the southern neighborhoods of the city, endowed with less infrastructure and urban services (Pérez Medina 2010). During the second half of the twentieth century Merida showed important changes in its demographic profile. From 1950 to 1960 its population increased from 142,858 to 170,834 inhabitants, a growth that produced an increase in the city's extension from 575 to 3631 hectares. From 1960 to 1970 the population increased by 20% and reached 212,097 inhabitants (Fuentes 2005). The vertiginous growth began in the seventies; in 1980 the population reached 400,142 inhabitants. From 1970 to 2000, the size of city's population increased by more than 350% due to the arrival of Maya people from communities within the state, that is, due to rural-to-urban migration (INEGI 1971, 2001). This migratory flow has been explained by several socioeconomic factors, the most important being the adverse conditions experienced in neighboring rural zones, the fall of the *henequen* agro-industry and reduced financial support from the government to agricultural activities. The estimation of the size of the Maya population residing in Merida depends on the criteria used. Lizama Quijano (2012) reports that by 2005 65,000 people (11% of the Merida's population) were Maya speakers.

Motul is located in the north-central region of the state, a former area of henequen production, and is approximately 40 km from Merida. The Maya population have settled in this city for centuries. During the colonial period Motul experienced a very slow rate of growth and its population depended on its ancestral agricultural system. During the boom years of henequen there was much greater demographic and economic growth. However, after the fall of the henequen agro-industry (1970–1980) and its associated emigration, the economic activities in the region diversified towards small-scale livestock holdings and agricultural production of citrus, and more recently manufacturing operations and gastronomic tourism. Currently, Motul is a small-sized city and by 2015 was inhabited by about 36,000 people, 25% of whom were Maya speakers (INEGI 2016).

Despite their differences in size and economic dynamic, we believe Maya people from both cities share historical, economic, and sociocultural characteristics. In both locations, the Maya bring with them the colonial heritage of poverty and a present characterized by accelerated changes in their patterns of food consumption and physical activity.

The studied sample consisted of 260 child-mother-grandmother triads, 180 of who were from Merida and 80 from Motul. Participants were recruited in local primary schools attended by participant children. We selected schools located in zones of the city where Maya speakers and people with the lowest income were concentrated. In each school children were selected if they were between 6 and 8 years old, if their mother and maternal grandmother were present in the household and when maternal and paternal Maya surnames were present in the three generations. During fieldwork we collected several anthropometric measures and applied bioelectric impedance analyses to estimate body composition in members of three generations. Many grandmothers lived outside Merida, in communities from different regions of the state; we visited them accompanied by their daughters (children's mothers). In some cases, all three generations lived in the same household. During home visits we obtained information about the socioeconomic conditions of children's families and we interviewed mothers and grandmothers to know their living conditions during their childhood.

Briefly, eighty-eight percent of households were inhabited by nuclear families, consisted of parents (fathers and mothers) and offspring, all other families were headed by women. The average family size in the sample was 5.1 and the mean number of children was 2.4 (SD = 1.1), while around 42% of families had three or more children. The age of mothers and grandmothers was on average 33 (SD = 5.5) and 60 (SD = 8.5) years old. The median of total years of maternal education was 9, which on average correspond to secondary level. The level of father's education was very similar to that shown by mothers. Most grandmothers only reached the second year of primary education, although in many Yucatecan small towns primary schools offered only 3 or 4 years. Fifty-two percent of mothers were housewives with external employment of income generating activities, sixteen percent were housewives and small-scale merchants, and the remaining thirty-two percent were unskilled workers and owners of small business. Eighty-two percent of fathers were low-qualified employees in different businesses, fifteen percent were traders and self-employed technical workers, and only two percent owned small businesses with a workforce. Forty-eight percent of families were classified as *overcrowded* based on number of people per rooms of the house.

In the following discussion we firstly describe the growth and nutritional status of participants and then we present two examples of how conditions experienced by mothers and grandmothers during their growth and development shape the biological status of the youngest generations. Some of the results presented in Sects. 5.5 and 5.6 have been published by members of our research team in Azcorra et al. (2015, 2016a, b).

5.4 Growth and Nutritional Status of Children, Mothers, and Grandmothers

In general, participants from Merida and Motul showed similar anthropometric characteristics; therefore, the data presented in this chapter derives from the combined samples of children, mothers, and grandmothers from the two cities.

As reported in other social groups experiencing sociocultural and economic transitions, the Maya children and adults of our study show the paradoxical coexistence of deficit in linear growth and excess body weight. This is called the *nutritional dual burden* (NDB). In children this trend is clear in both sexes although boys seem to be slightly more affected in height and girls show greater excess in body mass index

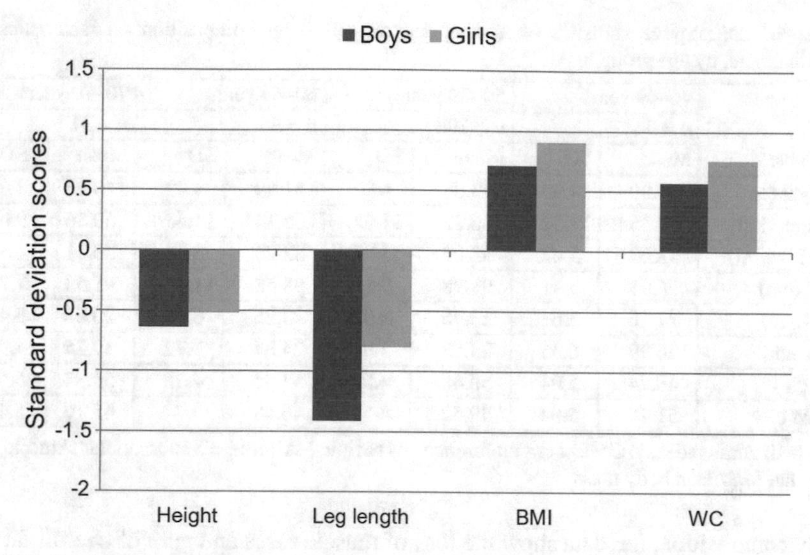

Fig. 5.1 Standard deviation-scores for height, leg length, body mass index (BMI), and waist circumference (WC), by sex group

Table 5.1 Nutritional status in children and mothers according to their height, body mass index (BMI), and waist circumference (WC)

	Stunting	Height below 150 cm	Exceeded BMI	Exceeded WC
Children (%)	12	–	37	32
Mothers (%)	–	65	86	55

Stunting: height-for-age below 5th percentile; exceeded BMI and WC: in children BMI-for-age and WC-for-age above 85th percentile (Frisancho 2008) and mothers BMI above 25 kg/m^2 as suggested by the World Health Organization and waist circumference above 88 cm as suggested by the National Cholesterol Education Program (NCEP) Adult Treatment Panel III (ATP III)

(BMI) and waist circumference (WC) (Fig. 5.1). Eleven percent of children met the criteria of low height-for-age, defined as a standard deviation score (*z*-score) < −1.6 for the age and sex compared to an international growth reference. In both boys and girls, linear growth deficit was more pronounced in leg length (LL = height − sitting height) than in sitting height (28% vs. 6%). On the other extreme of malnutrition, more than thirty percent of children were classified as having high BMI and WC-for-age. Mothers were of low stature (147.80 cm, SD = 4.9) and there was a high prevalence of exceeded BMI and WC (Table 5.1). According to their BMI, 42% of mothers met the criteria for overweight (25–29 kg/m^2) and 44% for obesity (above 30 kg/m^2).

The analysis and interpretation of anthropometric characteristics of grandmothers is challenging due to the anatomical and physiological changes observed in old age (Elia 2001; Kennedy et al. 2008). The age of grandmothers in this study ranged from 40 to 84 years old. When participants were grouped according to their age, we observed that height, weight and BMI decreased with age (Table 5.2). In terms of

Table 5.2 Descriptive statistics of anthropometric and body composition characteristics of grandmothers, by age group

	40–49 years		50–59 years		60–69 years		70–79 years	
	n = 34		*n* = 100		*n* = 95		*n* = 31	
Variable	Mean	SD	Mean	SD	Mean	SD	Mean	SD
Height (cm)	144.99	4.95	143.66	4.89	141.99	4.42	140.16	4.67
Weight (kg)	69.45	11.82	66.22	11.82	65.14	11.63	60.36	9.07
BMI (kg/m²)	33.03	5.42	32.02	5.07	32.23	5.14	30.71	4.38
WC (cm)	96.95	9.41	97.98	9.89	98.58	11.42	96.50	7.75
TS (mm)	27.16	5.67	23.96	6.02	23.26	6.50	20.61	5.49
SS (mm)	26.99	6.36	23.58	5.98	24.87	7.72	19.25	4.90
BF (%)	48.74	5.04	50.48	4.58	53.41	4.75	54.84	5.42
LBM (%)	51.25	5.04	49.52	4.59	46.58	4.74	45.16	5.42

BMI body mass index, *WC* waist circumference, *TS* tricipital skinfold, *SS* subscapular skinfold, *BF* body fat, *LBM* lean body mass

body composition, the data show the loss of muscle mass and gain of overall fat tissue as age increases. Ninety-two percent of grandmothers had a height below 150 cm, which corresponds to the nearest whole centimeter of −1.65 *z*-score and fifth percentile of the Comprehensive Growth Reference of Frisancho (2008), based on the NHANES III survey (McDowell et al. 2009). However, this finding should be treated with caution due to reduction in height in individuals after the age of 40 years reported in longitudinal studies (Minaker 2011). BMI also has limitations as a measure of adiposity in geriatric subjects (McTigue et al. 2006; Donini et al. 2012); some evidence suggests that older people show a greater proportion of fat mass than younger people having the same BMI (Baumgarter et al. 1995). Waist circumference seems to be a stronger predictor of body fat and health-risk factors than BMI in old subjects (McTigue et al. 2006; Donini et al. 2012). In fact, there is some agreement in the use of WC as a measure of abdominal obesity when studying the elderly (Gutiérrez-Fisac et al. 2004; Alemán-Mateo et al. 2009; Wyka et al. 2010; Rosas-Carrasco et al. 2012). With respect to this measure, 84% of grandmothers had a WC above 88 cm, indicating an increased risk for abdominal obesity.

Sixty-three percent of grandmothers reported having been diagnosed with some sort of chronic disease. Type 2 diabetes mellitus and hypertension were the diseases with the highest percentages (27% and 34% respectively). Hypercholesterolemia and hypertriglyceridemia (24%) were also diseases with high percentages in grandmothers.

The coexistence of chronic undernutrition (stunting in children and short stature in adults) and overnutrition in the same individual represents the worst scenario of nutritional dual burden (NDB); both conditions increase the risk of chronic diseases in the short and long term. Overnutrition in NDB can be established according to BMI or WC of individuals. In our sample only one child (0.4%) showed the coexistence of low height-for-age and exceeded BMI or WC. However, the prevalence of this condition was high in mothers: 59% were shorter than 150 cm and had a BMI above 25 kg/m² and 55% were shorter than 150 cm and had a WC greater than 88 cm.

Mother-child correlations were slightly higher in LL ($r = 0.288$) than in height ($r = 0.247$) and sitting height ($r = 0.193$). In terms of body mass, correlations in weight and BMI were very similar ($r = 0.280$ and $r = 0.289$, respectively) and relatively lower in WC ($r = 0.231$). A brief inspection of the nutritional status of mother-child dyads shows that stunting in children and short stature in mothers was present in twelve percent of mother-child pairs and in thirty-three percent of dyads both members presented excessive BMI. The coexistence of maternal overnutrition (BMI above 25 kg/m^2) and stunting in children (NDB at household level) was present in ten percent of the mother-child pairs.

5.5 The Effect of Living Conditions Experienced by Mothers and Grandmothers During Their Childhood on Grandchildren's Linear Growth

As we have shown, chronic growth deficits in height-for-age among Maya children are an unresolved problem in Yucatan. Several nongenetic factors may contribute to the presence of low height-for-age in children of populations experiencing disadvantaged living conditions. These factors include maternal short stature (as result of their own pre and postnatal growth), low birth weight, inadequate early feeding practices, and a high frequency of infectious diseases during the first years of life. Is the growth of the Maya children also shaped by intergenerational factors? To answer this question, we analyzed whether living conditions experienced by mothers and grandmothers during their childhood are associated with children's height and LL. We focused not only on height because as we showed in the previous section children's LL seems to be a more sensitive indicator of ecological conditions than height.

The biosocial backgrounds of mothers and grandmothers were analyzed through their current height and LL measures (z-score values) and socioeconomic parameters that describe the household conditions during their own growth. As in other populations with high levels of material poverty, the identification of useful socioeconomic parameters among the Maya is challenging because of the high degree of homogeneity in their living conditions. Interviews allowed us to identify that family size (number of people in the household) and the materials of construction of the house were two sensitive indicators of living conditions during grandmothers' childhood. We constructed a home quality index defined by three types of grandmother's dwellings: (1) those where perishable materials (dirt floor and cardboard or metal sheets for walls and ceilings) constitute the majority of construction, (2) those with traditional building materials, which include dirt floor, palm leaf ceilings, and daub walls, and (3) those where durable materials (cement floor and cement blocks in walls and ceilings) were mostly used during construction. For us, grandmothers who lived in a house built with durable materials belonged to families who experienced relatively better living conditions. In the case of children's mothers, we identified that her father's employment stability was the most important variable to describe their childhood living conditions. Most of the mother's fathers worked, upon their arrival in Merida, as masons and laborers in construction activities. We suggest the more frequent the father's job loss, the lower the childhood living conditions of the mother. The physical and emotional stress on children from parental employment insecurity, as well as housing and food insecurity due to poverty, is becoming an increasingly important topic in the literature on the causes of both stunting and overweight. Chapter 3 in this book discusses some of this literature.

To investigate the intergenerational influences of biosocial background of mothers and grandmothers on children's growth, we analyzed height and LL (both in *z*-score values) using adjusted multiple linear regression models. In each model, we used maternal height or LL, paternal job loss during childhood, grandmother's height < 150 cm, home quality index, and family size as predictor variables.

As expected, maternal height positively predicted the growth of children. Grand-maternal short height was not associated with outcome variables. After controlling for the mother's and grandmother's height, our analyses showed that, compared to the category of perishable materials of construction, grandmothers' better-quality dwellings were associated with higher values of children's height and LL. Contrary to what we expected, greater grand-maternal family size during childhood was positively associated with children's leg length. Our analyses also showed that paternal job loss during maternal childhood did not influence the children's linear growth.

How can these results be interpreted? Firstly, our results show that the quality of housing construction seems to be a useful indicator of the living conditions of mothers and grandmothers during their childhood. A common feature among the Maya who reside in urban and semi-urban areas is that they live in housing literally built by the family itself. As time passes and the economic conditions of the family improve, the house is upgraded with permanent materials replacing perishable ones (García Gómez 2007). The home is the first environment experienced during infancy and childhood; in this space children eat, sleep, play, and live with their parents, other caregivers, and siblings. Therefore, characteristics and quality of this space are important in the health and emotional state of individuals during their first years of life.

Second, our findings suggest that poor ecological conditions experienced by recent maternal ancestors are associated with adverse growth trajectories in descendants. We showed in the previous section that stronger associations in maternal-child dyads were found in LL; in this analysis we found that grandmother's family size was positively associated with LL, but not with height. Altogether, these results support the hypothesis that LL relative to total height is a more sensitive indicator of the quality of the environment (Leitch 1951; Bogin and Varela-Silva 2010) and intergenerational influences for growth status than is sitting height or total height.

Grand-maternal family size was another intergenerational predictor of children's linear growth. Our results showed that grandmothers who lived in larger families during their childhood predicted better growth status for their grandchildren. This finding makes sense in the context of the demographic characteristics of subsistence, or partially self-subsistent, social groups. In general, rural societies dependent on agricultural activities for self-consumption take advantage of the family size to increase their production of staples (Kramer 2005). This finding allows us to suggest that those grandmothers who belong to bigger families in their childhood experienced better ecological conditions during growth, which in turn has a positive intergenerational influence on the growth of their grandchildren.

5.6 Short Maternal Height and Child's Adiposity

Body composition is a phenotypic trait highly susceptible to variations in ecological conditions. Observational and prospective studies have shown the impact of early nutrition during gestation (Law et al. 1992; Fall et al. 1995; Curhan et al. 1996; Barker 1998; Ravelli et al. 1999; Martorell et al. 2001) and the first years of postnatal life (Schroeder et al. 1999) on the development of *adiposity* in adulthood. Stunted juveniles and early adolescents have been shown to have greater central adiposity than normal-height children (Bogin and Sullivan 1986; Hoffman et al. 2007). Hoffman et al. (2007) found that stunted children had more truncal fat mass, both absolutely (kg) and proportionally (%) and increased gain of truncal fat mass. Additionally, it has been found that stunted boys accumulate more body fat and gain less lean mass (kg and %) than non-stunted boys when they reach puberty (Martins et al. 2004).

These findings have been interpreted as an evidence for programming, a concept coined by Lucas (1991) and defined as a *process whereby a stimulus or insult at a critical period of development has lasting or lifelong significance…* on phenotype of the organism. However, the term programming may imply that once the program is stablished, during a critical stage of development, the resulting trajectory of organism has very little flexibility in the context of highly dynamic ecosystems (Gluckman and Hanson 2004). Alternatively, Bateson proposes the term induction to refer to a range of processes influencing the developing organism (Bateson 2001).

Some physiological processes, most occurring at a metabolic level, have been proposed to explain how early malnutrition increases the risk of higher adiposity in later life: (1) impaired fat oxidation (Frisancho 2003; Hoffman et al. 2000b), (2) alterations in the metabolism of cortisol and insulin (Sawaya et al. 2004), and (3) changes in appetite regulation and increased susceptibility to high fat diets (Hoffman et al. 2000a; Sawaya et al. 1998). Chapter 3 in this book discusses some of this literature.

According to the *Intergenerational Influences Hypothesis*, such metabolic adjustments can be transmitted from one generation to another. To test this hypothesis, we analyzed the association between maternal height and children's body fatness. We recognize that maternal height is an imperfect measure for nutritional history since it is the product of a complex interaction between environment and parent's genetics component. However, as has been shown by several studies, the parent-child correlations in linear growth tend to be low in populations living under disadvantaged circumstances, where other environmental factors tend to induce more variation in phenotype (Mueller 1976). In this study we hypothesized that children of shorter mothers will show higher levels of adiposity in comparison to children of taller mothers.

We used three parameters of children's adiposity: fat mass index (FMI = kg/m^2), WC and sum of triceps, subscapular and suprailiac skinfolds (SumSkf). FMI and WC were used as measures for total body fat and central adiposity, respectively. SumSkf was considered to be an estimate of subcutaneous fat. FMI was estimated from a bioelectric impedance analysis using the equation developed by Ramírez et al. (2012) based on a sample of Mexican school children from several ethnicities, including the Maya. The influence of maternal height on offspring adiposity was analyzed adjusting for the effect of children's birth weight and socioeconomic conditions in the child home environment. We built a socioeconomic index (SEI) based on mother's education, father's occupation, and household crowding status. The mother's education was classified according to three categories: 1 = junior high school or less, 2 = high school, and 3 = university, while the father's occupation was classified as 1 = employee, 2 = self-employed, and 3 = employer. Households were categorized according to their crowding status in 0 = crowded and 1 = not crowded. The SEI was calculated by adding socioeconomic variable response categories. Separate regression models were performed for boys and girls since intergenerational influences can produce different effects according to offspring sex. Figure 5.2 shows the relationships between maternal height and children's adiposity parameters adjusted for children's height, BW and household SEI. The results obtained showed differences by children's sex; in both boys and girls maternal height was inversely associated with children's FMI, WC, and SumSkf. The association was significant statistically for boys but not for girls. We found no significant differences in SEI, maternal height, BMI, children's body fatness, and birth weight between boys and girls.

These findings support partially our hypothesis that short maternal stature may induce higher levels of adiposity in offspring. We argue that the stature of children's mothers in this study is the result of unfavorable nutritional, social, economic, and political conditions experienced by women during their own prenatal growth, childhood, and adolescence. How can intergenerational influences detected only in boys be interpreted? From a developmental programming perspective, evidence coming from animal models shows that male offspring tend to be more vulnerable to prenatal and perinatal adversities, such as deficiencies in maternal diet during pregnancy (Kwong et al. 2000; Woods et al. 2005; Sinclair et al. 2007; Micke et al. 2010) and lactation (Zambrano et al. 2005). In this line of analysis some evidence suggests that female sheep (Wilcoxon et al. 2003) and rat (Vickers et al. 2011) conceptuses are more protected in the presence of suboptimal intrauterine conditions through placental adaptations. From an evolutionary point of view, our results might be interpreted as evidence of how resources are preferentially directed towards female offspring in adverse contexts to increase reproductive success in the group.

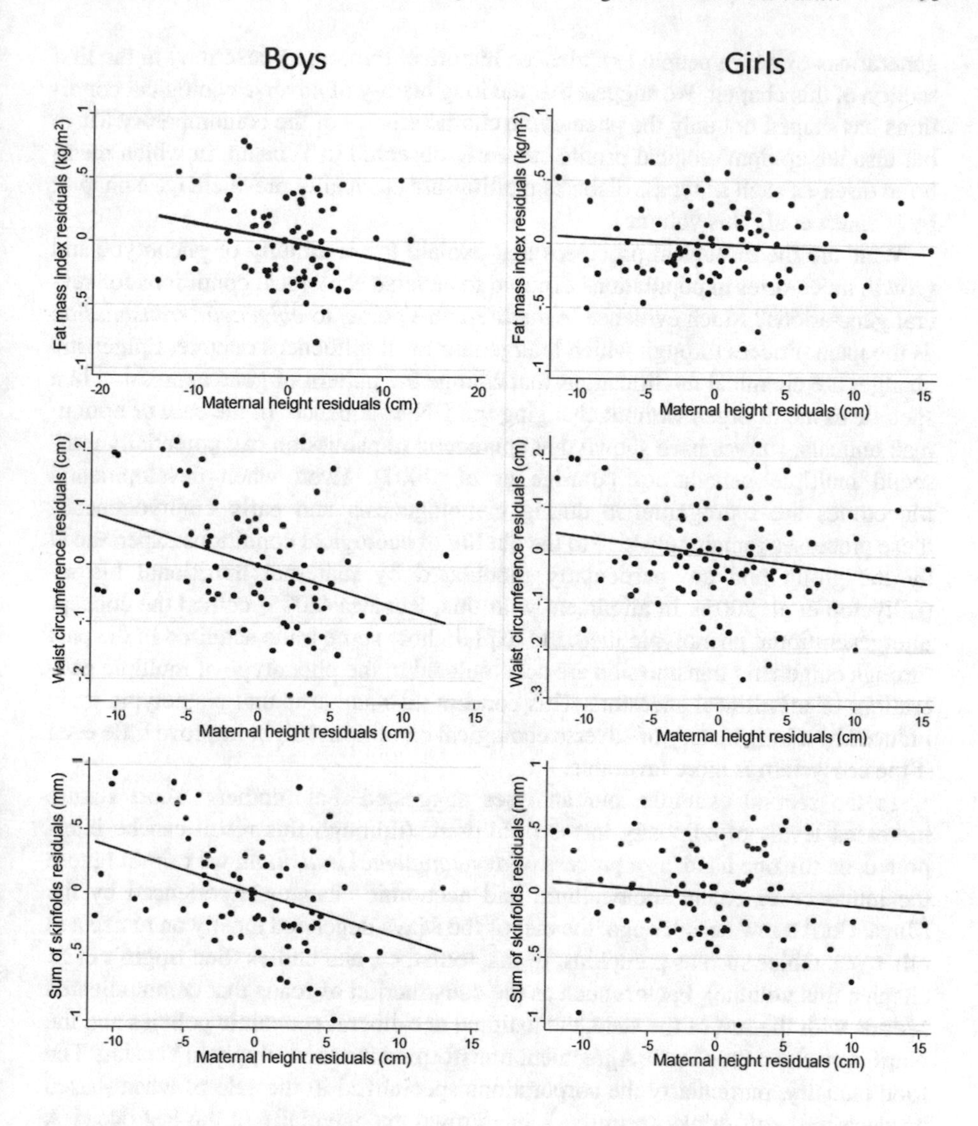

Fig. 5.2 Scatterplots and regression of children's FMI, WC, and SumSkf on maternal height by children's sex. Models adjusted for children's age, height and birth weight and household socio-economic index

5.7 Final Commentaries and Conclusions

The data and analyses presented in this chapter allow us to interpret the phenotype and nutritional status of the contemporary Maya as the historical result of the social and political conditions experienced in the last four centuries. Even considering variations in life span and reproductive patterns, it is possible that at least 15 generations of Maya people experienced historical trauma, as described in the first section of this chapter. We suggest that the long history of adverse ecological conditions has shaped not only the phenotypic characteristics of the contemporary Maya but also the epidemiological profile currently observed in Yucatan, in which metabolic diseases such as type 2 diabetes mellitus are especially prevalent (See chapter by Méndez et al., this volume).

What are the biological processes that explain the continuity of phenotype and growth trajectories in populations exposed to adverse ecological conditions for several generations? Much evidence in the literature points to *epigenetic transmission* as the main process through which intergenerational influences operate. Epigenetic changes are chemical modifications that change the pattern of gene expression in a specific tissue or organ without changing the DNA sequence. In the case of nonhuman animals, studies have shown that epigenetic transmission can potentially transcend multiple generations (Burdge et al. 2007). Even when developmental trajectories are reprogrammed during gametogenesis and early embryogenesis, these processes remain sensitive to the quality of ecological conditions experienced by the group and are particularly modulated by maternal nutritional history (Lillycrop et al. 2005). In alignment with this, Kuzawa (2005) coined the concept intergenerational phenotypic inertia to explain how some traits acquired in the past through epigenetic transmission are accumulated in the phenotype of multiple generations of matrilineal ancestors. This concept also suggests that phenotypic traits induced by a long history of adverse ecological conditions might improve little even if the ecosystem is more favorable.

In the second example, our analyses suggested that mothers' short stature increased levels of adiposity in their children. Although this result can be interpreted, on the one hand, as a process of *developmental induction*, we cannot ignore the influence of recent sociocultural and economic changes experienced by the Maya. Until a few decades ago, the diet of the Maya depended mostly on maize and other vegetables such as pumpkins, beans, tomatoes, and chilies (See Bogin's et al. chapter, this volume). Factors such as the construction of roads that communicated Merida with the rest of the state, the national neo-liberal economic policies and the North American Free Trade Agreement transformed the food supply in Yucatan. The food industry, particularly the corporations specialized in the sale of wheat-based products and soft drinks (*refrescos*), has grown exponentially in the last decades. While this process is observed in several parts of Mexico, Yucatan shows a particular high trend of consumption. The Maya have experienced an abrupt change, called "coca colonization" by Leatherman and Goodman (2005), in their diet and it is possible that we have not yet observed all the consequences of this pattern. Other factors such as rural-urban migration, the decrease in the family size, and the modest increase in income have intensified the exposure to a diet which is energy-dense, but nutritionally poor (See chapters of Bogin et al. and Goodman and Leatherman, this volume).

Several research questions arose during this study. Some of these questions could help to understand the current biological status of contemporary Maya people and their determinants. Our study was designed to investigate the intergenerational influences on the postnatal growth of the younger generation. Our understanding of the factors associated with prenatal growth among the Maya and the influence of socioecological factors during pregnancy is still limited (Azcorra et al. 2016a, b; Azcorra and Méndez 2018) and represents a priority for future research.

So far, in recent decades the Maya have chosen to migrate to the Mexican Caribbean, Merida city or the United States, to escape the poverty in which they grow up. These strategies partially alleviate the socioeconomic adversity of the Maya families but intensifies the globalization processes that increase the risks of chronic degenerative diseases due to dietary unbalances and sedentary lifestyles. It is clear that official social welfare programs implemented in Mexico over the last few decades have failed to reduce the levels of poverty and inequality among the minority ethnic groups including the Maya. We believe that Maya people will not overcome the traces of historical trauma if structural changes in the economic and political systems are not carried out in the short term. Meanwhile we will continue to be in debt to the Maya, an historical debt for the technologies, art, mathematics, and science that they developed before the Conquest and our current debt of gratitude for their participation and guidance of our research.

References

Alemán-Mateo H, Lee SY, Javed F et al (2009) Elderly Mexicans have less muscle and greater total and truncal fat compared to African-Americans and Caucasians with the same BMI. J Nutr Health Aging 13:919–923

Ancona Ricalde EM (2017) Conformación de la población en Progreso, Yucatán: Familias, redes sociales y laborales de extranjeros residentes en el puerto, 1870–1910. PhD Dissertation. Centro de Investigación y de Estudios Superiores en Antropología Social

Azcorra H, Dickinson F, Bogin B et al (2015) Intergenerational influences on the growth of Maya Children: the effect of living conditions experienced by mothers and maternal grandmothers during their childhood. Am J Hum Biol 27:494–500

Azcorra H, Dickinson F, Datta Banik S (2016a) Maternal height and its relationship to offspring birth weight and adiposity in 6- to 10-year-old Maya children from poor neighbourhoods in Merida, Yucatan. Am J Phys Anthropol 161:571–571

Azcorra H, Méndez N (2018) The influence of maternal height on offspring's birth weight in Merida, Mexico. Am J Hum Biol 30(6):e23162

Azcorra H, Vázquez-Vazquez A, Mendez N et al (2016b) Maternal Maya ancestry and birth weight in Yucatan, Mexico. Am J Hum Biol 28:436–439

Barker DJP (1998) Mothers, babies and health in later life. Churchill Livingstone, Edinburgh

Bateson P (2001) Fetal experience and good adult design. Int J Epidemiol 30:928–934

Baumgarter RN, Heymsfield SB, Roche AF (1995) Human body composition and the epidemiology of chronic disease. Obes Res 3:73–95

Bogin B, Sullivan T (1986) Socioeconomic status, sex, age, and ethnicity as determinants of body fat distribution for guatemalan children. Am J Phys Anthropol 69:527–535

Bogin B, Varela-Silva MI (2010) Leg length, body proportion, and health: a review with a note on beauty. Int J Environ Res Public Health 7:1047–1075

Bracamonte P (2007) Una deuda histórica. Ensayo sobre las causas de pobreza secular de los mayas yucatecos. Centro de Investigaciones y de Estudios Superiores en Antropología Social – Miguel Ángel Porrúa, México, DF

Burdge GC, Slater-Jefferies J, Torrens C et al (2007) Dietary protein restriction of pregnant rats in the F_0 generation induces altered methylation of hepatic gene promoters in the adult male offspring in the F_1 and F_2 generations. Br J Nutr 97:435–439
Cook S, Borah W (1978) Essays in population history: Mexico and the Caribean, vol 3 Vol. University of California Press, Berkeley
Curhan GC, Chertow GM, Willett WC et al (1996) Birth weight and adult hypertension and obesity in women. Circulation 94:1310–1315
Donini LM, Savina C, Gennaro E et al (2012) A systematic review of the literature concerning the relationship between obesity and mortality in the elderly. J Nutr Health Aging 16:89–98
Elia M (2001) Obesity in the elderly. Obes Res 9:244S–248S
Fall CHD, Osmond C, Barker DJP et al (1995) Fetal and infant growth and cardiovascular risk factors in women. BMJ 310:428–432
Farriss NM (1984) Maya society under colonial rule. The collective enterprise of survival. Princeton University Press, Princeton, NJ
Frisancho AR (2003) Reduced rate of fat oxidation: a metabolic pathway to obesity in the developing nations. Am J Hum Biol 15:522–532
Frisancho AR (2008) Anthropometric standards: an interactive nutritional reference of body size and body composition for children and adults. The University of Michigan Press, Ann Arbor, MI
Fuentes JH (2005) Espacios, actores, prácticas e imaginarios urbanos en Mérida, Yucatán, México. Universidad Autónoma de Yucatán, México
García Bernal MC (1978) Población y encomienda en Yucatán bajo las Austrias. Consejo Superior de Investigaciones Científicas. Sevilla, España
García Gómez C (2007) Producción social de la vivienda popular. In: García C, Bolio E (eds) Autoproducción de vivienda en Mérida. Zonas urbanas en proceso de consolidación. Universidad Autónoma de Yucatán, Mérida, pp 111–133
Gluckman PD, Hanson M (2004) The fetal matrix: evolution, development and disease. Cambridge University Press, Cambridge
Gracey M, King M (2009) Indigenous health part 1: determinants and disease patterns. Lancet 374:65–75
Gutiérrez-Fisac JL, López E, Banegas JR et al (2004) Prevalence of overweight and obesity in elderly people in Spain. Obes Res 12:710–715
Hoffman DJ, Martins PA, Roberts SB et al (2007) Body fat distribution in stunted compared with normal-height children from the shantytowns of Sao Paulo, Brazil. Nutrition 23:640–646
Hoffman DJ, Roberts SB, Verreschi I et al (2000a) Regulation of energy intake may be impaired in nutritionally stunted children from the shantytowns of Sao Paulo, Brazil. J Nutr 130:2265–2270
Hoffman DJ, Sawaya AL, Verreschi I et al (2000b) Why are nutritionally stunted children at increased risk of obesity? Studies of metabolic rate and fat oxidation in shantytown children from Sao Paulo, Brazil. Am J Clin Nutr 72:702–707
INEGI (1971) IX Censo General de población 1970. Aguascalientes: Instituto Nacional Estadística, Geografía e Informática. http://www.inegi.org.mx/est/conteidos/proyectos/ccpv/cpv1970/default.aspx. Accessed 15 Jul 2018
INEGI (2001) XII Censo General de Población y Vivienda 2000. Aguascalientes: Instituto Nacional de Estadística, Geografía e Informática. http://www.inegi.org.mx/est/contenidos/Proyectos/ccpv/cpv2000/default.aspx. Accessed 15 Jul 2018
INEGI (2016) Encuesta Intercensal 2015. Panorama sociodemográfico de Yucatán. Instituto Nacional dc Estadística y Geografía, México. http://www3.inegi.org.mx/sistemas/Panorama2015/Web/Contenido.aspx#Yucatán31000. Accessed 15 July 2018
Katz F (1974) Labor conditions on haciendas in Porfirian Mexico: some trends and tendencies. Hisp Am Hist Rev 54:1–47
Kelley JCH (1991) Contrast in somatic variables among traditional and modernized Maya females. Int J Anthropol 6:159–177
Kennedy RL, Malabu U, Kazi M et al (2008) Management of obesity in the elderly: too much and too late? J Nutr Health Aging 12:608–621

Kramer KL (2005) Maya children: helpers at the farm. Harvard University Press, Cambridge
Kuzawa CW (2005) Fetal origins of developmental plasticity: are fetal cues reliable predictors of future nutritional environments? Am J Hum Biol 17:5–21
Kuzawa CW, Gravlee CC (2016) Beyond genetic race: biocultural insights into the causes of racial health disparities. In: Zuckerman MK, Martin DL (eds) New directions in biocultural anthropology. Wiley Blackwell, Hoboken, NJ, pp 89–105
Kwong WY, Wild AE, Roberts P et al (2000) Maternal undernutrition during preimplantation period of rat development causes blastocyst abnormalities and programming of postnatal hypertension. Development 127:4195–4202
Law CM, Barker DJP, Osmond C (1992) Early growth and abdominal fatness in adult life. J Epidemiol Community Health 46:184–186
Leatherman TL, Goodman AH (2005) Coca-colonization of diets in the Yucatan. Soc Sci Med 61:833–846
Leitch I (1951) Growth and health. Br J Nutr 5:142–151
Lillycrop KA, Phillips ES, Jackson AA et al (2005) Dietary protein restriction of pregnant rats induces and folic acid supplementation prevents epigenetic modification of hepatic gene expression in the offspring. J Nutr 135:1382–1386
Lizama Quijano JJ (2012) Del pueblo a la urbe. El perfil maya de la blanca Mérida. Centro de Investigaciones y de Estudios Superiores en Antropología Social, México, DF
Lovell WG, Lutz CH (1996) "A Dark Obverse": Maya Survival in Guatemala: 1520–1994. Geograph Rev 86:398–407
Lucas A (1991) Programming by early nutrition in man. Ciba Found Symp 156:38–50
Martins PA, Hoffman DJ, Fernandes MTB et al (2004) Stunted children gain less lean body mass and more fat mass than their non-stunted counterparts: a prospective study. Br J Nutr 92:819–825
Martorell R, Stein AD, Schroeder DG (2001) Early nutrition and later adiposity. J Nutr 131:874S–880S
McCollough JM (1982) Secular trend for stature in adult male Yucatec Maya to 1968. Am J Phys Anthropol 58:221–225
McDowell MA, Fryar CD, Ogden CL (2009) Anthropometric reference data for children and adults: United States, 1988–1994. Vital Health Stat 11(249):1–68
McTigue KM, Hess R, Ziouras J (2006) Obesity in older adults: a systematic review of the evidence for diagnosis and treatment. Obesity 14:1485–1497
Micke GC, Sullivan TM, Gatford KL et al (2010) Nutrient intake in the bovine during early midgestation causes sex-specific changes in progeny plasma IGF-I, liveweight, height and carcass traits. Anim Reprod Sci 121:208–217
Minaker KL (2011) Common clinical sequelae of aging. In: Goldman L, Schafer AI (eds) Cecil medicine, 24th edn. Elsevier Saunders, Philadelphia, PA
Mueller WH (1976) Parent child correlations for stature and weight among school aged children: a review of 24 studies. Hum Biol 48:379–397
Peniche Moreno P (2016) El cólera morbus en Yucatán. Medicina y salud pública 1833–1853. Centro de Investigaciones y Estudios Superiores en Antropología Social – Miguel Ángel Porrúa, Ciudad de México
Pérez Medina S (2000) Segregación y desequilibrios urbanos en Mérida. Cuadernos de Arquitectura de Yucatán. Universidad Autónoma de Yucatán, México, DF
Pérez Medina S (2010) Segregación, recreación y calidad de vida en Mérida. Universidad Nacional Autónoma de México, México, DF
Ramírez E, Valencia ME, Bourges H et al (2012) Body composition prediction equations based on deuterium oxide dilution method in Mexican children: a national study. Eur J Clin Nutr 66:1099–1103
Ransom M (2006) La era del henequén. In: Barrera Rubio A, Gubler R (eds) Los mayas de ayer y hoy. Memorias del Primer Congreso Internacional de Cultura Maya, Consejo Nacional para la Cultura y las Artes, vol II, 1st edn. Instituto Nacional de Antropología e Historia, México, DF, pp 998–1007

Ravelli ACJ, van der Meulen JHP, Osmond C et al (1999) Obesity at the age of 50 y in men and women exposed to famine prenatally. Am J Clin Nutr 70:811–816
Rosas-Carrasco O, Juárez-Cedillo T, Ruiz-Arregui L et al (2012) Overweight and obesity as markers for the evaluation of disease risk in older adults. J Nutr Health Aging 16:14–20
Roys RL (1957) The political geography of the Yucatan Maya. Carnegie Institution of Washington, Washington, DC
Sawaya AL, Grillo LP, Verreschi L et al (1998) Mild stunting is associated with higher susceptibility to the effects of high fat diets: studies in a shantytown population in Sao Paulo, Brazil. J Nutr 128:415S–420S
Sawaya AL, Martins PA, Grillo LP et al (2004) Long-term effects of early malnutrition on body weight regulation. Nutr Rev 62:127S–133S
Schroeder DG, Martorell R, Flores R (1999) Infant and child growth and fatness and fat distribution in Guatemalan adults. Am J Epidemiol 149:177–185
Sinclair KD, Allegrucci C, Singh R et al (2007) DNA methylation, insulin resistance, and blood pressure in offspring determined by maternal periconceptional B Vitamin and methionine status. PNAS 104:19351–19356
Siniarska A, Wolanski N (1999) Living conditions and growth of Maya Indian and non-Maya boys from Yucatan in 1993 in comparison with other studies. Int J Anthropol 14:259–288
Sotero MM (2006) A conceptual model of historical trauma: implications for public health practice and research. J Health Dispar Res Pract 1:93–108
Stephens C, Nettleton C, Porter J et al (2005) Indigenous peoples' health-why are they behind everyone, everywhere? Lancet 366:10–13
Turner JK (1909) México bárbaro. Quinto Sol, México, DF
Varela-Silva MI, Azcorra H, Dickinson F et al (2009) Influence of maternal stature, pregnancy age, and infant birth weight on growth during childhood in Yucatan, Mexico: a test of intergenerational effects hypothesis. Am J Hum Biol 21:657–663
Varela-Silva MI, Dickinson F, Wilson H et al (2012) The nutritional dual burden in developing countries – how is it assessed and what are the health implications? Coll Antropol 36:39–45
Vickers MH, Clayton ZE, Yap C et al (2011) Maternal fructose intake during pregnancy and lactation alters placental growth and leads to sex-specific changes in fetal and neonatal endocrine function. Endocrinology 152:1378–1387
Walters K, Simoni J (2002) Reconceptualizing native women's health: An "Indigenist" stress-coping model. Am J Public Health 93:520–524
Wells A (1985) Yucatan's gilded age: haciendas, henequen and international harvester, 1860–1915. University of New Mexico Press, Albuquerque
Whitbeck LB, Adams GW, Hoyt DR et al (2004) Conceptualizing and measuring historical trauma among American Indian people. Am J Com Psychol 33:119–130
Wilcoxon JS, Schwartz J, Aird F et al (2003) Sexually dimorphic effects of maternal alcohol intake and adrenalectomy on left ventricular hypertrophy in rat offspring. Am J Physiol 285:E31–E39
Williams D, Neighbors H, Jackson J (2003) Racial/ethnic discrimination and health: findings from community studies. Am J Public Health 93:200–208
Wolanski N, Dickinson F, Siniarska A (1993) Biological traits and living conditions of Maya Indian and non-Maya girls from Merida, Mexico. Int J Anthropol 8:233–246
Woods LL, Ingelfinger JR, Rasch R (2005) Modest maternal protein restriction fails to program adult hypertension in female rats. Am J Physiol 289:R1131–R1136
Wyka J, Biernat J, Kiedik D (2010) Nutritional determination of the health status in Polish elderly people from an urban cnvironment. J Nutr Health Aging 14:67–71
Zambrano E, Martinez-Samayoa PM, Bautista CJ et al (2005) Sex differences in transgenerational alterations of growth and metabolism in progeny (F2) of female offspring (F1) of rats fed a low protein diet during pregnancy and lactation. J Physiol 566:225–236

Chapter 6
A Critical Biocultural Perspective on Tourism and the Nutrition Transition in the Yucatan

Thomas Leatherman, Alan H. Goodman, and J. Tobias Stillman

Abstract Biocultural anthropology is a broad and holistic approach to the study of human biology within social and cultural contexts. A critical biocultural approach furthers these considerations by paying particular attention to how historical and political economic forces help shape social contexts and biological variation, as well as the research process itself. In this chapter we first discuss the emergence of a critical biocultural anthropology and then provide a case study of the political ecology of food and nutrition in the Yucatan following the development of the tourism industry in the 1970s and 1980s.

Tourism-led economic development has had a complex biocultural impact on Mayan communities. On the one hand, tourism has brought jobs and expanded cash-generating activities, roadways, markets, and other aspects of infrastructural development necessary to broader economic growth. Yet, most jobs are low-wage and precarious, and rural communities linked to tourism are experiencing major transformations in social relations and disrupted agricultural systems, as well as dietary shifts associated with increasingly delocalized and commoditized food systems. We find increases in stature and a reduction in dental enamel hypoplasias indicating improved nutritional well-being in children, but broader patterns of growth reveal a "dual burden" of persistent stunting coupled with overweight and obesity in the same communities. Regional patterns throughout the Yucatan reflect high rates of obesity in both children and adults, as well as high rates of Type 2 diabetes, indicating a widespread nutrition transition among the Yucatec Maya.

T. Leatherman (✉)
Department of Anthropology, University of Massachusetts, Amherst, Amherst, MA, USA
e-mail: tleatherman@anthro.umass.edu

A. H. Goodman
School of Natural Sciences, Hampshire College, Amherst, MA, USA
e-mail: agoodman@hampshire.edu

J. T. Stillman
Save the Children Foundation, Washington, DC, USA

H. Azcorra, F. Dickinson (eds.), *Culture, Environment and Health in the Yucatan Peninsula*, https://doi.org/10.1007/978-3-030-27001-8_6

6.1 Introduction

An important theme within political economic and human ecology research over the past five decades has been documenting the impacts on rural agrarian communities of their rapid transition into societies and economies dominated by capitalist economic relations. Biological and biocultural anthropologists have more recently contributed by evaluating changes in diets, activity patterns, nutrition, and health status. A common theme was how shifts toward cash cropping and wage labor and a greater reliance on market foods over locally produced foodstuffs often led to reduced dietary diversity, impaired nutritional status, and worse health outcomes (Dewey 1989). In the context of low incomes and high food prices many rural producers reliant on market foods often consumed worse diets that they had as basic subsistence food producers. In the past several decades more studies have focused on how economic transitions are associated with a nutrition transition toward calorie rich and nutrient poor diets, greater obesity, and the frequent appearance of "dual burden" households, or the co-occurrence of stunting and obesity in the same community, households, and even within individuals from infancy to adulthood (Delisle 2008; Doak et al. 2005). Indeed, dual burden has become a frequent, even expected, nutritional outcome of poorer populations experiencing economic growth accompanied by nutrition transitions.

These patterns are prevalent in extensive research throughout Latin America and the Yucatan (Dickinson et al. 1993; Gurri 2015), and have been linked to greater urbanization and involvement in capitalist markets (Dickinson et al. 1993; Gurri 2015; Leatherman and Goodman 2005b). The dominant form of economic transition in the Yucatan over the past four decades is tourism-led economic development (Wilson 2008). Tourism is often heralded as a "soft path" form of economic growth that has the potential for job creation, greater foreign exchange, and with less of an environmental, social, and cultural footprint from other forms of development (e.g., industry). Yet, tourism, like other forms of development, has contradictory and ambiguous effects (Stonich 1998; Stronza 2001; Pi-Sunyer and Thomas 1997; Wilson 2008). For example, some argue for the benefits associated with job creation, positive impacts on unemployment and under-employment, and the generation of foreign exchange, while others point to the precarity and low wages of predominantly unskilled service jobs, and negative impacts on agrarian economies and community social relations. But most studies fail to take into account internal variation within countries, regions, or communities (Wilson 2008); the winners, losers and others who may benefit in some dimensions and lose in others. Thus, an assessment of impacts leads us to a political economic perspective (Wilson 2008).

In addition to the economic and social impacts of tourism, a major concern is also with its impacts on environments, food systems, nutrition and health (e.g., Daltabuit and Leatherman 1998; Stonich 1998; Stronza 2001). This requires a biocultural perspective attentive to the political economy of tourism-led capitalist development and the political ecology of food systems. Addressing these concerns calls for an approach that can integrate perspectives from human ecology and adaptability with political economy; in short, a critical biocultural approach to human biology and health (Goodman and Leatherman 1998a; Leatherman and Goodman 2011). As we describe below, critical biocultural approaches can take many forms. In the case of rural agrarian populations such as the Maya of the Yucatan, where the sort of social and economic transformations involved in tourist economies include changes to local production systems, social and familial relations, food systems, diet nutrition and health—we have framed it as a study in the political ecology of food systems and health.

In the section that follows we outline the development of a critical biocultural approach derived from earlier work and then provide a case study of the political ecology of food systems, diet, and nutrition among rural Mayan communities differentially integrated into the tourist economy.

6.2 A Brief History of Critical Biocultural Anthropology

As medical and nutritional anthropology were growing as sub-disciplines of anthropology in the 1970s, a biocultural framework based in an ecological model emerged that framed human health and nutrition at the intersection of physical and social environments, culture, social organization, and human biology (Armelagos et al. 1992; Jerome et al. 1980). This model was used to examine specific human-environment interactions involving nutrition, disease or other biological indicators of stress and adaptation (Goodman et al. 1988), and served as a framework for examining social, nutritional, and epidemiologic transitions (Armelagos et al. 2005). The promise of such an integrative model led many to conclude that anthropology had achieved a theoretically coherent integration of biological, ecological, and cultural domains (for a longer analysis see Goodman and Leatherman 1998b; Leatherman and Goodman 2011). Yet, because they framed human ecosystems as closed, homeostatic systems, unconnected to regional and global processes, these models were increasingly seen as flawed. Obviously, such closed models do not work within the Yucatan which has been enmeshed in global (colonial and post-colonial) politics and economies for the past five centuries.

Along with the closed framing of ecosystems, the concept of adaptation came under fire from evolutionary biology (Levins and Lewontin 1985) and anthropology (Armelagos et al. 1992; Singer 1989), as tautological, teleological, reductionist, and victim-blaming. Critiques of the "small but healthy" hypothesis exposed the faulty logics of designating short stature in resource poor environments as an "adaptation" and the perils of "quitting early" and failing to link short stature to other functional indicators of health (Martorell 1989). An overriding critique of the adaptation concept was that it viewed people as passively responding to autonomous, external environmental forces rather than recognizing their role in constructing those environments (Lewontin 1995; Leatherman and Goodman 2005a; Smith and Thomas 1998). As a result, it also failed to evaluate how the costs of adaptive response might generate new problems and constraints for human biology. In short, by seeing responses to change as adaptive or maladaptive according to a cost-benefit calculation it missed the nuance and ambiguity embedded in social and biological responses to changing environments. Yet, transitions of all sorts, including the integration of agrarian societies into capitalist economies, are filled with ambiguity. This is certainly the case for the social and biological impacts of a developing tourist economy in the Yucatan.

In the 1960s and early 1970s, human biologists were largely concerned with understanding adaptations to physical and biotic extremes. The initial assumption was that under stable, extreme conditions, human genetic adaptations would emerge and could be identified by investigators. However, two decades of research showed that human populations exhibited many more ontogenetic or developmental responses than genetic responses to environmental stressors (Frisancho 1993; Smith 1993). Thus, biological plasticity was recognized as the key to understanding the human adaptive process. An example of this shift in thinking is evident in the way a "thrifty phenotype" hypothesis has gained favor over the "thrifty genotype" hypothesis which suggested a genetic basis for rising rates of obesity as part of a pattern of metabolic disease in native American populations (Neel 1962). Newer approaches such as the developmental origins of adult health and disease (DOHaD) posit ways in which environmental stressors shape fetal metabolism and adult health and disease by producing "thrifty phenotypes" prone to increased fat storage, increased adiposity, and obesity (Hales and Barker 1992).

It also became clear to many biocultural researchers that groups living in challenging physical environments were often also living in social environments with limited access to means of production, economic opportunities, political power, health care, and education. The resulting stressors with origins in relations of power, such as food insecurity and malnutrition, invariably had a greater impact on biology and health than did physical stressors such as high altitude and cold temperatures (e.g., Greksa 1986; Leonard 1989; Thomas 1998). Thus, rather than searching for adaptations, biocultural research in the 1980s and 1990s increasingly became oriented toward documenting biological compromise or dysfunction in impoverished environments and the biological impacts of social and economic change (Thomas 1998). Social environments took precedence over physical environments and measures of stressors expanded to include psychosocial stressors and their impact on health conditions such as hypertension and immune suppression (e.g., Blakey 1994; Dressler and Bindon 2000; Goodman et al. 1988; McDade 2002). Yet, while it became relatively common to associate biological variation with some aspect of socioeconomic variation, it was rare that the context or roots of the socioeconomic variation were addressed. For example, research on "modernizing" populations documented how devastating such changes can be on human biology and health but provided little or no information about processes of modernization. The socioeconomic conditions, workloads, and environmental exposures that contribute to diminished health thus became seen as *natural and even inevitable* aspects of changing environments, rather than contingent on historical processes and social and economic relations (Leatherman and Thomas 2001).

Emerging interests in the biological dimensions of poverty and inequality, and the need to frame these analyses in rich social and economic contexts, demanded new models and approaches with biological and biocultural anthropology. Specifically, these interests drew biocultural researchers to integrate perspectives from anthropological political economy with those from ecology and adaptability. Goodman and Leatherman (1998b) suggested a number of thematic shifts to biocultural approaches that came to characterize a more "critical biocultural" orientation, including an expanded view of environmental and historical contexts and contingencies, attention to the social relations that structure (and are structured by) human-environment interactions, and granting greater agency to humans in constructing their environments; but also recognizing constraints on agency, what Levins and Lewontin (1985) called "conditional rationality" in coping with social and environmental problems.

Critical biocultural approaches share much with the structural violence perspective (Farmer 2004) commonly used in anthropology and studies of global health to consider the systemic ways in which societal structures harm or disadvantage some individuals (Rylko-Bauer et al. 2009). There is also shared ground with the eco-social approach in public health, which "seeks to embrace a social production of disease perspective while aiming to bring in comparably rich biological and ecological analysis" to answer "who and what drives current and changing patterns of social inequalities in health" (Krieger 2001, p. 672). All three attempt to integrate history, political economy, and biology in social and environmental contexts that are "geographically broad" and "historically deep" (Farmer 1996, p. 274). These perspectives also focus on the health inequities and limitations to human agency that result from social and structural inequalities. They specifically focus on how the many forms of social inequalities literally become embodied in biology and health; how lived realities "get under the skin" along multiple pathways of embodiment. In short, the goal of a critical biocultural approach is to link structures of inequality, constrained agency, and pathways to embodiment within historically and ethnographically grounded contexts, lived experience, and local biologies.

6.3 Trends in Biocultural Analyses

In recent decades, biocultural anthropologists have increasingly focused on the health and nutritional consequences of social and ecological vulnerabilities. Research projects and approaches have taken on many forms and foci, including social inequalities and health, populations in transition, bio-psychological stress and response, and the biological consequences of race and racism (Leatherman and Goodman 2011). These are illustrative of studies that go beyond standard measures of socioeconomic status (SES) to study vulnerabilities along multiple axes including race, gender, income, and occupation. It is now well accepted that social inequalities underlie health disparities across different contexts.

Biocultural anthropologists have contributed to these observations through grounded research on the interactions among social inequalities, livelihoods, food security, nutrition, and illness. Such work includes the reproduction of poverty and poor health in Peru (Leatherman and Jernigan 2014) and the interactions among economic vulnerability, food security, diets, and nutrition in the USA, Costa Rica, and southern Africa (see respectively Crooks 1998; Himmelgreen and Romero-Daza 2009; Hampshire et al. 2009). These examinations move beyond simple correlations of SES and nutrition and health outcomes in their efforts to track pathways through which inequalities and health are mutually constitutive and reinforcing.

A deeper appreciation of history makes clear that humans are constantly in transition—from prehistoric shifts in foraging to food production, to conquest and colonization, and integration into capitalist economies. Armelagos et al. (2005) frame this history in terms of epidemiologic transitions in disease patterns resulting from evolutionary, historical, and political-economic processes associated with social change, and apply these transitions across time and space. Goodman and Armelagos (1985), for example, argued that political hierarchies and resource extraction from the peripheries to the center of pre-capitalist social formations in prehistory played a key role in declining health in rural areas. Using historic records and modern epidemiological health surveys, Santos and Coimbra (1998) have researched the health effects of colonization on indigenous populations in Brazil through a series of historical events from initial contacts, to economic booms and busts (rubber and timber), to more recent migrations of settlers into the Amazon.

Transitions to market-based economies is a historic transition that can have negative, positive, and uneven effects on nutrition and health (Leatherman 1994; Kennedy 1994; Dewey 1989; Pelto and Pelto 1983). Gross and Underwood (1971) were among the first to document the economic and nutritional impacts of a shift from subsistence to sisal production in northeast Brazil. More recently the Tsimane' Amazonian Panel Study (TAPS) (Leonard and Godoy 2008) has investigated the biocultural impacts of market integration in the Bolivian Amazon (Godoy et al. 2009; Reyes-García et al. 2009). When shifts to markets occur in contexts of tourism, as in the Yucatán, complex transformations are set into motion in social relations, values, and economics as well as diet, nutrition, and health.

In the following section we present a case study in tourism-led social and economic change that is part of a recent nutrition transition in Latin America. The nutrition transition to which we refer is a trend observed in countries experiencing recent and ongoing development leading to significant changes in food availability, access, and cost (Popkin 2001; Popkin and Gordon-Larsen 2004). The most common pattern is a shift away from more locally grown foods in favor of imported and processed foods which can often be purchased in higher quantities at less cost than local produce. In many cases, this transition has led initially to improved food security and nutritional status and is then followed by emerging overnutrition and increasing rates of overweight, obesity, and related metabolic illnesses. In this case study, we consider how tourism-led economic development led to patterns of change in social relations, food production, diet, and nutrition in Mayan communities in the Yucatán Peninsula

6.4 The "Coca-Colonization" of Diet in the Yucatán

Our case study centers on changing patterns of agriculture, diet, and nutrition among Mayan children and adults of the Yucatán Peninsula in Southern Mexico. These rapid changes are an example of a nutritional transition, in this case, linked to international tourism and increasing dietary delocalization. Individuals, households, and communities differentially experience and respond to tourism-led development, and differently reflect changing food systems and values in their diets and nutrition.

Along with many other Native American groups, the Maya have experienced a recent rise in chronic diseases such as diabetes and obesity. Over a quarter century ago, Federico Dickinson and collaborators (1993) conducted nutritional and health surveys among women living in rural sisal producing regions, women from the coastal town of Progresso, and women migrating from the sisal region to the coast (sampling 72 women in each group). About 86% of the women were overweight of which about 50% were obese. Rates of hypertension and diabetes were also high. Moreover, they found these high rates of overweight and obesity in all three groups, suggesting that overweight and obesity was a widespread public health problem in the Yucatan. More recently, the 2012 National Survey of Nutrition and Health reported that 82% of adult women and 78.6% of men in the state of Yucatan were overweight or obese (Instituto Nacional de Salud Pública 2013). The rates of obesity in women and men (46.2% and 43.3%, respectively) was actually greater than the percent of overweight (35.8% and 35.3%, respectively). The rates for urban and rural samples were roughly equal. The biggest differences were found in rural zones where 43.3% of men were overweight and 34% obese, and 32.1% of rural women were overweight but 51% obese. Perhaps more disturbing, rates of overweight and obesity in children and adolescents are high, and above the national average. Forty-five percent of children (5–11 years) and 43.4% of adolescent boys and girls (ages 12–19 years) are overweight or obese. Moreover, 30% of adults have been diagnosed with diabetes (32% women and 28.1% men), with higher rates in those over 60 years of age (43.8%, vs. 21.9% in those 20–39 years, and 38.7% in those 40–59 years of age). Similarly, Loria et al. (2018) report that between 1962 and 2000 rural Maya adults underwent a 40-year transformation from conditions of malnutrition (such as pellagra) due to caloric and nutrient insufficiencies to one of high rates of obesity and diabetes.

This evidence of "overnutrition" or surplus of calorie consumption leading to overweight and obesity in the Yucatec Maya is an apparent contradiction as the Maya have frequently been depicted as economically marginal, impoverished, and undernourished. Indeed, studies of Maya child growth, a key indicator of community level nutrition, illustrated severe stunting and little change in growth between the 1930s and the early 1980s (Daltabuit 1988; Leatherman et al. 2010). However, by the mid to late 1990s, increases in both heights and weights were evident (Leatherman et al. 2010; Gurri et al. 2001). Despite these increases in growth, we found persistent indications of poor diet quality in rural Yucatecan Maya communities (Leatherman and Goodman 2005b and summarized below). Now the Maya are subject to what Dickinson has termed the "double-edged sword of malnutrition" and Doak and co-workers (2005) term the "dual burden", where there is a coexistence of undernutrition (i.e., stunting) and overnutrition (overweight and obesity) in the same community, household or individual. Varela-Silva and co-workers (2012) provide a thoughtful assessment of dual burden among Maya in Merida, the capitol of Yucatan, Mexico. How did the dual burden of undernutrition coupled with overnutrition emerge in the Yucatán?

One of the more notable aspects of tourism development in the Yucatán is the commoditization of food systems, the increased distribution and consumption of commercialized foods and especially "junk foods," which are calorie rich but deficient in micronutrients. Thus, a proximate answer to the above question is a trend toward the high fat and sugar consumption typical of western diets, i.e., a "westernization" of local diets. On the other hand, glossing processes of dietary change as westernization misses the changing local and regional dynamics of food systems and diets that are the result of global processes, including tourism-based social and economic change.

A "westernization" of Yucatecan diet is one form of a broader process of "dietary delocalization," a process wherein local peoples consume foods produced out of the region (Pelto and Pelto 1983). Since ancient times, trade, internal colonization, and migration have promoted the exchange of foods across regions. After 1492, such exchanges became global, and more recently dietary delocalization has been increasingly linked to the commoditization and commercialization of food systems and diets (Dewey 1989; Pelto and Pelto 1983). Shifts from locally produced to market and commercialized foods have been associated with greater dietary diversity and improved levels of nutrition in industrialized nations (Pelto and Pelto 1983). However, the increased commoditization of foodstuffs typically coincides with increasing market prices that stress the budgets of the poor majority and frequently leads to reduced dietary diversity.

An assessment of changing food systems, nutrition, and health in the context of tourism-based economic and social transformations among the Yucatec Maya begins with the identification of several points of contradiction. Child growth has improved somewhat, but nutritional deficiencies persist, and at the same time, adult obesity and diabetes rates are skyrocketing, portending increases in childhood rates of obesity and diabetes. Commoditization of food systems can lead to an overall increase in food availability, diversity, and consumption, but also can heighten inequalities in access and detrimentally affect nutrition and health. From our perspective, these contradictions reflect the manner in which social, cultural, and health impacts of tourism-led development are distributed unevenly and experienced unequally among Mayan communities, families, and individuals. Much of this variation is influenced by the manner and degree to which communities, households, and individuals articulate with local production systems and with the tourism-based economy.

6.5 Global Contexts and Local Transformations of Dietary Change

Throughout Latin America and much of the developing world, nations are turning to tourism as an efficient means for generating economic development and foreign capital. Mexico, and especially its Caribbean coast, is a major a destination for North American and European tourists. Mexico has been a leader in Latin America in developing a robust tourist economy. Over the past decade, tourism has been one of Mexico's highest sources of foreign exchange alongside oil and remittances from migrants (Wilson 2008, Estevez 2016). A recent report from Universidad Anáhuac in *La Economista* notes that 38 million tourists visited Mexico in 2017, an increase of over 9% from the preceding year, and generating 21 billion dollars in revenue (De La Rosa 2018). Following early success in Acapulco and Puerta Vallarta on the Pacific Coast, Mexico turned to the development of the Caribbean in the 1970s, spearheaded by FONATUR (National Fund for Tourism Development). This region, with Cancun as its epicenter, became the fastest growing tourism region, meeting national goals of stimulating migration and generating job creation and foreign exchange. The later point was less successful due to the fact that so much of the investment in the new tourist economy came from foreign transnational corporations (TNCs) and thus much of the profits were expatriated out of Mexico. For the Mexican government, they developed an "undeveloped" region; for Mayan communities, there was a new source of jobs; for the tourist economies, Mayans were a source of cheap labor.

Indeed, the former "Territory" of Quintana Roo in the Yucatán Peninsula, which became a state in 1974, has been transformed from one of the Mexico's most marginal regions, with few areas with paved roads and electrification, to a hotspot of global tourism. From the 1970s to the 1990s, Cancun grew from a fishing village of just over 400 inhabitants to the state of Quintana Roo's most important city with a population of over 600,000 people (INEGI 2012) and an estimated 700,000 by 2017 (Cancun Population 2018). This development has been a significant economic success for the Mexican government, a few Mexican capitalists, and many foreign investors. However, such rapid and totalizing development does not come without cultural and economic costs (Pi-Sunyer and Brooke 1997; Daltabuit and Leatherman 1998). As environmental resources, labor, and food become increasingly commoditized and symbols of prestige become increasingly western, Mayan culture, environments, and lives are inevitably changed. For example, with the growth of the tourist economy, households and communities have become increasingly dependent on the large urban centers for income, and the reliance on milpa agriculture and home-gardens has declined in some parts of the Yucatan. Krumrine (2016), for example, reports a sharp decline in milpa agriculture and even home gardens in recent years in Coba, a site of archeotourism and one of the sites of our earlier research. Some young families prefer not to pool resources and labor, or even share food and meal preparation, with their parents and in-laws. This reflects an erosion of familial relations, and more practically a scarcity of labor for milpa production and household domestic tasks, and greater social and economic insecurity for parental generations. As one woman told us, "we would starve without Cancun, but it will be the death of us."

Our concern here is with one form of change in particular, the commoditization of food systems and its link to dietary change, nutrition, and health. Decreased milpa production and a growth in commoditized food and labor markets has increased Mayan dependence on nontraditional and store-bought foods. Although a growth in markets and commoditization of foodstuffs may result in increased food availability and diversity, much of the population may be unable to afford market prices and consume a less diverse diet. Variation in diets is therefore likely to be found among households differing in levels of milpa production and success in the cash economy.

Here, we contrast food systems, diet, and nutrition in Mayan communities that differ in subsistence base and articulation with the tourist industry. The communities of *Akumal* and the recently built *Ciudad Chemuyil* are service villages to a popular resort on the Caribbean coast, that is to say that the villages were established largely as residences for those employed by the resort and other businesses associated with tourism. The local economy is based on wage work or commerce. Because these communities lack agricultural plots and have little land for home gardens, inhabitants are dependent on purchased food from local vendors. *Coba*, in contrast, has had direct local involvement in the tourist economy through archeotourism, focused on the nearby archaeological site of Coba, but residents try to maintain agricultural production and home gardens. Finally, *Yalcoba* is an inland farming community with little direct exposure to tourists, but substantial outmigration of men to Cancun on a weekly basis to work primarily in construction (Leatherman and Goodman 2005b).

6.6 Dietary Delocalization and Food Commoditization

While communities in the Yucatán are increasingly consuming foods from further and further away, the nature of changing food systems is markedly different for the coastal (Akumal and Ciudad Chemuyil) versus inland (Coba and Yalcoba) communities. In the coastal communities, a fully commercialized system is now in place. Foods are purchased from local stores and traveling vendors, often specializing in food stuffs from specific growing regions. In the two inland communities, products such as corn, beans, and squash from the *milpa* (agricultural fields) are harvested and available for consumption when in season. These are regularly supplemented by market foods and diets reflect these shifts. By the mid-1980s in Yalcoba, Daltabuit (1988) had already noted a shift from locally produced foods such as honey, tubers, *posoles* (stews), and wild meat, toward commercial foodstuffs such as rice and pasta, sodas, and snacks.

This trend was further advanced in the 1990s as even maize and beans, two key staples, were increasingly purchased from government subsidized stores

(CONASUPO) or the many small family owned-stores (*tiendas*). Local *tiendas* often also resell to the surrounding community small quantities of produce (e.g., tomatoes, potatoes, cabbage, carrots, onions, garlic and peppers) purchased from larger towns' markets. While a greater variety of foods are available, prices are high. As one resident of Coba noted "there are more foods available now, but no money to buy them" (Leatherman and Goodman 2005b, p. 838). Some viewed this period as a time of steadily decreasing food availability, while others saw it as a time of growth in opportunity and consumption.

The most dramatic aspect of commercialization of food systems in the region is the pervading presence of *Coca-Cola, Pepsi*, other sugar loaded drinks, and a variety of chips, cookies, candies, and locally called *comidas chatarras* ("junk foods"; snacks high in salt, sugar, and fats). What we saw in the Yucatan is part of a country-wide process (see, Bogin et al. 2014). Mexico had risen to the world's largest consumers of soft drinks, reaching an estimated 434 liters of Coca-Cola beverages per capita in 2015; a 60% increase over 2012. In 2012, Mexico led the world in consumption of Coca-Cola beverages, consuming 745 8-ounce servings, 240 more than Chile and 240 more than US – its two closest competitors (Statista, accessed December 5 2018). Indeed, the Mexican market is so important that it has been the site of an ongoing "Cola War" between Coke and Pepsi—a fight over the "stomach share" of the Mexican people. Coke's international company slogan is "an arms' length from desire"—to make Coke available at every corner in every town or village in every part of Mexico (Pendergrast 2000). Pepsi's slogan was the "power of one" marketing a Pepsi and chips or other snack food as a single purchase. Pepsi and Coke trucks make weekly visits to large and small stores in both Coba and Yalcoba to deliver sodas and snacks, as well as set up display cases, and provide coolers for the sodas. Upon entering almost any *tienda*, displays of snack food are the first things to catch one's eyes. Indeed, in Yalcoba we observed that three quarters individuals that entered a *tienda* purchased one or more snacks or high sugar drinks. As Coke and Pepsi's fight for greater stomach share intensifies, we expect to see an even greater penetration of soda and snack foods in the diets of Mexicans and the Yucatec Maya. Bogin et al. (2014) provide updated and additional information on the marketing and consumption of soft drinks in the Yucatan.

6.7 Diet and Nutrition

As shifts in foods systems are at the intersection of global and local economies, local diets and nutrition reflect actions and experience within these intersecting realities. We gathered information on demography, economic activities, diet, and nutrition from 80 coastal and inland community households. A dietary survey (including food frequency and 24-h recall instruments) provided information on the range of foods eaten, the source of foods, and their contribution to nutrient intakes (see Leatherman and Goodman 2005b for details on methods). We also estimated nutrient profiles in households from Yalcoba with steady employment in the tourist economy in or near Cancun, compared to those relying on their own subsistence production and irregular wage work to meet basic needs. This latter comparison is an effort to understand the dietary and nutritional consequences of differential involvement in local versus tourist economies.

The leading food categories in the diets of the communities we studied were tortillas, fats (oil and lard), sodas, snacks and sugar, beans, meat, and rice and pasta (Table 6.1). Together, these foods comprised 68 and 78% of calories consumed in the coastal, Akumal and Ciudad Chemuyil, and inland, Coba and Yalcoba, farming communities, respectively. Compared to inland communities, people in the coastal communities consume half the tortillas, but twice the fruits, between 1.5 to 4 times the meat, and three times the quantity of dairy and junk foods. The residents in Yalcoba who have steady employment in Cancun follow the coastal pattern. They consume just over half the tortillas, twice the meat, eggs, fruit, and dairy, and five times the bread as their counterparts who rely on farming and irregular wage work to meet their food needs. It appears that when it comes to diet in these research sites, you are where you work.

Together, sugar and junk food account on average for 16% of calories in the coastal communities, compared to 10.5 % in Coba and 8% in Yalcoba. Because our dietary surveys recorded primarily foods eaten in the household, and most sodas and snacks are consumed away from home, these are likely to be underestimates.

Table 6.1 Commonly consumed foods and macronutrients

	Coastal (*N* = 26)	Coba (*N* = 30)	Yalcoba (*N* = 24)
Food items			
Tortillas	23.0	44.0	46.0
Oil/lard	12.0	10.5	9.5
Sodas/snacks	12.0	4.5	4.0
Sugar	4.0	6.0	4.0
Beans	5.0	6.0	6.0
Meat	8.0	2.0	5.0
Rice/pasta	4.5	5.5	3.0
Bread/crackers	5.0	4.0	3.0
Fruits	6.0	3.0	3.0
Eggs	3.0	2.5	2.0
Dairy	3.5	1.0	1.0
Macronutrients			
Carbohydrate	62.5	69.3	68.6
Fat	29.2	23.3	23.2
Protein	11.6	10.6	11.0
Protein quality	79.0	67.0	73.0

Leatherman and Goodman (2005b). Coca-colonization of diets in the Yucatan. Soc Sci Med 61: 833–846

The top part of this table lists the percent contribution of commonly consumed foods to total caloric intake (median value) and the bottom part provides estimated contribution of macronutrients to total caloric intake, and protein quality scores (from Leatherman and Goodman 2005b)

Local distributors of soft drinks in Coba and Yalcoba in 1996 and 1998 reported weekly sales reflecting an average per-capita consumption of one cola per day and consumption is at least 50% greater in the coastal communities. For instance, in 1996, seventy-five school-aged children in Yalcoba reported average weekly intakes of 7.4 soft drinks (mostly Coke or Pepsi), 10.2 snack foods (e.g., chips or cookies), and 11.8 candies (e.g., lollipops). Previously, Daltabuit (1988) found that during a morning school break in Yalcoba, it is typical for children to buy a soft drink and a snack, accounting for about 350 calories, or one-quarter of an elementary school child's daily caloric intake. Chips and sweets were marketed in one-peso packages, a price most people and, perhaps especially, children can afford.

We found that the diets in all of the communities showed potential deficiencies in a number of micronutrients, including Vitamins A, B-2, B-12, and E, and especially the mineral zinc. The coastal communities exhibited a better micronutrient profile; only zinc was deficient in coastal diets. Our analyses detected several potential deficiencies in Coba and Yalcoba: Vitamin B2 or riboflavin, B-12 (cobalamins) and vitamin E in Yalcoba; and vitamins A, B2, and B12 in Coba. Yet, Yalcoba households with steady employment consumed more high-quality animal protein and better micro-nutrient profiles. Households without steady incomes that were reliant on irregular wage work and marginal *milpa* production experienced at least seasonal shortages in vitamin A and C, as well as deficiencies in many of the B vitamins and zinc. These vitamins are essential to immune health as well as growth and development and insufficient intake can lead to irregular and reduced growth. The potential micronutrient deficiencies described here can become more problematic in high maize diets and with increased consumption of soft drinks. Plant-based diets, high in fiber and phytates, are associated with increased requirements and low bioavailability of a number of micro-nutrients such as zinc, iron, calcium, and vitamin B-12 all of which are necessary for either hemoglobin production, bone health, growth, and brain development (Allen et al. 1992; Calloway et al. 1993). Thus, when the remaining "non-maize" calories come from sugar, soft drinks, and snack foods, micronutrient status inexorably worsens.

6.8 Ambiguous Impacts on Nutrition: Anthropometry and Linear Enamel Hypoplasias

As noted earlier, Federico Dickinson et al. (1993, p. 315) observed what they called the "double-edged sword" of malnutrition in rural populations and especially women in the Yucatán a quarter century ago, noting that they "spend most of their life under very strong nutritional stress: when they are young, malnutrition is highly prevalent, when they are older, obesity is quite common." Similarly, we found that children were growing better since the 1980s, but that stunting persisted and were coupled with high numbers of overweight and obese adults. These ambiguous impacts of tourism on nutritional status emerge from our research on childhood and adult anthropometry, supplemented with dental enamel hypoplasias (DEH) in school children.

In June of 1998 dental enamel defect and anthropometric data were gathered from 468 Mayan school children between the ages of 6 and 17. All of the children attended either primary or secondary school in Yalcoba and lived with their families in the surrounding community. According to a census of Yalcoba two years old by the time of our field work, the sample represented approximately 85 percent of the total population between the ages of 6 and 16 (Government of Mexico 1996). The results, therefore, should be fairly representative of the community as a whole. Any bias will tend towards children of slightly higher socioeconomic status who may have had better access to education, especially at the secondary school level.

Each child's weight, height, sitting height, upper arm circumference, and triceps skin fold thickness was recorded. Height-for-age z-scores were calculated in comparison to the NCHS growth standard (US Department of Health and Human Services 1987). Subsequently, the sample was grouped by age and heights were compared to a 1938 sample of 215 Yucatec Mayan children between the ages of 6 and 18 (Steggerda 1941) and a 1987 sample of 288 Yalcoba school children between the ages of 7 and 13 (Daltabuit 1988).

Each child was also examined for the presence of dental enamel defects on the buccal surface of their maxillary central incisors. The location and severity of each defect was recorded on a tooth diagram. The present analysis is confined to the permanent dentition and is presented as either linear enamel hypoplasia (LEH) or any hypoplastic defect (AHD). LEHs refer to defects recorded on the field diagrams as a clear or unambiguous linear enamel hypoplasia. Such defects were readily apparent to both observers and were typically bilateral. Due to their higher sensitivity, LEHs are used for all chronological analysis.

Using the methodology outlined in Goodman et al. (1987), each tooth was divided into 6 linear zones of equal width. The estimated timing of a maxillary central incisors crown development was divided by the number of zones (6) to arrive at an estimation of age at the formation of each zone. The observable enamel in the maxillary central incisors of our sample is estimated to represent a period between approximately 8 months and 4.5 years (Table 6.2). In many of the younger children

Table 6.2 Frequency of paired defects by developmental zone

	All defects		LEH[a]	
	%	*n*	%	*n*
Zone 1 (~8–15.5 months postpartum)	3.3	394	1.0	394
Zone 2 (~15.5–23 months postpartum)	21.0	396	7.6	396
Zone 3 (~23–30.5 months postpartum)	29.3	392	13.3	392
Zone 4 (~30.5–38 months postpartum)	29.1	385	16.9	385
Zone 5 (~38–45.5 months postpartum)	38.4	367	28.9	367
Zone 6 (~45.5–54 months postpartum)	10.0	349	5.2	349
Total (~8 months–4.5 years postpartum)	76.0	358	53.1	358

[a]*LEH* linear enamel hypoplasias

the permanent teeth had not fully erupted leaving a portion of the gingival zones unobservable. Additionally, in older children tooth caps or attrition sometimes obscured a zone. If any portion of a zone was unobservable, the entire zone was scored as missing. Analysis of the distribution of defects by zone was confined to children with fully scorable teeth ($n = 347$).

LEHs are used for temporal analyses. A permanent central incisor of a child whose sixth birthday was in June of 1998 (born in June 1992) and was observed at that time provided a record of physiological stress from March 1993 (8 months of age) to January 1997 (age of 4.5 years; or 1.5 years prior to measurement). At the other end, a child whose seventeenth birthday was in June 1998 and was observed at that time provided a record of physiological stress from March 1982 (8 months of age) to January 1986 (age 4.5 years). Thus, the sample age range of 6–17 years provides a record of stressful events ranging from March 1982 to January 1997.

Overall, 76 percent of children with fully scorable teeth presented with at least one paired AHD and 53 percent presented with at least one paired LEH (Table 6.2). Females have slightly lower paired defect rates than males. The highest defect rate occurs in Zone 5 (physiological disruption between ~38 and 45 months postpartum) while the second and third greatest rates occur in Zones 3 and 4 (physiological disruption between ~23 and 38 months postpartum). The lowest rate of defects occurs in the most incisal zone (Zone 1: physiological disruption between ~8 and 15 months postpartum), while the second lowest rate occurs in the most gingival zone (Zone 6: physiological disruption between ~45 and 54 months postpartum).

There is a clear relationship between age and paired defect prevalence. Children with no paired defects are significantly younger than children presenting with paired defects. Additionally, there is a steady decrease in defect rates in younger children when the sample is separated by year of birth (Fig. 6.1). This downward trend is particularly marked in combined developmental zones 3 through 6 (23–54 months postpartum).

There is also a relationship between defect rates and the year in which defects developed. In 1983, 38.5 percent of the children with developing teeth developed paired linear enamel hypoplasias, while in 1993, only 3.1 percent of the children with developing teeth developed paired linear enamel hypoplasias (Fig. 6.2). This strong secular trend is relatively smooth: the LEH rate declines almost every year over this eleven-year span. This decrease in the percent of children developing LEH mirrors the previously discussed decrease in prevalence by age.

Stature in children differed little between1938 and 1987 (+1.02 cm for males and −0.53 for females) (Fig. 6.3), but between 1987 and 1998 the mean height of Yalcoba children increased by 2.5 cm in males and 2.7 cm in females (Leatherman et al. 2010). Average weight in children between 1987 and 1998 increased by 1.8 kg for females and 2.0 kg for males, with the majority of the gains after age 12 (Leatherman et al. 2010). Child BMIs were similar to US reference values, and 13.4 % were considered overweight. Thus, modest increases in height and relatively greater increases in weight occurred during the same time span as the decreases in dental defects, and this further suggests improvement in nutrition and/or reduced disease in the decade between 1987 and 1998.

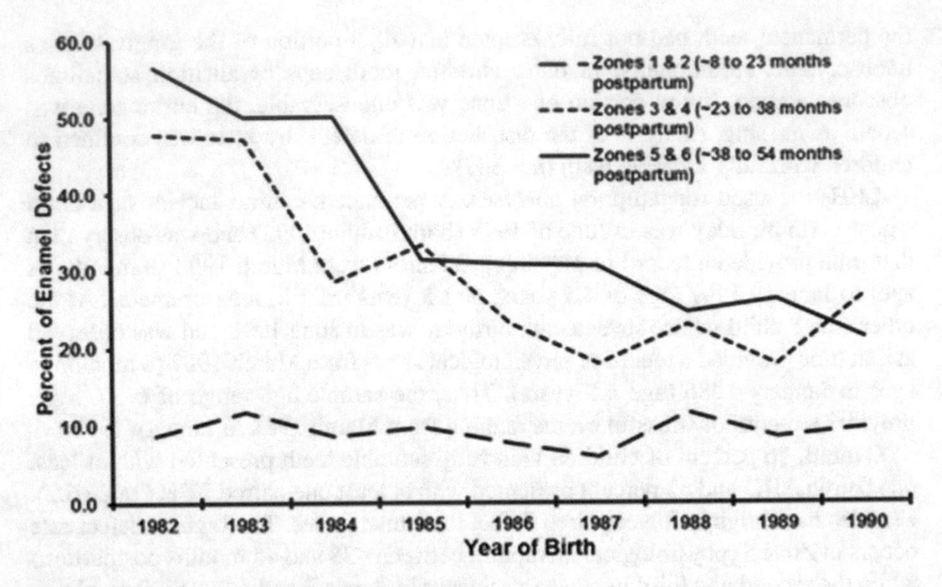

Fig 6.1 Percent of enamel defects by year of birth in children from Yalcoba

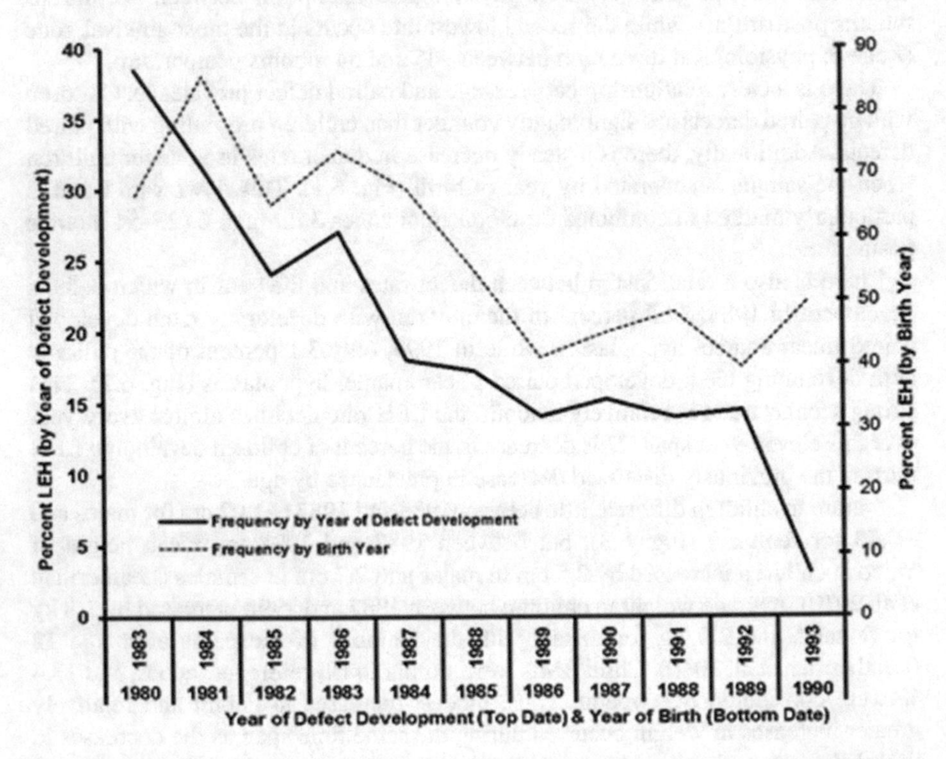

Fig 6.2 Percent of linear enamel hypoplasias by year of defect development and year of birth in children from Yalcoba

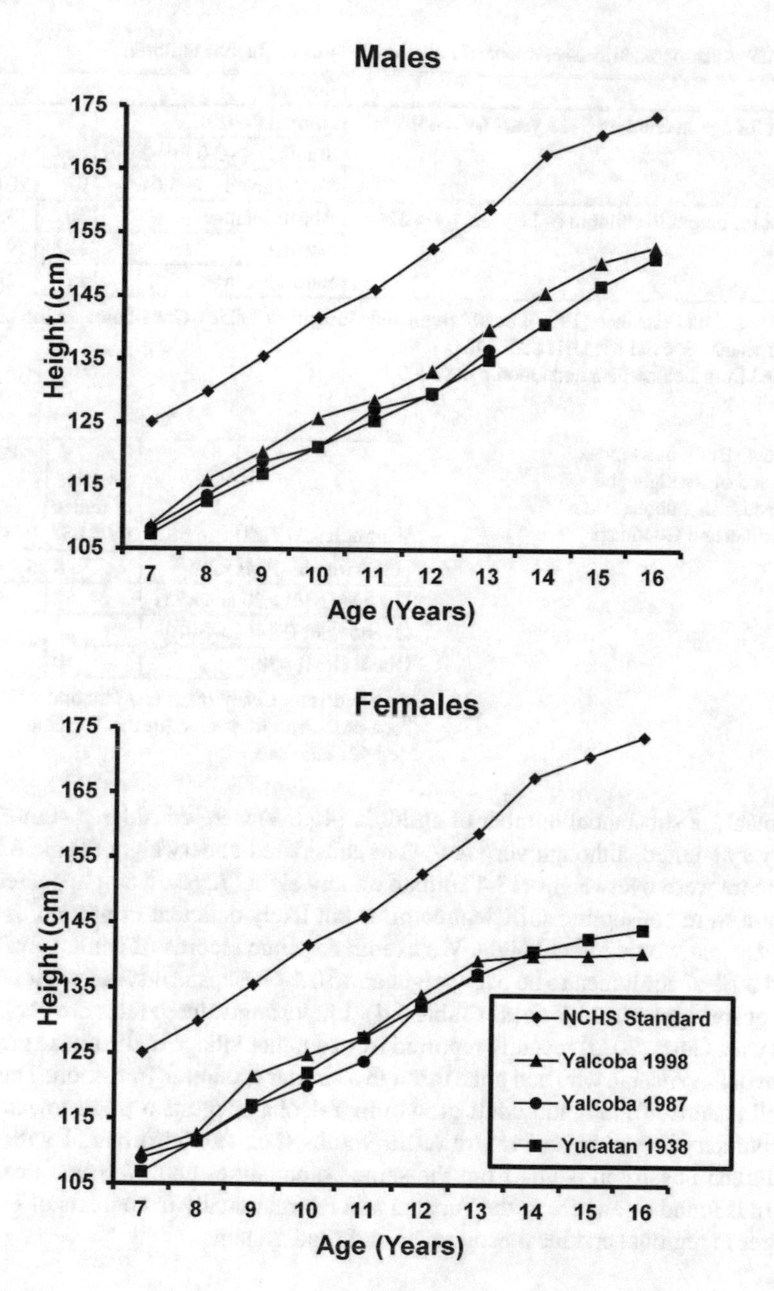

Fig 6.3 Stature for age in male and female children in Yalcoba, 1938–1998 (from Leatherman et al. 2010)

Table 6.3 Anthropometric assessment of nutritional status in Yalcoba children

	Z-score	*N*	Percent
Height for age in children 6–18 years (*N* = 456)	Normal (>−2.0)	157	34.4%
	Stunted (≤−2.0 to −2.99)	208	45.6%
	Very stunted (≤−3.0)	91	20.0%
Weight for height in children 6–11 years (*N* = 224)[a]	Above average	30	13.4%
	Normal	192	79.0%
	Below average	17	7.5%

[a]Levels based on Frisancho (1990); Leatherman and Goodman (2005b). Coca-colonization of diets in the Yucatan. Soc Sci Med 61: 833–846
Modified from Leatherman and Goodman (2005b)

Table 6.4 Body mass index (BMI) and overweight and obese adults in Yalcoba (from Leatherman and Goodman 2005b)

Weight level (BMI)	Percent of males (*N* = 83)	Percent of females (*N* = 214)
Underweight (BMI <20)	8	3
Normal (BMI ≥20 and <25)	52	33
Overweight (BMI 25–30)	30	44
Obese (BMI >30)	10	20

Modified from Leatherman and Goodman (2005b). Coca-colonization of diets in the Yucatan. Soc Sci Med 61: 833–846

Notably, a substantial number of children (45.6%) were considered stunted and 20% very stunted, although very few were considered underweight (Table 6.3). In fact, more were overweight (13.4%) than underweight (7.5%). So, it appeared that children were consuming sufficient calories but likely deficient in protein or other nutrients that would affect height. We examined clinic records of adult growth and found 30% of adult men to be overweight and 10% obese, and 44% of adult women to be overweight and 20% obese (Table 6.4). Interestingly, these rates are very similar to what Gurri (2015) recently reported from another village in the maize growing region of the Yucatan who had gone through a similar economic transition. Thus, the overall pattern of child and adult growth in Yalcoba suggests a trend towards the dual burden of chronic undernutrition in youths (i.e., stunting) paired with overweight and obesity in adults from the same communities and even families. This pattern is found elsewhere in the Yucatan and internationally in contexts of broader changes in regional and local economies and food systems.

6.9 Discussion and Conclusions

The case study outlined above provides a view into the a particularly dynamic period of social, economic, and nutrition transition following the development and spread of a tourist economy in the Yucatan Peninsula. Increased consumption of

Coca-cola and other *comidas chatarras* is emblematic of these broader changes. Coke and Pepsi in the Yucatán serve as iconic symbols of the commoditization of food systems and diets, and more broadly of global exchanges of goods and tastes. What was once a symbol of the west, of wealth, of whiteness, and of moments of respite amidst a long day of work in a hot and humid *milpa*, has become just part of everyday life.

The nutritional impact of dietary change in the Yucatan is filled with ambiguity. Mayan communities are no longer characterized by food scarcity and hunger, but by ample calories paired with micro-nutrient deficiencies resulting in dual burden households with stunted children and overweight and obese adults. A number of different scenarios has been proposed by which the dual burden of under- and over-nutrition can exist within a single individual, a single household, or community (Delisle 2008). Many scholars have argued that the dual burden may emerge as a result of a mismatch between environments and developmental programing (Hales and Barker 1992). Infants whose mothers experience marginal nutritional environments during their pregnancy or even before are born with a more "thrifty" metabolism, allowing them to quickly store fat. When populations undergo economic and nutritional transitions, like that which we are describing in the Yucatán, conditions are created for a significant mismatch that can increase the risk of obesity, diabetes, and cardiovascular disease for the population.

While we have not tested for the action of the thrifty phenotype as an underlying bio-social mechanism in the Yucatec Maya discussed here, our research does suggest that this dual burden is present in the Yucatán. Children in Yalcoba are growing taller but are still stunted compared to local and international growth standards (Leatherman et al. 2010; Leatherman and Goodman 2005b). This may reflect persistent micronutrient deficiencies since caloric intakes appear to be adequate. While obesity and diabetes have reached epidemic proportion in Mexican urban zones (Instituto Nacional de Salud Publica 2013; Arroyo et al. 1999), conditions are not quite as extreme among rural communities (Loria et al. 2018). However, an analysis of adult weights from Yalcoba collected from clinic records in 1998 found that about 40% of the men and 64% of the women were overweight and obese (Leatherman and Goodman 2005b). These levels fell short of those from urban centers in the Yucatán (Azcorra et al. 2013), but began to approach Mexico City estimates. Recent results from the 2012 Nutrition and Health Survey found roughly equal rates of overweight and obesity in rural and urban populations. Thus, it appears that the nutrition transition has been most pronounced in urban populations but has come to characterize rural communities as well.

There is little doubt that Mayan diets and nutritional status are increasingly shaped by the intersection of tourism and allied global and local processes underlying a nutrition transition. It is not clear to what extent junk food consumption is associated with weight gains in this and other communities in the Yucatán. It certainly is an important part of the overall shift in food commoditization and toward diets adequate or excessive in calories, but deficient in micronutrients such as zinc. As more caloric intake is met through sodas and snack foods, this certainly does not improve nutrition and may prove to be particularly detrimental. The "Cola Wars" and consumption of *comida chatarra* show no sign of slowing—and the emerging pattern of childhood undernutrition and adult overnutrition is a serious threat to well-being. Indeed, recent reports indicate that rates of overweight and obesity in the Yucatan are among the highest in Mexico, and especially among children and adolescents.

Mexico is the leading per-capita consumer of coke products (Case 2014) and has overtaken the US as a global leader in obesity (at 32.8%, FAO 2013). Government actions and policies demonstrate that they do believe that increasing rates of obesity are linked to the consumption of soft drinks and other junk foods. In response to the epidemic of obesity and diabetes, Mexico imposed a 1-peso (8 cent USA dollar) tax on each liter of soft drinks and other sugary beverages in January 2014, and by March of the same year potential declines in sales of 7% were noted (Case 2014). The hope is that rates of obesity will fall in time.

Finally, this case study presents a more nuanced examination of local-level changes and impacts that make up broader epidemiological and nutrition transitions. Individuals within human societies and cultures do not experience change equally but because of past experience and access to power and resources are more and less able to take advantage of opportunities and cope with potentially negative impacts. Those with greater access to power and resources and with more economic and cultural capital have better diets and nutritional status. An increase in the diversity of foods is generally associated with better dietary diversity and better nutritional status. But one must be able to afford diversity, and for those who cannot, diets based on market access can reduce dietary diversity and impair nutritional status. Complicating matters, acquired tastes and habits for processed sweet, salty, and fatty foods, clearly evident in the Yucatán turn dietary choices from micronutrient dense to calorically dense foods. The contingencies of past experience and present realities will continue to shape the nutrition and health transitions in these and other groups. The goals of a critical biocultural approach—linking structures of inequality, constrained agency, and pathways to embodiment within ethnographically grounded local contexts, lived experience, and local biologies—are well suited to mapping these transitions into the future.

References

Allen L, Backstrand JR, Stanek EJ et al (1992) The interactive effects of dietary quality on the growth of young Mexican children. Am J Clin Nutr 56(2):353–364

Armelagos G, Leatherman T, Ryan M et al (1992) Biocultural synthesis in medical anthropology. Med Anthropol 14:35–52

Armelagos G, Brown P, Turner B (2005) Evolutionary, historical and political economic perspectives on health and disease. Soc Sci Med 61:755–765

Arroyo P, Pardio P, Fernandez V et al (1999) Obesity and cultural environment in the Yucatan region. Nutr Res 57(5):S78–S82

Azcorra H, Varela-Silva MI, Rodriguez L, Bogin B, Dickinson F (2013) Nutritional status of Maya children, their mothers, and their grandmothers residing in the City of Merida, Mexico: Revisiting the leg-length hypothesis. American Journal of Human Biology 25(5):659–665

Blakey ML (1994) Psychophysiological stress and disorders of industrial society: a critical theoretical formulation for biocultural research. In: Forman S (ed) Diagnosing America: anthropology and public engagement. University of Michigan Press, Ann Arbor, pp 149–192
Bogin B, Azcorra H, Wilson HJ, Vázquez-Vázquez A, Avila-Escalante ML, Castillo-Burguete MT, Varela-Silva I, Dickinson F (2014) Globalization and children's diets: The case of Maya of Mexico and Central America. Anthropological Review 77(1):11–32
Calloway D, Murphy SP, Beaton GH et al (1993) Estimated vitamin intakes of toddlers: Predicted prevalence of inadequacy in village populations in Egypt, Kenya, and Mexico. Am J Clin Nutr 58:376–384
Cancun Population. (2018). Retrieved 2018-12-5, from http://worldpopulationreview.com/world-cities/cancun
Case B (2014) Soft-drink thirst quenched by Pena Nieto tax: Corporate Mexico. Bloomberg Business, March 2014. http://www.bloomberg.com/news/articles/2014-03-27/soft-drink-thirst-quenched-by-pena-nieto-tax-corporate-mexico. Accessed 10 Apr 2019
Crooks D (1998) Poverty and nutrition in Eastern Kentucky: the political economy of childhood growth. In: Goodman A, Leatherman T (eds) Building a new biocultural synthesis: political economic perspectives in biological anthropology. University of Michigan Press, Ann Arbor, pp 339–358
Daltabuit M (1988) Mayan women: work, nutrition and child care. Ph.D. Dissertation. University of Massachusetts, Amherst
Daltabuit M, Leatherman T (1998) The biocultural impact of tourism on Mayan communities. In: Goodman A, Leatherman T (eds) Building a new biocultural synthesis: political-economic perspectives in biological anthropology. University of Michigan Press, Ann Arbor, pp 317–338
Delisle H (2008) The double burden of malnutrition in mothers and the intergenerational impact. Ann NY Acad Sci 1136:172–184
Dewey K (1989) Nutrition and the commoditization of food systems in Latin America. Soc Sci Med 28:415–424
Dickinson F, Castillo MT, Vales L et al (1993) Obesity and women's health in two socioeconomic areas of Yucatan, Mexico. Coll Antropol 2:309–317
Doak CM, Adair LS, Bentley M et al (2005) The dual burden household and the nutrition transition paradox. Int J Obes 29:129–136
Dressler WW, Bindon J (2000) The health consequences of cultural consonance: Cultural dimensions of lifestyle, social support, and arterial blood pressure in an African American community. Am Anthropol 102:244–260
De La Rosa A (2018) Divisas y llegada turistas perfilan nuevo record. El Economista. January 10, 2018 ((https://www.eleconomista.com.mx/empresas/Divisas-y-llegada-de-turistas-perfilan-nuevo-record-20180110-0166.html). Accessed December 5, 2018.
Estevez D (2016) Remittances supercede oil as Mexico's main source of foreign income. Forbes. May 16, 2016. (https://www.forbes.com/sites/doliaestevez/2016/05/16/remittances-supersede-oil-as-mexicos-main-source-of-foreign-income/#6adb83891754). Accessed December 5, 2018.
Farmer P (1996) On suffering and structural violence: a view from below. Daedalus 125(1):251–283
Farmer P (2004) An anthropology of structural violence. Curr Anthropol 45:305–325
Frisancho AR (1993) Human adaptation and accommodation. University of Michigan Press, Ann Arbor
Frisancho AR (1990) Anthropometric standards for the assessment of growth and nutritional status. University of Michigan Press, Ann Arbor.
FAO (2013) The state of food and agriculture 2013: Food systems for better nutrition. Food and Agriculture Organization of the United Nations, Rome.
Godoy R, Reyes-Garcia V, Vadez V et al (2009) The relation between forest clearance and household income among native Amazonians: results from the Tsimane' Amazonian panel study, Bolivia. Ecol Econ 68:1864–1871
Goodman A, Armelagos GJ (1985) Disease and death at Dr. Dickson's Mounds. Nat Hist 94(9):12–18

Goodman AH, Leatherman TL (1998a) Building a new biocultural synthesis: political-economic perspectives on human biology. University of Michigan Press, Ann Arbor

Goodman AH, Leatherman TL (1998b) Traversing the chasm between biology and culture: an introduction. In: Goodman AH, Leatherman TL (eds) Building a new biocultural synthesis: political-economic perspectives on human biology. University of Michigan Press, Ann Arbor, pp 3–42

Goodman AH, Allen LH, Hernandez GP et al (1987) Prevalence and age at development of enamel hypoplasias in Mexican children. Am J Phys Anthropol 72:7–19

Goodman AH, Thomas RB, Swedlund A et al (1988) Biocultural perspectives on stress in prehistoric, historical, and contemporary population research. Am J Phys Anthropol 31:169–202

Greksa L (1986) Growth patterns of 9–20 year old European and Aymara high altitude natives. Curr Anthropol 27(1):72–74

Gross D, Underwood B (1971) Technological change and caloric costs: sisal agriculture in Northeastern Brazil. Am Anthropol 73(3):725–740

Gurri F (2015) The disruption of agricultural subsistence systems in rural Yucatan, Mexico may have contributed to the coexistence of stunting in children with adult overweight and obesity. Coll Antropol 39(4):847–854

Gurri F, Balam G, Moran E (2001) Well-Being changes in response to 30 Years of regional integration in Maya populations from Yucatan, Mexico. Am J Hum Biol 13:590–602

Government of Mexico, 1996. Census of Yalcoba. Unpublished.

Hales CN, Barker DJ (1992) Type 2 (non-insulin-dependent) diabetes mellitus: the thrifty phenotype hypothesis. Diabetologia 35:595–601

Hampshire KR, Panter-Brick C, Kilpatrick K, Casiday RE (2009) Saving lives, preserving livelihoods: Understanding risk, decision-making and child health in a food crisis. Social Science & Medicine 68(4):758–765

Himmelgreen D, Romero-Daza N (2009) Anthropological approaches to the global food crisis: Understanding and addressing the "Silent Tsunami". NAPA Bulletin 32:1–11

Instituto Nacional de Salud Pública (2013) Encuesta Nacional de Salud y Nutrición 2012. Resultados por entidad federativa. Yucatán. Instituto Nacional de Salud Pública, Cuernavaca, México. https://encuestas.insp.mx/

INEGI (2012). Censo General de Población y Vivienda 2010. Aguascalientes: Instituto Nacional de Estadística, Geografía e Informática.

Jerome NW, Kandel RF, Pelto GH (1980) An ecological approach to nutritional anthropology. In: Jerome NW, Kandel RF, Pelto GH (eds) Nutritional anthropology: contemporary approaches to diet and culture. Redgrave Publishing Co., Pleasantville, NY, pp 13–45

Kennedy E (1994) Health and nutrition effects of commercialization of agriculture. In: Braun J, Kennedy E (eds) Agricultural commercialization, economic development, and nutrition. The Johns Hopkins University Press, Baltimore, pp 79–99

Krieger N (2001) Theories for social epidemiology in the 21st century: an ecosocial perspective. Int J Epidemiol 30:668–677

Krumrine KJ (2016) Effects of diet and culture change on growth, development and nutrition among Yucatec Maya children. PhD Dissertation. Buffalo University. State University of New York

Leatherman TL (1994) Health implications of changing agrarian economies in the Southern Andes. Hum Organ 53:371–380

Leatherman T, Goodman A (2005a) Context and complexity in human biological research. In: McKinnon S, Silverman S (eds) Complexities: beyond nature & nurture. University of Chicago Press, Chicago, pp 179–195

Leatherman TL, Goodman A (2005b) Coca-colonization of diets in the Yucatan. Soc Sci Med 61:833–846

Leatherman TL, Goodman AH (2011) Critical biocultural approaches in medical anthropology. In: Singer M, Erickson P (eds) Companion to medical anthropology. Wiley & Blackwell, New York, pp 29–48

Leatherman TL, Jernigan K (2014) Introduction: biocultural contributions to the study of health disparities. Ann Anthropol Pract 38:171–186

Leatherman TL, Thomas RB (2001) Political ecology and constructions of environment in biological anthropology. In: Crumley C (ed) New directions in anthropology and environment. Alta Mira Press, Walnut Creek, pp 113–131

Leatherman TL, Goodman A, Stillman T (2010) Changes in stature, weight and nutritional status with tourism-based economic development in the Yucatan. Econ Hum Biol 8:153–158

Leonard WR (1989) Nutritional determinants of high-altitude growth in Nuñoa, Peru. Am J Phys Anthropol 80:341–352

Leonard W, Godoy R (2008) Tsimane' Amazonian panel study. Econ Hum Biol 6(2):299–301

Levins R, Lewontin R (1985) The dialectical biologist. Harvard University Press, Cambridge

Lewontin R (1995) Genes, environment, and organisms. In: Silver R (ed) Hidden histories of science. New York Review Book, New York

Loria A, Arroyo P, Fernandez V et al (2018) Prevalence of obesity and diabetes in the socioeconomic transition of rural Mayas of Yucatan from 1962 to 2000. Ethn Health 20:1–7. https://doi.org/10.1080/13557858.2018.1442560. [Epub ahead of print]

Martorell R (1989) Body size, adaptation and function. Hum Organ 48(1):15–20

McDade TW (2002) Status incongruity in Samoan youth: a biocultural analysis of culture change, stress, and immune function. Med Anthropol Q 16:123–150

Neel JV (1962) Diabetes mellitus: a "thrifty" genotype rendered detrimental by "progress". Am J Hum Genet 14(4):353–362

Pelto GH, Pelto PJ (1983) Diet and delocalization: dietary changes since 1750. J Interdiscip Hist 14:507–528

Pendergrast M (2000) For God, country and Coca-Cola: the unauthorized history of the great American soft drink and the company that makes it. Charles Scribner's Sons, New York

Pi-Sunyer O, Brooke TR (1997) Tourism, environmentalism and cultural survival in Quintana Roo, Mexico. In: Johnston B (ed) Life and death matters: human rights and the environment at the end of the millennium. Altamira Press, Walnut Creek, pp 187–212

Popkin BM (2001) The nutrition transition and obesity in the developing world. J Nutr 131(3):871S–873S

Popkin BM, Gordon-Larsen P (2004) The nutrition transition: worldwide obesity dynamics and their determinants. Int J Obes 28:S2–S9

Reyes-García V, Molina JL, McDade T et al (2009) Inequality in social rank and adult nutritional status: evidence from a small-scale society in the Bolivian Amazon. Soc Sci Med 69:571–578

Rylko-Bauer B, Whiteford L, Farmer P (2009) Global health in times of violence. School for advanced research advanced seminar series. SAR Press, Santa Fe, New Mexico

Santos R, Coimbra C (1998) On the (un)natural history of the Tupi-Monde Indians: Bioanthropology and change in the Brazilian Amazon. In: Goodman A, Leatherman T (eds) Building a new biocultural synthesis: political economic perspectives in biological anthropology. University of Michigan Press, Ann Arbor, pp 269–294

Singer M (1989) The limitations of medical ecology: the concept of adaptation in the context of social stratification and social transformation. Med Anthropol Q 10(4):218–229

Smith M (1993) Genetic adaptation. In: Harrison GA (ed) Human adaptation. Oxford University Press, New York

Smith G, Thomas RB (1998) What could be: biological anthropology for the next generation. In: Goodman AH, Leatherman TL (eds) Building a new biocultural synthesis: political-economic perspectives on human biology. University of Michigan Press, Ann Arbor, pp 451–473

Steggerda M (1941) Maya Indians of Yucatan. Carnegie Institution of Washington, Washington, DC, p 531

Stonich S (1998) Political ecology of tourism. Ann Tourism Res 25(1):25–54

Stronza A (2001) Anthropology of tourism: forging new ground for ecotourism and other alternatives. Annu Rev Anthropol 30:261–283

Statista. Annual per capita consumption of Coca-Cola Company's beverage products from 1991 to 2012, by country (in servings of 8-fluid ounce beverages). https://www.statista.com/statistics/271156/per-capita-consumption-of-soft-drinks-of-the-coca-cola-company-by-country/. Accessed Dec. 5, 2018.

Thomas RB (1998) The evolution of human adaptability paradigms: toward a biology of poverty. In: Goodman AH, Leatherman TL (eds) Building a new biocultural synthesis: political-economic perspectives on human biology. University of Michigan Press, Ann Arbor, pp 451–473

US Department of Health and Human Services (1987) Anthropometric reference data and prevalence of overweight, United States, 1976–80. In: Vital and Health Statistics Series 11, vol 238. National Center for Health Statistics, Washington, DC

Wilson TD (2008) Economic and social impacts of tourism in Mexico. Lat Am Perspect 35(3):37–52

Chapter 7
Effect of Salaried Work in Cities and Commercial Agriculture on Natural Fertility in Rural Maya Women from the Yucatan Peninsula, Mexico

Allan Ortega-Muñoz and Francisco D. Gurri

Abstract We examined 548 reproductive histories of peasant Maya women from the Maize and Citrus Regions of the State of Yucatan, Mexico, to explore the impact of commercial agriculture and circular migration on rural fertility patterns. Since the 1970s households from the Citrus Region combined subsistence with commercial agriculture, and those from the Maize Region combined subsistence agriculture with salaried work in the nearby cities of Merida and Cancun. We compared the Age-Specific Fertility Rates between cohorts that reproduced before and after economic development, and between populations. Age at desired fertility was determined to search for stopping behavior and Coale and Trussell (Popul Index 40(2):185–258, 1974; Popul Index 44(2):203–213, 1978) model of marital control was used to determine variations from natural fertility. Citrus Region women reduced fertility after development, had fewer children than those from the Maize Region, and most of them stopped reproducing after age 36. However, Total Fertility Rate remained high (5.08). Those in the Maize Region had higher fertility (7.24) and a natural fertility pattern. The importance of food production maintained a high premium on fertility in both regions. While commercial agriculture reduced desired fertility, salaried work in the cities made large families desirable to incorporate labor opportunities into the household's traditional survival strategy.

A. Ortega-Muñoz (✉)
Department of Physical Anthropology, Centro INAH Quintana Roo,
Instituto Nacional de Antropología e Historia, Chetumal, Quintana Roo, México

F. D. Gurri
Department of Sustainability Science, Environmental Anthropology and Gender Lab,
El Colegio de la Frontera Sur (ECOSUR), Campeche, México
e-mail: fgurri@ecosur.mx

H. Azcorra, F. Dickinson (eds.), *Culture, Environment and Health in the Yucatan Peninsula*, https://doi.org/10.1007/978-3-030-27001-8_7

7.1 Introduction

In this chapter, we compared the fertility patterns of two Maya agricultural populations from the State of Yucatan that went through different development processes. In one, the Citrus Region, subsistence agriculturalists became successful commercial citrus cultivators for the market. In the other, the Maize Region, circular migration to the cities of Merida and Cancun created a mixed economy where people began to depend on salaried work in the cities without abandoning food production at home. In each region, we also compared generations who reproduced before and after development.

According to Bongaarts and Cotts Watkins (1996) and Caldwell (1978), fertility reductions during the first demographic transition responded to an increase in intergenerational wealth transfers from parents to children and the institutionalization of social welfare for the elderly. Most rural populations in less developed countries (LDCs), however, depend on staple-based household agriculture, which centers on family labor (Bentley et al. 1993; Gurri 2010; Knodel et al. 2000; Lee and Kramer 2002; McC Netting 1993), and the elderly must depend on their adult children for survival (Pickard 2011; Rabell Romero and Murillo López 2016).

Nevertheless, there has been a secular reduction in fertility in rural areas around the world where government-controlled fertility programs have successfully introduced contraceptives (Cleland and Rutstein 1986; Neyer and Andersson 2008; Presser et al. 2006; Pullum et al. 1985). Contraceptive use, however, has only been effective where social and economic conditions make fertility reductions desirable (Becker 1991; Pritchett 1995). In rural areas the transformation of staple-based subsistence agriculturalists during the second half of the twentieth century into agricultural producers for the market incorporated them into a worldview which promotes low fertility (Bryant 2007; Caldwell and Schindlmayr 2003; Dixon-Mueller 2000). Tobin (1967) and Lee et al. (2000) argue that markets will substitute the institutionalization of social welfare for the elderly by allowing parents to save, reducing their dependence on older children to support them during old age. Boserup (1984) proposed that the application of modern technology would lessen desired fertility by reducing the demand for child labor. Gurri and Ortega-Muñoz (2015) agreed with Boserup and added that technology and production for the market further upset intergenerational wealth flows by extending a child's dependence period, and by substituting the importance of local kin-related networks for commercial partnerships outside the community. Fertility reductions in the Citrus Region were thus to be expected.

Roads connecting rural areas to developing urban centers, not only promoted commercial agriculture, they also facilitated population movement. Through these roads travelled the largest rural-urban migration in history (Gungwu 1997), depopulating the countryside and leading to fertility declines (Rosero-Bixby 1983). Efficient transportation systems also promoted circular migration between the cities and the countryside (Michaels 2008). Its influence on fertility patterns isn't clear. In part, because circular migration is invisible to census takers (Hugo 1982) and because not all situations are comparable (Saint-Maurice and Pintassilgo 2018). Yet, if salaried work in the cities does not accelerate the independence of the younger generation, intergenerational wealth transfers shouldn't be affected, and we would expect fertility to remain unchanged.

In both regions, Maya agriculturalists have depended for centuries on a diversified subsistence strategy centered on the *milpa*, a polyculture based on maize (*Zea mays*), different types of beans (*Phaseolus* sp.), squash (*Cucurbita* sp.), chili peppers (*Capsicum* sp.) and complimented with other cultigens (Fedick 1996; Redfield and Villa Rojas 1962; White 1999). In addition, and as part of the survival system, the Maya harvest fallow fields, hunt and gather wild and semi-domesticated plants, do backyard agriculture and forest management (Atran et al. 1993; Terán and Rasmussen 1994; Redfield and Villa Rojas 1962). Since the 1970s, vaccination campaigns, available cash to pay for private medicine and medical attention, home and sanitary improvements reduced the disease load and the infant mortality rate traditionally associated with pre-industrial agricultural populations (Gurri García et al. 2001). In addition, women have been subject of intensive birth control campaigns (Feldman et al. 2009; Pullum et al. 1985).

In the Citrus Region following Villanueva Mukul's (1994) classification, government sponsored water well drilling in the 1940s offered subsistence farmers the opportunity to cultivate citrus for the market. The development of Merida and Cancun in the early 1970s expanded their markets, turning citrus cultivation into the most important economic activity, and even promoted the development of a local leather and shoe industry. Farmers in the Citrus Region maintained control over the production and distribution of oranges and other citruses. They organized themselves to deliver their fruit to local and regional markets, and today they export to the USA (Sagarpa 2012). Next to the capital city of Merida, the Citrus Region was the fastest economic growing area in the state of Yucatan (Rosales González 1988, 2012).

The Maize Region (Villanueva Mukul 1994) also began to change in the early 1970s. Unlike the Citrus Region, the Maize Region did not develop locally. Starting in 1973 a modern highway and an efficient public transportation system connected local population centers to the growing cities of Merida and Cancun. Easy and rapid access to these cities allowed peasants to travel and incorporate to seasonal salaried work in Merida and Cancun without altering their traditional subsistence strategy. Food production in the *milpa* continued to be the center of their subsistence strategy, and determined their seasonal labor patterns and their division of labor by sex and age (Gurri García et al. 2001; Gurri and Moran 2002).

7.2 Materials and Methods

The towns and municipalities studied are shown in Fig. 7.1. The data collected forms part of two multidisciplinary human ecology projects. A team from the Anthropological Center for Research and Training (ASMRC) at Indiana University and the Centro de Investigación y Estudios Avanzados Unit Merida (Cinvestav-IPN) collected the

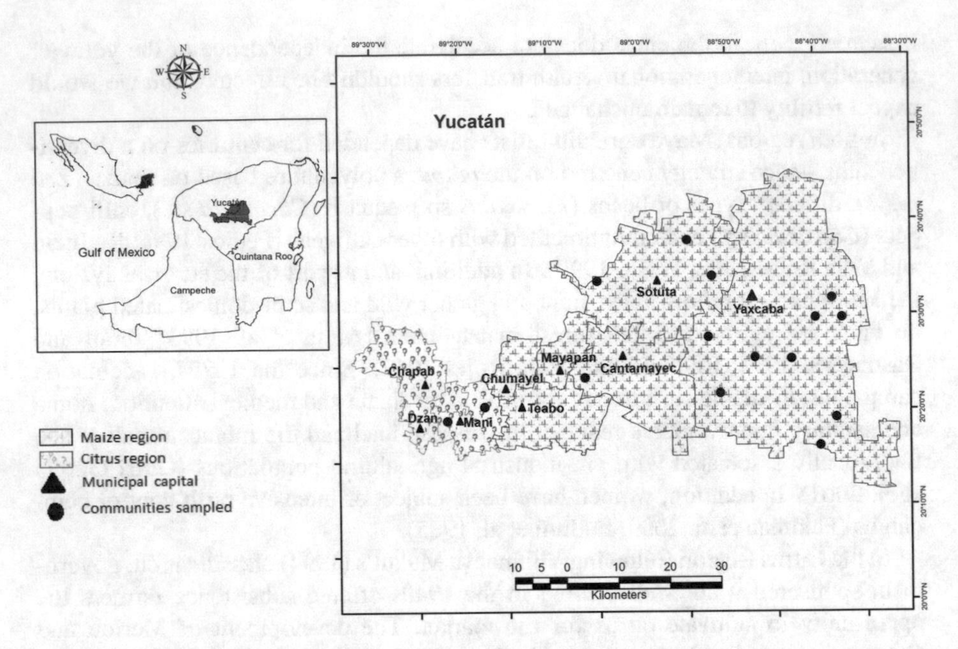

Fig. 7.1 Location Map of the Maize and Citrus Region in Yucatan, Mexico (map elaborated by Victor M. Ku Quej from ECOSUR)

Maize Region data during 1995–96. A second project with a combined team from the Center for the Study of Institutions, Population and Environmental Change (CIPEC), and El Colegio de la Frontera Sur (ECOSUR) collected the Citrus Region data in 1998. In both projects the second author trained and supervised all field and laboratory personnel. He also participated throughout the data collection periods.

In each project, enough domestic units were sampled to obtain a significant regional estimate of nutritional status in children under 10 years of age (see Gurri García et al. 2001, for a detailed description of sample estimation and collection in the Maize Region, and Balam et al. 1998, for the Citrus Region). Female team members applied a full reproductive history to all women 10 years of age and older present at the time of the visit. Women who did not speak Spanish were interviewed either by a Maya speaking technician or by a non-Maya speaker with the aid of an interpreter, also female. For the reproductive histories, recall was aided in several ways. Because data was collected at household level, all sons and daughters present were first included in the reproductive history and ordered by birth. Their birth dates were corroborated with official records such as vaccination cards and birth certificates. First pregnancy was then ascertained and then questions were asked about the second and so on until the reproductive history was complete. After a pregnancy had been recorded, the interviewer asked if anyone not mentioned had been born before and then proceeded to ask about the next birth. Date of birth was recorded for every live born child. Spontaneous abortions, when remembered, were recorded. According to Beall and Leslie (2014), asking about a woman's sequence of pregnancies and the outcome of each will improve recall. We believe the setting was also helpful. Every effort was made to distance the interviewee from friends and other family members. The later were interviewed or measured by other team members at the same time.

From the reproductive histories, we estimated and compared the *Age-Specific Fertility Rates* (ASFR) for each region for two *cohorts*: one born between 1920 and 1951, and another born between 1952 and 1979. In the Maize Region the first cohort consists of women who reproduced before the development of the cities of Merida and Cancun, and the second to those who reproduced during the 1970s and after when circular migration became important. The first Citrus Region cohort aggregates women who reproduced before and after the irrigation projects started. Nevertheless, agriculturalists did not come to depend on citrus cultivation until the 1970s during the oil industry growth in Tabasco and the growth of nearby Cancun and Merida, so as the Maize Region cohorts, we can think of them as women who reproduced before, and after economic development. For each cohort, ASFRs were estimated for 5 year intervals. The total person-years lived per cohort was divided over the number of children ever born to that specific interval.

To detect stopping behavior, women were asked how many children they desired to have. This gave us women who were satisfied with the number of children they had, those who had more than the desired number, those who still wanted more children, and those who left their fertility up to God. The age at which desired fertility was reached for each mother in the first category was recorded as the age of the mother when her last child was born, and for the second group, age at desired parity was used. Age of the mother during the survey was employed for those who knew they wanted more children and for those who had no family planning unless they were past reproductive age. The latter were ascribed an "Age at which desired fertility was reached" of 50 years. A life table analysis was carried out for mothers of each cohort in both regions, survival functions were compared statistically, and graphically using Kaplan Meier analysis to account for censored cases.

Coale and Trussell's (1978) age-specific ratio of marital to *natural fertility* was used to estimate the age-specific degree of control of marital over natural fertility known as m (Barclay et al. 1976; Wilson et al. 1988). As m increases, so does marital control. A function of the underlying total achieved fertility denoted as M in Coale and Trussell's formula (Coale and Trussell 1978) was used to make comparisons between regions and to other published values for natural fertility populations. Unlike m, the larger the value of M, the lesser the marital control. Thus, natural fertility populations will have higher values of M.

7.3 Results

A total of 278 women over the age of 10 were interviewed in five municipalities of the Maize Region and 270 in the three of the Citrus Region (Table 7.1). Age distributions by regions were compared with a Kolmogorv-Sminorf test for independent

Table 7.1 Women interviewed by date, region, and municipality, Yucatan, Mexico

Year of data collection	Region	Municipality	*n*
1995	Maize region	Cantamayec	55
		Chumayel	53
		Mayapán	47
		Sotuta	55
		Yaxcaba	68
		Total	278
1998	Citrus region	Chapab	71
		Dzan	99
		Maní	100
		Total	270
Total			548

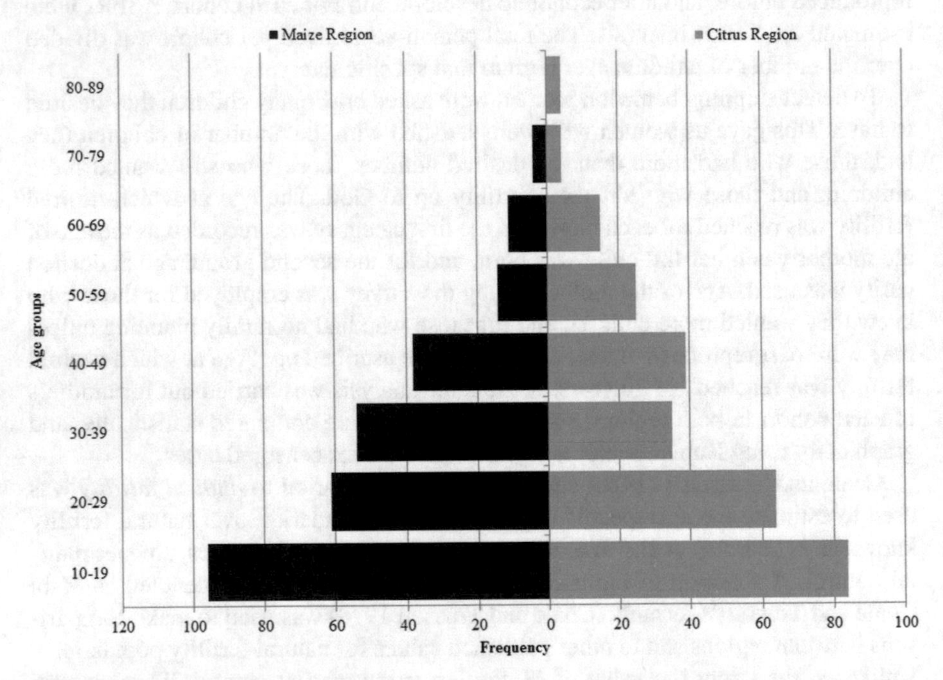

Fig. 7.2 Age composition of the women from the Maize and Citrus regions of Yucatan, Mexico

samples. They are not significantly different (KSz = 1.160, $\alpha = 0.136$). Their population pyramids are wide at the base and narrow at the top with a slight pinch between the ages of 25 and 35 years, particularly between 30 and 35 (Fig. 7.2).

The ASFR for Yucatec women are divided by region and cohort (Table 7.2). The ASFR data of Table 7.2 was graphed in Fig. 7.3. Total Fertility Rate (TFR) is greater in the Maize Region overall and in each generation (Table 7.2). Women from the

Table 7.2 ASFR and children ever born, by region and cohort, Yucatan, Mexico

	Citrus region			Maize region		
	Cohort			Cohort		
Age group	1920–1951	1952–1979	Total	1920–1951	1952–1979	Total
9–14	0.04	0	0.01	0.02	0.01	0.01
15–19	0.15	0.1	0.11	0.14	0.14	0.14
20–24	0.26	0.28	0.27	0.32	0.33	0.33
25–29	0.29	0.28	0.28	0.31	0.32	0.32
30–34	0.27	0.21	0.24	0.24	0.3	0.27
35–39	0.14	0.14	0.14	0.24	0.27	0.25
40–44	0.11	0.01	0.09	0.12	0.15	0.12
45–49	0.02	0	0.02	0.01	0	0.01
TFR	6.32	5.08	5.79	7.02	7.54	7.24
Children ever born						
	Citrus region			Maize region		
	Cohort			Cohort		
Age group	1920–1951	1952–1979	Total	1920–1951	1952–1979	Total
9–14	12	2	14	5	7	12
15–19	43	71	114	36	110	146
20–24	73	152	225	81	187	268
25–29	82	107	189	78	130	208
30–34	76	59	135	59	89	148
35–39	39	27	66	59	43	102
40–44	31	1	32	30	4	34
45–49	4	0	4	2	0	2
Total	360	419	779	350	570	920

ASFR age-specific fertility rates

Maize Region and women from the 1920 to 1951 cohorts in the Citrus Region had children throughout their reproductive lives. Most women from the 1952–79 cohorts in the Citrus Region, however, seem to stop having children after the 30- to 34-year age group and practically all of them stopped having children between the ages of 35 and 39 years. In addition, women from the younger cohorts in the Citrus Region start reproducing later (Fig. 7.3).

The responses obtained to the question of "How many children do you wish to have?" are shown in Table 7.3. Their responses are significantly different between regions ($\chi^2 = 137.664$, 2d.f. $\alpha < 0.0001$). Women in the Maize Region are not planning. Over 91% of them left their achieved fertility up to God. More than 80% of the Citrus Region Women, however, were planning or wanted to have less than 10 children and as many as 38.1% wanted between 0 and 4.

Means, medians, and 95% confidence intervals for age at which desired fertility was reached per region are shown in Table 7.4. Both average and median ages were significantly greater in the Maize Region than in the Citrus Region (Kaplan Meier analysis, $\varphi^2 = 32.425$ $\alpha = 0.001$). The survivorship curves were graphed in Fig. 7.4.

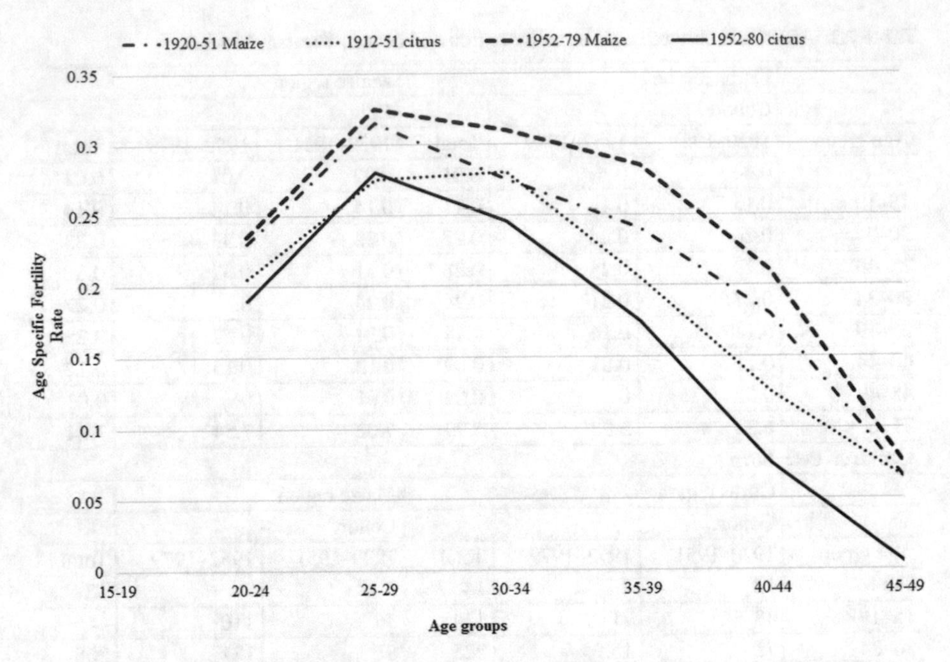

Fig. 7.3 Age-specific fertility rate by cohort and region for Maize and Citrus regions, Yucatan, Mexico

Table 7.3 Desired number of children in Yucatan, Mexico, by region

	Region		
Number of children	Maize	Citrus	Total
0 to 4	8.80%	38.10%	25.70%
5 to 10	0.00%	42.60%	24.60%
More than 10	91.20%	19.40%	49.60%
Total	113	155	268

Table 7.4 Age of mother (years) when desired fertility was reached in Yucatan, Mexico, by region

		95% Confidence interval			95% Confidence interval	
Region	Mean	Lower limit	Upper limit	Median	Lower limit	Upper limit
Maize	44.509	42.338	46.681	44.92		
Citrus	36.203	34.747	37.659	36.025	34.403	37.647
Global	38.606	37.319	39.894	39.381	36.986	41.776
$j^2 = 32.425$ $a = 0.00$						

The Maize Region's curve remained close to one until women were 36 or 37 years old. It did not drop sharply until age 45, and more than 40% of the women between 45 and 50 years old were still willing to have children. The survival curve in the Citrus Region, on the other hand, shows a different tendency. It begins to drop sharply before women are 25 years of age. By age 36, more than half of the women in the Citrus Region had reached their desired fertility.

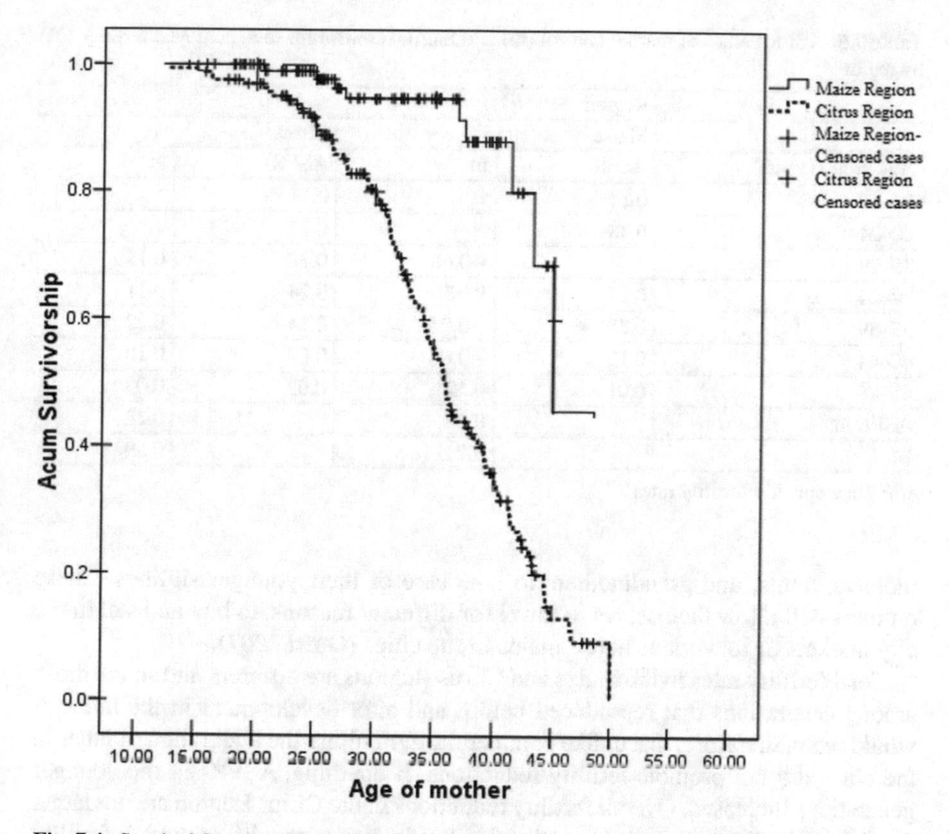

Fig. 7.4 Survival function for age at achieved desired fertility in rural Yucatan

Except for the last age group, 45–49, women from the Maize region show negative *m* values and one close to 0 (Table 7.5). Women in the Citrus Region have positive values at all ages and the 45 to 49 group has a positive value close to 0. Overall underlying fertility *M* are 0.71 for the Maize Region and 0.59 for the Citrus Region.

7.4 Discussion

Not all the indicators used in this paper are non-age biased estimators of fertility. It was important, therefore, that these populations were comparable, and they are. The population pyramids are similar, and there aren't statistical differences in the age distribution between each population. Their pyramids show two growing populations with a pinch in mothers whose eldest child is around 10 years old. Studies in Calakmul (Campeche) and Yucatan suggest that mothers start to depend on their 10–11 year old daughters who start to abandon school at that age (Gurri 2010). The child's activities are various, but one of their main activities is to help their

Table 7.5 ASFR values of marital control (m) and marital control for total achieved fertility (M) by region

	Region			
	Maize		Citrus	
Age group (years)	ASFR	m	ASFR	m
15–19	0.14		0.11	
20–24	0.33		0.27	
25–29	0.32	−0.06	0.28	0.13
30–34	0.27	0.02	0.24	0.11
35–39	0.25	−0.04	0.14	0.22
40–44	0.12	−0.01	0.09	0.10
45–49	0.01	0.12	0.02	0.03
ASFR m		0.01		0.12
M		0.71		0.59

ASFR age-specific fertility rates

mothers, aunts, and grandmothers to take care of their younger siblings. These mothers will allow themselves to travel for different reasons, to buy and sell in the city markets or to work as house maids in the cities (Gurri 2007).

Total fertility rates in the Maize and Citrus Regions are different and so are those among generations that reproduced before and after development in the latter. It would seem, therefore, that unlike commercial agriculture the integration of labor in the cities did not promote fertility reductions. If anything, ASFRs in the younger generations increased. Overall fertility reductions in the Citrus Region are not large. Total Fertility Rate is similar to urban fertility in countries with very high fertility rates (Roser 2018). Citrus Region fertility before development and Maize Region fertility before and after is higher and similar to other pre-transition rural populations (Ortega Muñoz 2012; Roser 2018).

There are noticeable changes in the reproductive strategies of the Citrus Region population after development. According to Wilson et al. (1988) there are two reproductive patterns. One, in which women reproduce throughout their reproductive life span, and one in which they stop reproducing while they are still fertile. In the former, variations in achieved fertility will depend on the age that women start reproducing and inter-birth spacing (Cleland and Rutstein 1986; Feldman et al. 2009). In the latter, they will depend on the age at which desired fertility is reached (Quilodrán 1980). The first demographic transition from hunter-gatherers to agriculturalists was obtained by reductions in the inter-birth interval (Armelagos et al. 1991). Stopping behavior is credited for reducing fertility during the second demographic transition in capitalist industrial urban centers (Bongaarts 2001; Caldwell and Caldwell 2003; Notestein 1945; Rodríguez 1996; Zavala de Cosío 2014).

Deviations from natural fertility may be observed in the way that the probability of having children is reduced with age, and it may be independent of total achieved fertility. In this chapter, we looked at five different ways of determining if changes in fertility patterns occurred. The first one compared the curves of the different

ASFRs. In natural fertility populations, ASFRs remain high for almost the entire reproductive life span and reduces gradually until menopause shown graphically as a convex curve (Coale and Trussell 1974, 1978; Juárez et al. 1989). Inter-population fertility differences will be reflected in the height of the curves. With stopping behavior, the curve will tend to be concave particularly when women stop having children at an early age. Usually, births will concentrate at the ages of highest reproduction between 20 and 29 and they will drop dramatically after that (Juárez et al. 1989; Quilodrán 1980).

The Maize Region and the predevelopment Citrus Region have natural fertility curves. Unlike secular trends in all of Mexico, ASFRs in the Maize Region after development are higher than they were before. This, however, should not be too surprising. Salaries in the cities are not high enough for workers to support their families on their own. In the Maize region, salary in the cities is only part of a diversified survival strategy that depends on the coordinated labor of men, women, and children of all ages (Daltabuit 1992; Gurri 2007). The incorporation of salaried work in the cities, therefore, will provide an incentive for larger families to take advantage of the new source of income.

Differences in total fertility between both Maize Region and predevelopment Citrus Region are a consequence of lower ASFRs throughout the reproductive life span in the later region, but particularly in the 20- to 24-year-old and the 35- to 39-year-old groups. This suggests that even before development, younger women there reproduced later, older women had fewer children and inter-birth spacing was greater than in the Maize Region. Women who reproduced after local development in the Citrus Region started to reproduce even later and appear to stop reproducing after age 36. The drop-in fertility is not drastic, and the curve is more convex than that of populations with fewer children where women stop reproducing long before Citrus Region women do (Juárez et al. 1989; Quilodrán 1980).

Desired number of children can give us an idea of a woman's acceptance of conscious family planning, and actual fertility varies between 1 and close to 0 in most countries and close to 2 in Africa (Günther and Harttgen 2016). Our estimations of desired fertility underestimate its real value due to ex-post rationalization of number of children already born (Pritchett 1995). This is probably why women would report desiring as many as ten children. The same method was applied in both regions, however, and both populations are comparable in age so that this bias would have an equal effect in the Maize and Citrus Regions, and the differences are telling. Independently of desired children, most Citrus Region women expressed a desired family size. In the Maize Region most women left their fertility up to God. This does not mean that they did not have an expected number, or that they didn't use contraceptives (Cleland and Rutstein 1986) but it most likely precludes stopping behavior.

Stronger evidence that there is a change in reproductive pattern in the Citrus Region is presented by the age at which desired fertility was reached. The survival functions for each region are significantly different. The curves combine all generations and suggest that most women in the Citrus Region have reached desired fertility by age 35 while Maize Region women continue to reproduce well into their 40s.

A way to measure deviation from natural fertility is provided by Coale and Trussell (1978) model of fertility control by age. In this model, as fertility rates decline more steeply than the standard determined by the underlying overall fertility (M) with age, *m* will increase. When there is little or no marital control *m* values will be negative. Thus, as populations depart from natural fertility, *m* values will be greater. As expected, Maize Region populations show a natural fertility pattern. Women in the Citrus Region, on the other hand, show positive numbers.

We started this chapter expecting that women in the Citrus Region would show changes in total fertility rate and fertility pattern. They did; reductions to five children per woman, however, are way above Mexico's TFR of 2.9 children per woman in 1994 or 2.4 in 2000 (Consejo Nacional de Población 2001). In addition, while 37 years of age at last birth is below natural fertility, it is still well above the 20 to 29 years of age group of most populations showing stopping behavior (Juárez et al. 1989; Knodel 1987; Quilodrán 1980; Suchindran and Koo 1992). The population dependent on circular migration and salaried work in the cities did not change its fertility patterns. They show a natural fertility curve, do not seem to want to limit their fertility and their TFRs and *M* are similar to very high fertility populations throughout history (Coale and Trussell 1974, 1978).

It is important to remember that both Citrus and Maize Region populations continue to depend on food production for survival. The *milpa* in both regions is essential and it continues to rule the peasant's cycle and household labor distribution even if it is no longer the principal source of income (Gurri 2007; Gurri García et al. 2001). Staple-based food production requires family labor and strong local non-monetary reciprocity (McAllister et al. 2012; McC Netting 1993; Lee and Kramer 2002; Gurri and Ortega-Muñoz 2015). Thus, while commercial agriculture may have encouraged fertility reduction by reducing the age of achieved fertility in the Citrus Region it is clear that large families continue to be important.

Salaries for unqualified labor are too low to satisfy household needs. Domestic units must therefore depend on food production at home in order to incorporate salaried income into their survival strategy (Daltabuit 1992). To do this, the Maya had to orchestrate the labor of larger families (Gurri 2007), promoting high fertility. In the Maize Region, food production allowed families to invest their wages in non-food items. These included cement floors and antibiotics. It also attracted resident private physicians and pharmacies resulting in improved well-being (Gurri García et al. 2001).

Both Citrus and Maize regions have been bombarded with intensive birth control campaigns for more than 50 years and development has changed the life of the Maya. Nevertheless, fertility continues to be high. Perhaps it is time to consider that while commercial agriculture may help reduce fertility, it will not substitute institutional care for the elderly and as long as it does not provide a truly viable alternative to food production, fertility will remain high. It is also unlikely that circular migration and economic growth in the cities can reduce fertility, for as long as the salary for unqualified labor continues to be as low as it is in Mexico, workers will continue to depend on food production.

Acknowledgements The authors wish to thank the women from the Maize and Citrus Region who opened their homes to the research teams. It is hard for us to imagine a more open and friendly population than the Maya of these two regions. We also wish to thank Drs. Emilio Moran and Gilberto Balam Pereira who were principal investigators and mentors to the second author. The Instituto Nacional Indigenista de México, the State of Yucatan, and National Science Foundation grant SBR#9420727 funded research in the Maize Region. The project Effects of Land Use Change on Landscape, Land Tenure, Social and Economic Organization from the Center for Institutions, Population, and Environmental Change (CIPEC) at Indiana University financed research in the Citrus Region. Any errors or misrepresentations are the sole responsibility of the authors.

References

Armelagos GJ, Goodman AH, Jacobs KH (1991) The origins of agriculture. Population growth during a period of declining health. Popul Environ 13(1):9–22

Atran S et al (1993) Itza Maya tropical agro-forestry. Curr Anthropol 34(5):633–700

Balam B et al (1998) Un estudio de los cambios sociales y agroecológicos en Yucatán mediante entrevistas familiares y el sistema de percepción remota: influencia de los cambios ecológicos en la salud y nutrición de dos tipos de población campesina de Yucatán, México. Cinvestav, México

Barclay GW et al (1976) A reassessment of the demography of traditional rural China. Popul Index 42(4):606–635

Beall CM, Leslie PW (2014) Collecting women's reproductive histories. Am J Hum Biol 26(5):577–589. https://doi.org/10.1002/ajhb.22543

Becker GS (1991) A treatise on the family, enlarged edition. Harvard University Press, Cambridge

Bentley GR, Goldberg T, Jasienska G (1993) The fertility of agricultural and non-agricultural traditional societies. Popul Stud J Demog 47(2):269–281

Bongaarts J (2001) Fertility and reproductive preferences in post-transitional societies. Popul Dev Rev 27.(Supplement: Global Fertility Transition:260–281

Bongaarts J, Cotts Watkins S (1996) Social interactions and contemporary fertility transitions. Popul Dev Rev 22(4):639–682

Boserup E (1984) Technical change and human fertility in rural areas of developing countries. In: Schutjer WA, Stokes CS (eds) Rural development and human fertility. MacMillan Publishing Co., New York, pp 23–33

Bryant J (2007) Theories of fertility decline and the evidence from development indicators. Popul Dev Rev 33(1):101–127

Caldwell JC (1978) A theory of fertility: from high plateau to destabilization. Popul Dev Rev 4(4):553–578

Caldwell JC, Caldwell BK (2003) Pretransitional population control and equilibrium. Popul Stud J Demog 57(2):199–215

Caldwell JC, Schindlmayr T (2003) Explanations of the fertility crisis in modern societies: a search for commonalities. Popul Stud J Demog 57(3):241–263

Cleland J, Rutstein S (1986) Contraception and birthspacing. Int Fam Plan Perspec 12(3):83–90

Coale AJ, Trussell TJ (1974) Model fertility schedules: variation in the age structure of childbearing in human populations. Popul Index 40(2):185–258

Coale AJ, Trussell TJ (1978) Technical note: finding the two parameters that specify a model schedule of marital fertility. Popul Index 44(2):203–213

Consejo Nacional de Población (2001) La población de México en el nuevo siglo. Consejo Nacional de Población, México, DF

Daltabuit M (1992) Mujeres mayas: Trabajo nutrición y fecundidad. Instituto de Investigaciones Antropológicas. Universidad Nacional Autónoma de México, México, DF

Dixon-Mueller R (2000) Women and the labour market in changing economies: demographic issues. Conference Proceeding of International Union for the Scientific Study of Population, p 1–14

Fedick SL (1996) An interpretative kaleidoscope: alternative perspectives on ancient agricultural landscapes of the Maya lowlands. In: Fedick SL (ed) The managed mosaic. Ancient Maya agriculture and resource use. University of Utah Press, Salt Lake City, pp 107–131

Feldman BS, Zaslavski AM, Ezzati M et al (2009) Contraceptive use, birth spacing, and autonomy: an analysis of the "Oportunidades" program in rural Mexico. Stud Family Plann 40(1):51–62

Gungwu W (1997) Global history and migrations. Westview Press, Boulder, CO

Günther I, Harttgen K (2016) Desired fertility and number of children born across time and space. Demography 53(1):55–83. https://doi.org/10.1007/s13524-015-0451-9

Gurri FD (2007) Globalización y cambios en la calidad de vida en familias campesinas de Yucatán, México. Ciencia 58(2):60–68

Gurri FD (2010) Smallholder land use in the southern Yucatan: how culture and history matter. Reg Environ Chang 10(3):219–231

Gurri García FD, Balam Pereira G, Moran EF (2001) Wellbeing changes in response to 30 years of regional integration in Maya populations from Yucatan, Mexico. Am J Hum Biol 13:590–602

Gurri FD, Moran EF (2002) Who is interested in commercial agriculture?: subsistence agriculture and salaried work in the city among Yucatec Maya from the State of Yucatan. Cult Agric 24(1):42–48

Gurri FD, Ortega-Muñoz A (2015) Impact of commercial farming on household reproductive strategies in Calakmul, Campeche, Mexico. Am J Hum Biol 27(6):758–766

Hugo GJ (1982) Circular migration in Indonesia. Popul Dev Rev 8(1):59–83. https://doi.org/10.2307/1972690

Juárez F, Quilodrán J, Zavala de Cosío ME (1989) De una fecundidad natural a una controlada: México 1950–1980. Estudios Demográficos y Urbanos 4(1):5–51

Knodel J (1987) Starting, stopping, and spacing during the early stages of fertility transition: the experience of German village populations in the 18th and 19th centuries. Demography 24:143. https://doi.org/10.2307/2061627

Knodel J, Friedman J, Si Anh T et al (2000) Intergenerational exchanges in Vietnam: family size, sex composition, and the location of children. Popul Stud J Demog 54(1):89–104

Lee RD, Kramer KL (2002) Children's economic roles in the Maya family life cycle: Cain, Caldwell and Chayanov revisited. Popul Dev Rev 28(3):475–499

Lee RD, Mason A, Miller T (2000) Life-cycle saving and the demographic transition: the case of Taiwan. Popul Dev Rev 26:194–219

McAllister L, Gruven M, Kaplan H et al (2012) Why do women have more children than they want? Understanding differences in women's ideal and actual family size in a natural fertility population. Am J Hum Biol 24(6):786–799. https://doi.org/10.1002/ajhb.22316

McC Netting R (1993) Smallholders, householders: farm families and the ecology of intensive, sustainable agriculture. Stanford University Press, Stanford

Michaels G (2008) The effect of trade on the demand for skill: evidence from the interstate highway system. Rev Econ Stat 90(4):683–701

Neyer G, Andersson G (2008) Consequences of family policies on childbearing behavior: effects or artifacts? Popul Dev Rev 34(4):699–724

Notestein FW (1945) Population. The long view. In: Schultz TW (ed) Food for the world. The University of Chicago Press, Chicago, pp 36–56

Ortega Muñoz A (2012) Una frontera en movimiento. Migración, fecundidad e identidad en el sur de Quintana Roo y norte de Honduras Británica (Belice), 1900–1935. Instituto Nacional de Antropología e Historia, El Colegio de México, México, DF

Pickard L (2011) The supply of informal care in Europe. ENEPRI Research Report no. 94. CEPS, Brussels

Presser HB, Hattori MLK, Parashar S et al (2006) Demographic change and response: social context and the practice of birth control in six countries. J Pop Res 23(2):135–163

Pritchett LH (1995) Desired fertility and the impact of population policies. Popul Dev Rev 20(1):1–55

Pullum TW, Casterline JB, Juárez F (1985) Changes in fertility and contraception in Mexico, 1977–1982. Int Fam Plan Perspect 11(2):40–47

Quilodrán J (1980) Algunas características de la fecundidad rural en México. Demogr Econ 14(4):397–410

Rabell Romero C, Murillo López S (2016) Redes sociales de apoyo a necesidades de la vida cotidiana de las familias. Coyuntura Demográfica 10:42–47

Redfield R, Villa Rojas A (1962). [1934]) Chan Kom. A Maya village. The University of Chicago Press, Chicago

Rodríguez G (1996) The spacing and limiting components of the fertility transition in Latin America. In: Guzmán JM, Singh S, Rodríguez G et al (eds) The fertility transition in Latin America. Claredon Press, Oxford University Press, Oxford, NY, pp 27–47

Rosales González M (1988) Oxcutzcab, Yucatán 1900–1960. In: Campesinos, cambio agrícola y mercado. Colección Regiones de México. Instituto Nacional de Antropología e Historia, México, DF

Rosales González M (2012) Estrategias de intervención para el desarrollo local: experiencias en comunidades mayas del sur de Yucatán. Península VII(1):79–101

Roser M (2018) Fertility rate. Published online at OurWorldInData.org. https://ourworldindata.org/fertility-rate. Accessed 9 Sept 2018

Rosero-Bixby L (1983) Determinantes de la fecundidad en Costa Rica, Notas de Población. Revista Latinoamericana de Demografía 11(32):79–122

SAGARPA (2012) Plan rector del sistema producto. Cítricos. Estado de Yucatán. http://dev.pue.itesm.mx/sagarpa/estatales/EPT%20COMITE%20SISTEMA%20PRODUCTO%20CITRICOS%20YUCATAN/PLAN%20RECTOR%20QUE%20CONTIENE%20PROGRAMA%20DE%20TRABAJO%202012/PR_CITRICOS_YUCATAN 2012.doc.pdf. Accessed 8 Oct 2018

Saint-Maurice A, Pintassilgo SC (2018) Ethnic differences in results of fertility and mother's health care: Portuguese population and Cape Verdeans living in Portugal. J Popul Res 35(2):131–150

Suchindran CM, Koo HP (1992) Age at last birth and its components. Demography 29(2):227–245

Terán SC, Rasmussen CH (1994) La milpa de los mayas: La agricultura de los mayas prehispánicos y actuales en el noreste de Yucatán. Danida, Mérida, Yucatán, México

Tobin J (1967) Life cycle saving and balanced economic growth. In: Fellner W (ed) Ten economic studies in the tradition of Irving fisher. Wiley Press, New York, pp 231–256

Villanueva Mukul E (1994) La formación de las regiones en la agricultura. El caso de Yucatán. Maldonado Editores, Instituto Nacional Indigenista, Universidad Autónoma de Yucatán, CEDRAC, Mérida, Yucatán, México

White CD (1999) Reconstructing ancient Maya diet. The University of Utah Press, Salt Lake City

Wilson C, Oeppen J, Pardoe M (1988) What is natural fertility? The modelling of a concept. Popul Index 54(1):4–20

Zavala de Cosío ME (2014) La transición demográfica en México (1895–2010). In: Rabell C (ed) Los mexicanos: un balance del cambio demográfico. Fondo de Cultura Económica, México, DF, pp 80–114

Chapter 8
Agricultural Transformation and Ontogeny in Rural Populations from the Yucatan Peninsula at the turn of the Century: Studying Linear Enamel Hypoplasias and Body Composition in Adolescents

Francisco D. Gurri

Abstract This chapter associates early developmental stress, as evidenced by the presence or absence of linear enamel hypoplasias (LEH), with fat accumulation in adolescents using body mass index (BMI) in two agricultural populations from the Yucatan Peninsula with different degrees of dependence on store foods. Adolescent males with LEHs had lower BMIz scores while the reverse was true for females. BMI variation in males depended on their ability to satisfy an activity schedule with a high-energy demand that placed individuals who had early developmental problems at a disadvantage. Women on the other hand, while busy for longer hours than men, had more stationary activities and did not participate in sports. Systemic problems in utero and early childhood would therefore lead to a thrifty phenotype and, because they are stationary, these women tended to accumulate body fat faster than those without LEHs. That this happened both in Calakmul with a high fiber traditional diet and in Central Yucatan with a store-bought high carbohydrate diet suggests that labor allocation differences in agricultural populations in the Yucatan Peninsula put women at a disadvantage that may lead those with a thrifty phenotype to become overweight. That this process is much more evident in Central Yucatan than in Calakmul suggests that development will exacerbate the consequences of traditional gender inequalities already present in the agricultural household adaptive strategy.

F. D. Gurri (✉)
Department of Sustainability Science, Environmental Anthropology and Gender Lab, El Colegio de la Frontera Sur (ECOSUR), Campeche, México
e-mail: fgurri@ecosur.mx

H. Azcorra, F. Dickinson (eds.), *Culture, Environment and Health in the Yucatan Peninsula*, https://doi.org/10.1007/978-3-030-27001-8_8

8.1 Introduction

During the twentieth century, development initiatives promoting commercial agriculture and salaried work transformed subsistence agriculturalist around the world. According to most case studies of the time, childhood undernutrition followed in its wake (Adams 1974; Baer 1987; Clark 1980; Daltabuit Godás 1992; Dewey 1979, 1981; Fleuret and Fleuret 1980; Hoorweg et al. 1996; King 1971; Little 1991; Schoepf and Schoepf 1987). By the 1970s it became clear that for several reasons profits from commercial crops, cattle, or labor were seldom enough for the local populations to buy adequate amounts of high-quality foods to replace the nutrients present in traditional diets (Adams 1974; Baer 1987; Cattle 1976; Dewey 1979; Dewey 1981; Daltabuit Godás 1992; Fleuret and Fleuret 1991; King 1971; Nietschmann 1972; Schoepf and Schoepf 1987; Reategui 1976; Rawson and Valverde 1976; Williams 1973). This, they discovered, was particularly damaging to pregnant women and to children under the age of five years (Eder 1978; Gross 1975; Haas and Harrison 1977; Rappaport et al. 1971; Wilmsen 1978.

During this century, it was suggested that these new nutrient-poor and calorie-rich unbalanced diets that were substituting staple-based nutritional regimes were not only affecting child growth, they were also making adults gain weight. Thus, while childhood undernutrition continued to increase in rural areas, it began to be accompanied by adult obesity (Arroyo et al. 2007; Malina et al. 2007; Doak et al. 2005). Caballero (2005) christened this phenomenon as "the dual burden of the nutrition transition" which they originally observed in urban populations.

Recent studies suggested that the new calorie-rich obesogenic environments may not be the only determinant of metabolic syndrome in adults. Children growing up in poor nutritional environments will develop a thrifty phenotype (Hales and Barker 2001, 2013; Wells 2011). They will be shorter and develop permanent metabolic changes that will make them less efficient in the oxidation of body fat (Frisancho 2009). These children will tend to accumulate fat, will become overweight easier than children growing under better conditions, and will have a tendency to develop metabolic syndrome when exposed to obesogenic environments (Hoffman et al. 2000). Through the alteration of subsistence staple-based agricultural regimes, globalization may thus be generating ecological changes that affect ontogeny by reducing the nutritional quality of the diet in utero and early childhood provoking the presence of thrifty phenotypes. At the same time, it may be favoring the accumulation of body fat in adolescents and adults by increasing their dependence on an obesogenic calorie-rich store-bought diet.

In the Yucatan Peninsula, globalization has altered staple-based subsistence strategies to different degrees generating different double burden profiles between agriculturalists (Gurri 2015). Except for the touristic coast by the Caribbean, there is a developing gradient from north to south, and in the north from east to west. As one travels south, and west, agricultural communities become more dependent on the ancestral staple-based survival system. The purpose of this chapter is to contrast nutritional ecology differences in two regions of this developmental gradient. The first is composed of nine municipalities from the central part of the peninsula referred to as the Citrus-Maize region of the State of Yucatan (Villanueva Mukul 1994). The second one, refereed to here as Calakmul, is composed of populations from two municipalities straddling the border of the states of Campeche and Quintana Roo, next to the border with Guatemala. The former has been shown to have greater childhood stunting and greater overweight and obesity than the latter (Gurri 2015). In particular, I am interested in detecting under which circumstances will ecological relationships lead to the accumulation of body fat in adolescents whose growth may have been affected during early childhood.

Linear enamel hypoplasia (LEH) will be used as proxy of environmental conditions in utero and during early childhood. LEHs will form at this time when negative ecological conditions affect tooth development (Goodman and Rose 1990). Because LEHs can be observed as long as the tooth is present in the mouth, ecological conditions during early childhood may be inferred at any age. Increased frequencies of individuals in a population with at least one hypoplasia in the mouth have thus been interpreted as a consequence of negative environmental conditions in utero and during the first years of life (Blakey et al. 1994; Belcastro et al. 2007; Manzi et al. 1999; Maunders et al. 1992; Goodman et al. 1984, 1991, 1987; Gurri et al. 2001; Zhou and Corruccini 1998; Fenton 1998; van Gerven et al. 1995). Because the stressors that disrupt early enamel formation adversely affect critical periods of child development, enamel defects predict long-term health status. Studies of developmental defects of enamel (DDE) showed that they predict stunted growth in late childhood and adolescence (Masterson et al. 2018; Rugg-Gunn et al. 1997; Santos and Coimbra 1999), celiac disease in later life (Pastore et al. 2008; Rashid et al. 2011), and longevity (Armelagos et al. 2009).

The body mass index (BMI) estimated as weight/height^2 provides an accurate measure of overall fat mass (Farewell et al. 2018; Garrow and Webster 1985). Furthermore, it is as strongly correlated with several metabolic complications as are a number of better estimates of body fatness (Hariri et al. 2013; Freedman and Ford 2015; Sun et al. 2010; Steinberger et al. 2005). The relationship between body fat and BMI varies by age, sex, and between Black, White, Asian, and Hispanic populations (Camhi et al. 2011; Deurenberg et al. 1998; Rush et al. 2009). BMI comparisons seeking ecological relationships should be made between comparable populations or controlling for the effects of sex age and ethnicity.

In this chapter male and female adolescents with and without hypoplasia from the Maize-Citrus region of the State of Yucatan were compared to those from Calakmul in Campeche and Quintana Roo. We expected to find significant BMI differences between individuals with LEHs and those without in the obesogenic environment of central Yucatan, but not in the more traditional populations from Calakmul.

8.2 Background

In the Yucatan Peninsula, the staple-base system is known as the milpa. It has been the dominant household adaptive system since the Formative Period (2000 BCE to 150 CE) (Coe 1984), and it changed very little with the Conquest (Mariaca-Méndez 2015). It is a diversified polyculture that includes maize (*Zea mays*), different types of beans (*Phaseolus* sp.), squash (*Cucurbita* sp.), chili peppers (*Capsicum* sp.), and other cultigens that will vary locally. The milpa is complimented by harvesting fallow fields, hunting and gathering, backyard agriculture and forest management (Atran et al. 1993; Mariaca-Méndez 2015; Terán and Rassmussen 1992). In addition, it continues to regulate most community and family rituals as well as religious beliefs (Tuz Chi 2013).

In 1847, the Maya from the entire Peninsula rose up against all white Mexicans in a war later known as the Caste War (Barabas 1989; Reed 1964). The political and economic geography of the state of Yucatan until the 1970s would be shaped by the fortunes of this war (see Chap. 2). Thus, from the blood and ashes of the Caste War arose a divided Yucatan: a white Yucatan to the east, north-east, a Maya Yucatan to the center, south and east of the peninsula, and large amounts of abandoned tropical forests to the south. These regions, kept separate by war, developed independent social, political, and economic organizations.

To the west, north-west, white hacienda owners confiscated Maya communal land and incorporated the peasants into sisal (*Agave fourcroydes*) plantations (Murguía 1981; Villanueva Mukul 1985, 1994; Wells and Joseph 1996). In the south, south-west, central Yucatan, and southeast close to the border with Belize, the Maya planted their milpas. Finally, further south and south west close to the border with Guatemala, mature tropical forests covered long abandoned ancient Maya sites and served as refuge to a few Maya communities (Ferré 1996; Haenn 2005), and later as temporary home to workers from lumber companies dedicated to the extraction of precious wood, and rubber (Fort 1979; Haenn 2005) (see Chap. 2).

Starting in the 1970s two different processes transformed subsistence agriculturalists from the central Maya region, and a land distribution under agrarian reform reopened the southern forests to agriculture. In the central part of the Peninsula, the south and southeastern part of today's state of Yucatan, a government plan, named *Plan Chaak*, followed later by *Plan Tabi* encouraged Maya agriculturalists to replace maize dry cultivation with irrigated citrus commercial agriculture in the 1940s and 50s. Commercial citrus cultivation was slow to start, but eventually, in the early 1970s, expanding markets due to a growing urban population in Merida, development of oil extraction in Tabasco, and a booming tourist industry in the Caribbean, encouraged the increase in the amount of land dedicated to commercial citrus cultivation (Lazos 1995; Rosales González 1988). In the 70s and 80s, for example, as much as 33% of communally held land in the Municipio of Akil was dedicated to commercial citrus and horticultural production (INEGI 1991).

Further west, where no irrigation programs followed, a modern highway, and an efficient public transportation system provided peasants with access to the developing urban centers of Merida and Cancun in 1973 (Daltabuit Godás 1992; Gurri and Balam 1992). This road increased the number of peasants seeking salaried work in the cities, and transformed the economy of the maize region from an agricultural economy (Press 1975; Re Cruz 1996; Redfield 1950; Redfield and Villa Rojas 1934; Terán and Rasmussen 1994; Terán and Rassmussen 1992; Villanueva Mukul 1994; Villa Rojas 1985) to a mixed economy. In the Central Maize region most families have at least one member working in Mérida or Cancún during the week (Gurri and Balam 1992). Prior to 1985 only men migrated into the cities. The number of women migrants, however, has risen continually since then (Gurri and Balam 1992). Migration especially since 1980 has become circular. Peasant households are placing more emphasis on wage labor and participation in the cash economy without abandoning slash and burn agriculture in their home town (Gurri and Balam 1992).

In central Yucatan, population growth has forced agriculturalists to reduce fallow times and forest cover with disastrous consequences for the milpa system (Gurri and Moran 2003; Sohn et al. 1999). Peasants today must use herbicides which reduce the milpa's variability and there has been a reduction in the availability of wild plants and animals. In addition, since the 1980s the diet has become ever more dependent on cheap carbohydrate-rich store-bought foods (Daltabuit Godás 1992; Daltabuit Godás and Leatherman 1998; Leatherman and Goodman 2005). In the municipalities with citrus cultivation, meat and party foods from the market supplement their diet, while agriculturalists further east will usually only have enough money to complement their diets with cheap pastas, junk food, and sodas bought at the local stores (Balam Pereira and Gurri García 1994; Leatherman and Goodman 2005).

During the early 1970s colonists from all over Mexico occupied the forests from the southern Yucatan Peninsula in what turned out to be the countries' last agrarian land distribution. Today, households in southern Yucatan combine subsistence agriculture with jalapeño pepper (*Capsicum annum*) production for the market. Colonists do not have direct access to the market. All their production is sold to intermediaries that drive into the area to purchase their peppers. Intermediaries decide how much of the harvest they will buy, and set the prices. Urban labor markets are not easily accessible either. There is no circular migration, and those who migrate usually go to the US (Gurri García et al. 2008). In Calakmul, diet depends mostly on the traditional survival system. Agriculturalists harvest a diversified milpa and gather roots from abandoned fallows (Alayón-Gamboa and Gurri-García 2007). Their backyards are at the same time a source of income and food (Alayón-Gamboa and Gurri-García 2008), and their cuisine is heavily dependent on animals and vegetables hunted and gathered in the surrounding tropical forest (Flores Medina and Gurri García 2005; Gurri et al. 2001).

8.3 Materials and Methods

The data in this chapter was collected during four different research projects in the Yucatan Peninsula (see Fig. 8.1). Central Yucatan, the Citrus-Maize region, was studied between 1992 and 1998 while the author worked for the Centro de Investigación y Estudios Avanzados del Instituto Politécnico Nacional (Cinvestav-IPN) unit Merida in Mexico, and later as a Ph.D. student at Indiana University (IU). Fieldwork for the Calakmul project took place in 1999 when the author worked at El Colegio de la Frontera Sur (ECOSUR)—Campeche.

IU's Human Subjects' Committee preapproved the projects that took place while the author was a student at IU. Cinvestav and ECOSUR, like most Mexican research institutions and universities at the time, did not have anything equivalent. As with the IU projects, however, the Mexican projects were presented to state, municipal, community, and health authorities for review and authorization. The latter were the only ones in those days with a formal human subjects' committee equivalent. The same procedures were followed in every project. In each case, verbal informed consent was obtained from every household head who participated.

Enough households were selected in each project, to measure sufficient children to obtain representative estimates of nutritional status in children under 10 years of

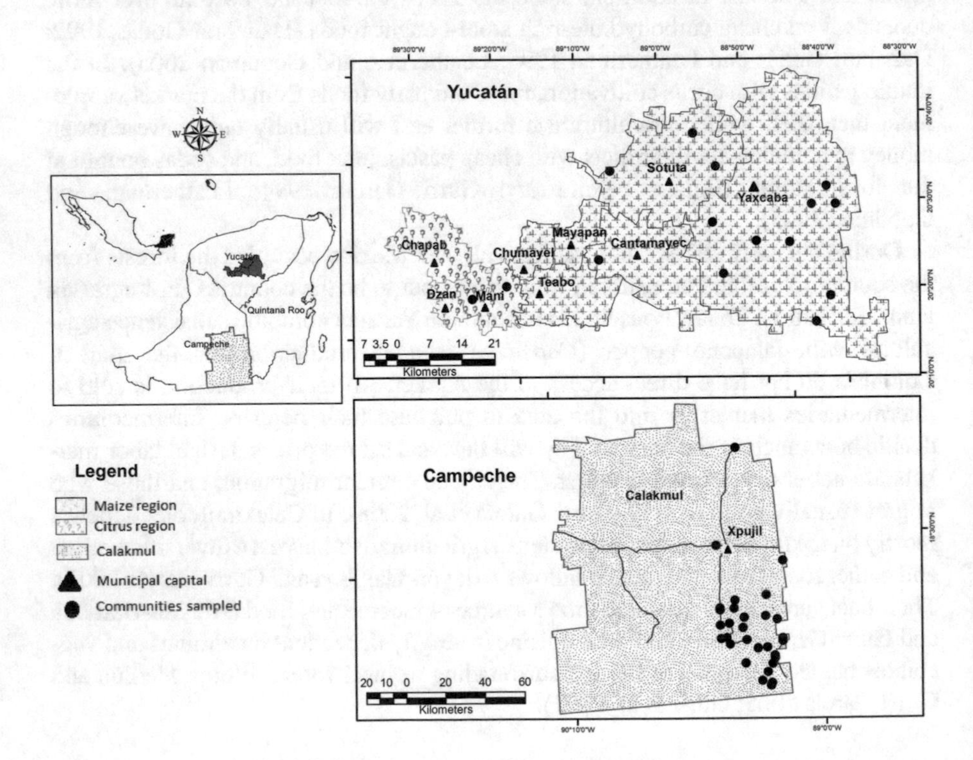

Fig. 8.1 Location map of the Maize and Citrus Regions in Yucatan, and Calakmul in Campeche, Mexico (map elaborated by Victor M. Ku Quej, ECOSUR)

age. In Yucatan number of households sampled was determined by a stratified sampling strategy by municipality and number of houses per community were selected at random (Gurri et al. 2001). In Calakmul, where the municipality boundaries are not clear, the number of households chosen was large enough to obtain a nutritional status estimate from the southernmost region of the States of Campeche and Quintana Roo (Gurri 2010). All individuals present in the household at the time of our visit were interviewed, measured, and examined for LEHs. Only adolescents from 10 to 20 years were analyzed for this chapter. A stratified sample size estimation for proportions (Cochran 1977) using 0.49 for Calakmul and 0.38 for Yucatan as an estimate of relative LEH frequency in adolescents (estimated from this sample) was used to ensure that sample size was the optimum required to make comparisons between sexes and populations with a desired precision of 0.05 and a 0.95 confidence interval. The sex age distributions of each region were tested with a Mann–Whitney μ test to make sure their populations were comparable and no adjustments had to be made for age or sex.

The permanent upper central incisors, the lower central incisors, and mandibular canines were examined for LEH in individuals 10 years and older. The first and the last have been recommended for population surveys as being the most hypoplastic teeth (Duray 1992; Goodman and Armelagos 1985; Goodman and Rose 1990). The lower central incisors were also included because of the frequent loss of the upper central incisors in Yucatec populations. Inspection of these teeth was done under indirect sunlight next to the door of the house, and with the aid of a flashlight held by an assistant.

Adolescents were measured for height barefoot, standing straight following the recommendations of Frisancho (1990) with a Martin Type Anthropometer. They were weighed in a portable bath scale without shoes and with as little clothing as possible. One anthropometrist performed the measure and repeated it aloud to a second one who confirmed the number and wrote it down. In each study, anthropometrists standardized before going into the field using the author as the "gold standard." In each region tests were performed to control for interobserver error (Ulijaszek and Kerr 1999). No error measurements were made between studies. Since the author was the gold standard in each, measurement error could be attributable to unconscious changes in the author's measurement technique, so little measurement error between groups should be expected. Nevertheless, the reader may want to keep this in mind when interpreting the significance levels of intergroup variance.

BMI was estimated as weight in kgs/height in cms^2. BMI was transformed into BMIz scores using WHO standards (de Onis et al. 2007). Differences in BMIz scores between sexes and regions were tested with a three-way ANOVA for individuals with and without hypoplasias. The factorial model was built with BMIz scores as dependent variable and presence or absence of LEH, Region and Sex as fixed factors. Hypoplasia differences between sexes and regions were tested with a χ^2.

8.4 Results

Figure 8.1 is a location map showing the municipalities and communities from the Citrus-Maize region and Calakmul where the studies took place. Table 8.1 shows sample size distribution by municipality. Oton P Blanco is part of the state of Quintana Roo and its communities were analyzed here as part of Calakmul. The population's age-sex distribution is shown in Fig. 8.2 for each region. The population pyramids are not significantly different (Mann-Whitney μ = 135639.5, $\alpha = 0.866$). Neither are the age distributions between male and female adolescents (Mann-Whitney μ = 146,695, $\alpha = 0.066$).

Table 8.2 shows the LEH relative frequency distribution for males and females in the Citrus-Maize region and in Calakmul. LEH frequencies in Southern Campeche were significantly higher than in Yucatan ($\chi^2_{1d.f.} = 14.416$, $\alpha < 0.0001$). No significant differences were found between males and females ($\chi^2_{1d.f.} = 0.038$, $\alpha = 0.845$).

Table 8.3 displays descriptive statistics for BMIz scores. They include mean BMI for reference only. The F values for the three-way ANOVA comparisons for BMIz scores may be found in Table 8.4. The factorial Model is significant. As expected, BMIz scores are significantly higher in females than in males (0.06 vs. −0.19, respectively) and BMIzs are significantly higher in Yucatan than in Calakmul (0.02 vs. −0.14). There aren't any significant BMIz differences between individuals with and without LEHs, but there is a significant interaction effect for BMIz between LEH and sex. Men with and without LEHs have similar BMIz scores. Women with LEHs have significantly higher BMIzs than women without, and unlike women, adolescent males with hypoplasias have lower BMIzs (Fig. 8.3). This is true for both

Table 8.1 Sample size by research site, municipality, and sex

		Sex		
Research site	Municipality	Male	Female	Total
Citrus-Maize Region	Cantamayec	36	34	70
	Chapab	17	23	40
	Chumayel	35	37	72
	Dzan	22	27	49
	Maní	33	51	84
	Mayapán	31	28	59
	Sotuta	36	54	90
	Teabo	15	25	40
	Yaxcabá	44	78	122
	Total	269	357	626
Calakmul	Calakmul	153	233	386
	Otón P. Blanco	27	23	50
	Total	180	256	436
Total		449	613	1062

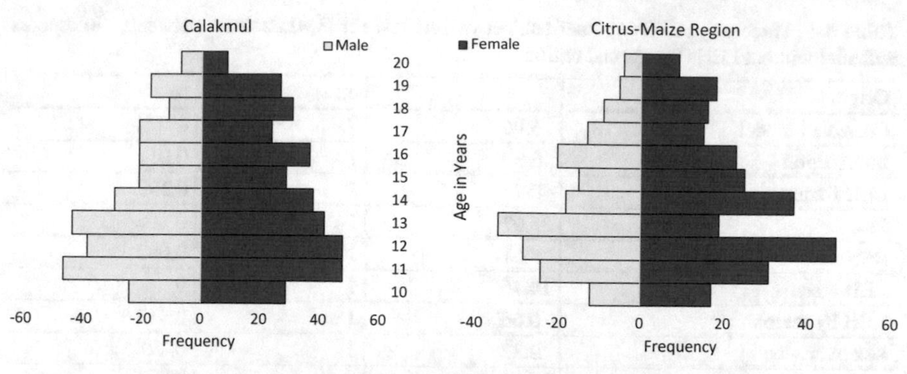

Fig. 8.2 Population Pyramids for the Maize-Citrus region, Yucatan and Calakmul, Campeche

Table 8.2 LEH frequencies by sex and region

	Region			
Sex	Citrus-Maize	Calakmul	Total	*n*
Male	0.38	0.52	0.43	449
Female	0.38	0.48	0.42	613
Total	0.38	0.50	0.43	1062
n	626	436	1062	

$\chi^2_{1d.f.} = 14.416$, $\alpha = <0.0001$ for LEH differences between regions
$\chi^2_{1d.f.} = 0.038$, $\alpha = 0.845$ for LEH differences between sexes

Table 8.3 BMI, BMIz scores for adolescents with and without LEHs by sex and region[a]

		Region:								
		Citrus-Maize Region			Calakmul			Total		
		Sex			Sex			Sex		
LEH:		Male	Female	Total	Male	Female	Total	Male	Female	Total
Without LEH	Mean BMI	18.87	19.43	19.19	18.38	19.80	19.24	18.70	19.57	19.21
	Mean BMIz	−0.01	−0.05	−0.04	−0.27	−0.08	−0.15	−0.10	−0.06	−0.08
	S.D. BMIz	0.96	0.96	0.96	1.06	0.96	1.00	1.00	0.96	0.97
	N	168	221	389	87	132	219	255	353	608
With LEH	Mean BMI	19.47	21.03	20.37	18.42	20.14	19.40	18.97	20.61	19.91
	Mean BMIz	−0.17	0.30	0.10	−0.47	0.14	−0.13	−0.31	0.22	−0.01
	S.D. BMIz	1.01	0.84	0.94	0.87	0.92	0.95	0.96	0.88	0.95
	N	101	136	237	93	124	217	194	260	454
Total:	Mean BMI	19.10	20.04	19.63	18.40	19.97	19.32	18.82	20.01	19.51
	Mean BMIz	−0.07	0.08	0.02	−0.37	0.03	−0.14	−0.19	0.06	−0.05
	S.D. BMIz	0.98	0.93	0.95	0.97	0.95	0.97	0.99	0.94	0.96
	N	269	357	626	180	256	436	449	613	1062

[a]WHO 2005 Standards were used for BMIz score transformations

Table 8.4 Three-way ANOVA: Inter-subject effects test for BMIz scores[a] between adolescents with and without LEHs by sex and region

Origin	*F*	d.f.	*α*
Corrected model	6.96	7	0
Intersection	6.50	1	0.011
LEH frequency	0.73	1	0.392
Sex	25.67	1	0
Region	9.51	1	0.002
LEH ∗ sex	14.76	1	0
LEH by region	0.66	1	0.417
Sex by region	2.32	1	0.128
LEH by sex by region	0.13	1	0.716

Model: Intersection + LEH + Sex + Region + (LEH∗Sex) + (LEH∗Region) + (Sex∗Region) + (LEH∗Sex∗Region)
[a]WHO 2005 Standards were used for BMIz score transformations

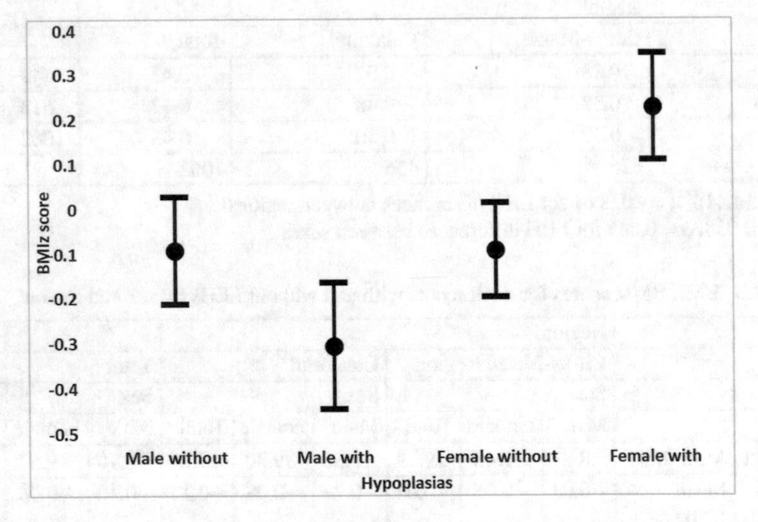

Fig. 8.3 BMIz scores for male and female adolescents with and without LEHs from both regions of the Yucatan Peninsula

regions. In Calakmul, however, male adolescents have sigificantly lower BMIz, than males in the Citurs-Maize region. Female adolescents without LEHs, have similar BMIz scores in both regions (–0.05 in Citrus-Maize Region and –0.08 in Calakmul). Only females with LEHs from the Citrus-Maize region have significantly higher BMIz scores than females without in either region (Fig. 8.4).

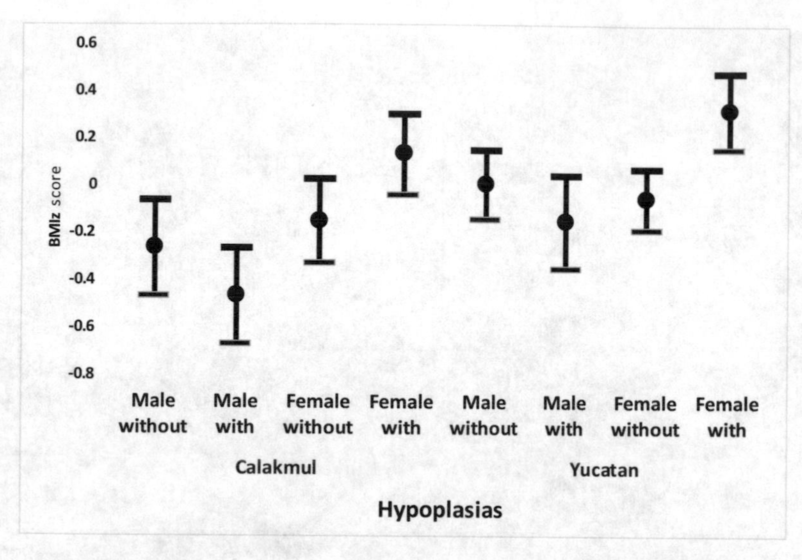

Fig. 8.4 BMIz scores for male and female adolescents with and without LEHs Per Region

8.5 Discussion

Use of BMI to measure fat accumulation is common, but as mentioned in the introduction, its meaning may vary by age, sex, and ethnic group. The samples used for this investigation, however, are comparable. They have similar sex-age distributions and are ethnically alike. The Citrus-Maize region population is Yucatec Maya and more homogeneous than that of Calakmul. The latter includes other Maya groups such as Chol and Mam, other non-Maya Amerindian groups such as the Zoque-Popoluca and admixed populations from different parts of Mexico (Gurri-García et al. 2001; Gurri 2007). Thus, while genetic differences have been encountered between Blacks, Whites, and Asians (Camhi et al. 2011; Deurenberg et al. 1998; Rush et al. 2009), we are unaware of any evidence that suggests different patterns of fat accumulation between Amerindian populations or between them and admixed "Hispanic."

In this analysis, LEH frequencies were significantly higher in Calakmul than in central Yucatan. Increased systemic perturbations during growth will augment LEH frequencies (Goodman et al. 1991; Dummer 1977; Kreshover 1960; Pindborg 1982, 1970; Sweeney et al. 1969, 1971; Sarnat and Schour 1942; Rose et al. 1985). In addition, the number of lesions the investigator will be able to detect will depend on the severity and duration of the disruptive event (Hutchinson and Larsen 1988; Ensor and Irish 1995). Thus, as growth disruptions become fewer, and the population's ability to recuperate from them increases, not only shall overall LEH frequencies be reduced, but the visible LEHs will concentrate in the middle most vulnerable part of the tooth (Gurri García 2011; Santos and Coimbra 1999; Lanphear 1990; Goodman et al. 1984, 1987; Goodman and Armelagos 1985). Development in the

Fig. 8.5 *Aljibe* for rainwater collection at José María Morelos y Pavón in Othon P. Blanco, Quintana Roo, Mexico (Calakmul Region)

Citrus-Maize region, indeed in the entire state of Yucatan, has positively modified the ecological conditions leading to higher LEH frequencies (Gurri et al. 2001). Changes since the 1970s include community health centers, affordable pharmacies, resident physicians, successful vaccination campaigns, and rehydration therapy for patients with diarrhea. At the household level, cement has substituted dirt floors and tubed water is now available. Accordingly, not only were there reductions in LEH between individuals growing before the 1970s and those who grew up after that, but also the LEH within tooth distribution in the Maize region changed from platykurtic to leptokurtic with the majority of the LEHs located in the middle half of the tooth (Gurri et al. 2001).

Calakmul, on the other hand, is Mexico's last frontier (Gurri 2010, 2015). There are a small number of health centers and the population depends on them for medicine and health care. There are few resident physicians and those who attend the clinic are usually young medical students doing their residency between Monday and Friday. Overall living conditions are harsher than in the Citrus-Maize region in Yucatan. Most houses have dirt floors, water is a scarce commodity, and at the time of this research, the communities' didn't have electricity. Most families depended on rainwater collected in large communal deposits known as "aljibes" and private water tanks (Figs. 8.5, 8.6, and 8.7). Higher LEH frequencies, as well as more platykurtic within tooth LEH distributions (Gurri Garcia 2011), are to be expected.

Fig. 8.6 *Aljibe* for rainwater collection at El Campanario, Calakmul, Campeche

Fig. 8.7 Homemade water tank installation for rainwater collection in Calakmul, Campeche

The diet in Calakmul is produced locally and is probably better balanced than that of central Yucatan. The latter have substituted fresh water from fruits in their backyards with carbonated sodas, reduced legumes and game traditionally obtained from and around the milpa, and have increased intake of sugars, fat, and "junk foods" such as animal crackers (Leatherman and Goodman 2005; Gurri García et al. 2018; Otero Prevost et al. 2017; Gurri et al. 2001). The higher BMIs in adolescents from the Citrus-Maize region found here as well as greater stunting in children under 10 years of age and increased prevalence of overweight and obesity in adults, particularly in women (Gurri 2015) are the consequences of this dietary shift.

In the association between LEH presence and BMIz scores, we found two different patterns: one for males and one for females suggesting gendered dependent micro environmental conditions during growth. Both patterns will reflect the result of the interaction between early development, environmental conditions during growth and adolescent phenotype. Although BMIz differences between males with and without LEHs were not significant, their tendency is similar to the one found by researchers in isolated traditional populations with low calorie diets and little modern medical attention. In these populations, individuals with DDE will have lower BMI and poorer health (Masterson et al. 2018; Rugg-Gunn et al. 1997; Santos and Coimbra 1999; Armelagos et al. 2009). The suggestion is that early systemic problems will affect the entire growth process, thus show up in health indicators during adolescence, and of course follow into adult life.

The results from the male pattern in these populations from the Yucatan Peninsula suggest that, more than diet, body composition may be related to energy expenditure. Individuals in Calakmul consume fewer calories than those in the Citrus-Maize region. Nevertheless, fat accumulation in adolescents in response to early systemic stress, as evidenced by the presence of LEHs, is similar. Like in most agricultural populations, energy demands on adolescent males are high. Teen-age boys accompany their parents to the fields; participate in logging, road building, and construction within and outside the community. Their recreational activities are also physical, they include soccer and basketball under the tropical sun (Alayón-Gamboa and Gurri-García 2007; Gurri García et al. 2018). As in the populations reviewed by Masterson et al. (2018) higher energy demands may be having a greater impact on children with an early age disadvantage.

Overweight as estimated by BMI in adult females in the Citrus-Maize region is as high as 64.6% and, while lower in Calakmul, it is still a high 54.1% (Gurri 2015) suggesting that unlike for men, females' micro-environment is obesogenic. Women with LEHs facing this environment have significantly higher BMIs, especially in the central Yucatan where dependence on a store food diet is greater. That teenage women with LEHs will have significantly greater BMIz suggests that those adolescents who suffered early systemic stress will accumulate fat easier than those without DDE when consuming an obesogenic diet.

Labor and recreational activities are divided by sex amongst agriculturalists. As mentioned above, male labor is energy intensive, and they usually participate in sports. Women take care of the household and their backyard. They leave their houses to go to the store or the mill, and in some families, particularly in Calakmul, they carry their husband's and children's meal to the field (Alayón-Gamboa and Gurri Garcia 2007). Even when teen age women work in the cities they do so in a domestic capacity as maids for upper middle class and rich Mexicans in the cities of Merida and Cancun (Gurri-García 2007). As for rest and recreation, Alayón-Gamboa and Gurri (2007) found that while adolescent boys play soccer or basketball, women tend to recover their energy by increasing their hours of sleep.

In Yucatan, rural development has reduced the number and severity of events that may lead to early systemic stress. However, it has partially substituted the balanced diet that evolved around the milpa with a calorie-rich store-bought diet (Leatherman and Goodman 2005; Gurri García et al. 2018; Otero Prevost et al. 2017). This has been unable to protect male children with lower adaptability from the consequences of an energy-demanding schedule. The consequences of a less demanding schedule, however, are clearly undesirable.

Early malnutrition or growth problems lead to a thrifty phenotype. In communities dependent on store foods, where children have a low-energy demanding schedule, higher BMIz scores since adolescence, obesity, and metabolic syndrome in adults may be expected (Hales and Barker 2001, 2013; Wells 2011). This is the case for urban children in Popkin's (1994) first urban nutrition transition and clearly the case for women in these agricultural populations. That BMIz scores are significantly higher in central Yucatan than in Calakmul, and that frequency of overweight adult women in the Citrus-Maize region is 10% higher than in Calakmul (Gurri 2015) suggests that development has exacerbated what may already be a traditional consequence of inequality in agricultural populations.

8.6 Conclusion

In the Citrus-Maize region and in Calakmul, development affected ontogeny. The impact was more severe in the former where store food items are more common. In both regions, the gendered division of labor and recreational activities that regulate the combination of diet and energy expenditure during growth mediated the nature of the impact. High-energy demands in childhood and adolescence prevented increases in BMI in males exposed to early developmental stress. Women with LEHs, on the other hand, showed significant BMI increases. The more obesogenic the diet, the greater the BMI increase in those who had suffered early developmental stress. Thus, while economic development may have improved conditions conducive to well-being in early childhood, it created an environment that increased fat accumulation in women with a thrifty phenotype. Health and nutritional intervention during early life will pay high dividends in the prevention of overweight and obesity in adults of rural populations where traditional diets have been altered by modernization.

Acknowledgements The author wishes to thank the householders from the State of Yucatan and Calakmul who allowed him and his team to visit and learn from their existence. I also thank Drs. Gilberto Balam and Emilio Moran. They were my mentors and principle investigators in the Maize-Region projects. Fieldwork in Yucatan was financed by the Instituto Nacional Indigenista de México, and NSF grant SBR#9420727; In Campeche, by the Consejo Nacional de Ciencia y Tecnología (CONACYT), Mexico grant 29264-H.

References

Adams RN (1974) Some observations on the inter-relations of development and nutrition programs. Ecol Food Nutr 3(2):85–88. https://doi.org/10.1080/03670244.1974.9990366. Accessed 12 Sept 2018

Alayón-Gamboa JA, Gurri FD (2007) Impact of seasonal scarcity on energy balance and body composition in peasant adolescents from Calakmul, Campeche Mexico. Am J Hum Biol 19:751–752

Alayón-Gamboa JA, Gurri-García FD (2007) Flujo y balance energético en los agroecosistemas de campesinos tradicionales y convencionales del sur de Calakmul, Campeche, México. In González Jácome CA, del Amo Rodríguez S, Gurri García FD (eds) Los nuevos caminos de la agricultura; procesos de conversión y perspectivas. México: Plaza y Valdez S.A. de C.V., pp. 243–260

Alayón-Gamboa JA, Gurri-García FD (2008) Home garden production and energetic Sustainability in Calakmul, Campeche, Mexico. Hum Ecol 36(3):395–407. Available at: http://link.springer.com/10.1007/s10745-007-9151-4. Accessed 18 Sept 2018

Armelagos GJ et al (2009) Enamel hypoplasia and early mortality: bioarcheological support for the Barker hypothesis. Evol Anthropol 18(6):261–271

Arroyo P et al (2007) Obesidad, morfología corporal y presión arterial en grupos urbanos y rurales de Yucatán. Sal Pub Mex 49:274–285. Available at: http://www.scielo.org.mx/pdf/spm/v49n4/04.pdf. Accessed 18 Sept 2018

Atran SCA et al (1993) Itza maya tropical agro-forestry. Curr Anthropol 34:633–700

Baer R (1987) Nutritional effects of commercial agriculture. Urban Anthropol Stud Cult Syst World Econ Dev 16(1):39–61

Balam Pereira G, Gurri García F (1994) A physiological adaptation to undernutrition. Ann Hum Biol 21:483–489

Barabas AM (1989) Utopias Indias. Movimientos sociorreligiosos en México. Grijalbo, México, D. F

Belcastro V et al (2007) Antiepileptic drugs and MTHFR polymorphisms influence hyperhomocysteinemia recurrence in epileptic patients. Epilepsia 48(10):1990–1994. Available at: https://s3.amazonaws.com/academia.edu.documents/44772963/Antiepileptic_Drugs_and_MTHFR_Polymorphi20160415-957-1ohapx.pdf?AWSAccessKeyId=AKIAIWOWYYGZ2Y53UL3A&Expires=1537562780&Signature=mL9%252BSJdfucYrxVUf4lzpfUGgtXc%253D&response-content-disposition=inl. Accessed 21 Sept 2018

Blakey ML et al (1994) Frequency and chronological distribution of dental enamel hypoplasia in enslaved African Americans: a test of the weaning hypothesis. Am J Phys Anthropol 95(4):371–383. Available at: http://www.ncbi.nlm.nih.gov/pubmed/7864059. Accessed 24 Sept 2018

Caballero B (2005) A nutrition paradox: underweight and obesity in developing countries. N Engl J Med 352:1514–1516

Camhi SM et al (2011) The relationship of waist circumference and BMI to visceral, subcutaneous, and total body fat: sex and race differences. Obesity 19(2):402–408. Available at: https://www.ncbi.nlm.nih.gov/pmc/articles/PMC3960785/pdf/nihms560719.pdf. Accessed 21 Sept 2018

Cattle D (1976) Dietary diversity and nutritional security in a coastal Miskito indian village, eastern Nicaragua. In: Helms MW, Loveland F (eds) Frontier adaptations in lower Central America. Institute for the Study of Human Issues, Philadelphia, pp 117–130
Clark WF (1980) The rural to urban nutritional gradient: application and interpretation in a developing nation and urban situation. Soc Sci Med Part D: Med Geogr 14(1):31–36. Available at: https://www.sciencedirect.com/science/article/pii/0160800280900295. Accessed 18 Sept 2018
Cochran WG (1977) Sampling techniques. John Wiley & Sons, London
Coe MD (1984) The Maya. Thames Hudson, London
Daltabuit Godás M (1992) In: Universidad Nacional Autónoma de México (ed) Mujeres mayas: trabajo, nutrición y fecundidad, México, D.F
Daltabuit Godás M, Leatherman TL (1998) In: Goodman AH, Leatherman TL (eds) Building a new biocultural synthesis: political-economic perspectives on human biology. University of Michigan Press, Ann Arbor, pp 317–338Building a new biocultural synthesis: political-economic perspectives on human biology
Deurenberg P et al (1998) Body mass index and percent body fat: a meta analysis among different ethnic groups. Int J Obes Rel Metabol Dis 22(12):1164–1171. Available at: http://www.ncbi.nlm.nih.gov/pubmed/9877251. Accessed 21 Sept 2018
Dewey KG (1979) Agricultural development, diet and nutrition. Ecol Food Nutr 8(4):265–273. https://doi.org/10.1080/03670244.1979.9990576. Accessed 21 Sept 2018
Dewey KG (1981) Nutritional consequences of the transformation from subsistence to commercial agriculture in Tabasco, Mexico. Hum Ecol 9(2):151–187
Doak CM et al (2005) The dual burden household and the nutrition transition paradox. Int J Obes 29(1):129–136. Available at: www.nature.com/ijo. Accessed 21 Sept 2018
Dummer PMH (1977) Severe enamel hypoplasia in a case of intestinal lymphangiectasia: a rare protein-losing enteropathy. Oral Surg Oral Med Oral Pathol 43(5):702–706. Available at: https://www.sciencedirect.com/science/article/pii/0030422077900548?via%3Dihub. Accessed 21 Sept 2018
Duray S (1992) The intertooth pattern of enamel defect occurrence. In: Paper presented at the sixty first annual meeting of the AAPA. Dental Anthropology Newsletter, Las Vegas Nevada, p 14
Eder JF (1978) The caloric returns to food collecting: disruption and change among the Batak of the Philippine tropical forest. Hum Ecol 6(1):55–69. Available at: http://link.springer.com/10.1007/BF00888566. Accessed 21 Sept 2018
Ensor BE, Irish JD (1995) Hypoplastic area method for analyzing dental enamel hypoplasia. Am J Phys Anthropol 98(4):507–517. https://doi.org/10.1002/ajpa.1330980410. Accessed 21 Sept 2018
Farewell CV et al (2018) Prenatal stress exposure and early childhood BMI: exploring associations in a New Zealand context. Am J Hum Biol 30(4):e23116. Available at: http://www.ncbi.nlm.nih.gov/pubmed/29476576. Accessed 21 Sept 2018
Fenton WT (1998) In: University of Arizona (ed) Dental conditions at Grasshopper Pueblo: evidence for dietary change and increased stress. Ph.D. dissertation, Tucson
Ferré D (1996) Geografía y sociedad en la reserva de la Biósfera de Calakmul. Voz Común 30:6–7
Fleuret P, Fleuret A (1980) Nutrition, consumption, and agricultural change. Hum Organ 39(3):250–260. https://doi.org/10.17730/humo.39.3.53332403k1461480. Accessed 21 Sept 2018
Fleuret P, Fleuret A (1991) Social organization, resource management, and child nutrition in the Taita Hills, Kenya. Am Anthropol 93(1):91–114. https://doi.org/10.1525/aa.1991.93.1.02a00050. Accessed 21 Sept 2018
Flores Medina VI, Gurri García FD (2005) Recetario regional de Calakmul. El Colegio de la Frontera Sur, Consejo Nacional para la Ciencia y las Artes, Instituto Nacional de Antropología e Historia, Campeche, Campeche. Available at: http://bibliotecasibe.ecosur.mx/sibe/book/000035579. Accessed 21 Sept 2018
Fort O (1979) La colonización ejidal en Quintana Roo: Estudio de caso. Instituto Nacional Indigenista, México. Available at: http://bibliotecasibe.ecosur.mx/sibe/book/000016930

Freedman D, Ford E (2015) Are the recent secular increases in the waist circumference of adults independent of change in bmi? Am J Clin Nutr 101(3):425–431
Frisancho R (1990) Anthropometric standards for the assessment of growth and nutritional status. University of Michigan Press, Ann Arbor
Frisancho AR (2009) Developmental adaptation: where we go from here. Am J Hum Biol 21(5):694–703. https://doi.org/10.1002/ajhb.20891. Accessed 21 Sept 2018
Garrow JS, Webster J (1985) Quetelet's index (w/h2) as a measure of fatness. Int J Obes 9:147–153
van Gerven DP et al (1995) The health and nutrition of a medieval Nubian population: the impact of political and economic change. Am Anthropol 97(3):468–480. Available at: https://www.jstor.org/stable/pdf/683267.pdf?refreqid=excelsior%3A8545e021bd5b629b3201cac8a2dd7882. Accessed 24 Sept 2018
Goodman AH, Armelagos GJ (1985) The chronological distribution of enamel hypoplasia in human permanent incisor and canine teeth. Arch Oral Biol 30(6):503–507
Goodman AH, Rose JC (1990) Assessment of systemic physiological perturbations from dental enamel hypoplasias and associated histological structures. Yearbook Phys Anthropol 33:59–110. https://doi.org/10.1002/ajpa.1330330506. Accessed 24 Sept 2018
Goodman AH et al (1984) Indications of stress from bones and teeth. In: Cohen MN, Armelagos GJ (eds) Paleopathology at the origins of agriculture. Academic Press, New York, pp 13–49
Goodman AH et al (1987) Prevalence and age at development of enamel hypoplasias in Mexican children. Am J Phys Anthropol 72(1):7–19
Goodman AH, Martinez C, Chavez A (1991) Nutritional supplementation and the development of linear enamel hypoplasias in children from Tezonteopan, Mexico. Am J Clin Nutr 53(3):773–781
Gross DR (1975) Protein capture and cultural development in the Amazon basin1. Am Anthropol 77(3):526–549
Gurri FD (2010) Smallholder land use in the southern Yucatan: how culture and history matter. Reg Envir Chang 10(3):219–231
Gurri FD (2015) The disruption of subsistence agricultural systems in rural Yucatan, Mexico may have contributed to the coexistence of stunting in children with adult overweight and obesity. Coll Antropol 10(6):723–733
Gurri FD, Balam G (1992) Regional integration and changes in nutritional status in the central region of Yucatan, Mexico: a study of dental enamel hypoplasia and anthropometry. J Hum Ecol (Spain) 3(2):417–432
Gurri García F (2011) Inferences from variation in linear enamel hypoplasias in three agricultural populations from the Yucatan peninsula, Mexico. Paper presented at the 80th annual meeting of the American association of physical anthropologists. Minneapolis, Minnesota, USA
Gurri García FD et al (2008) Distribución e impacto de las remesas en la dinámica de familias campesinas del sur de Campeche. El Colegio de la Frontera Sur, Campeche
Gurri García FD et al. (2018) Impact of community development and abandonment of traditional subsistence strategies on the diet of Maya Populations from the maize region of Yucatan, Mexico; Linking human biology and population dynamics with socio-envirnmental transformations in the Yucatan. Paper presented at the 56° Congreso Internacional de Americanistas (ICA). Salamanca, España
Gurri FD, Moran EF (2003) Who is interested in commercial agriculture?: subsistence agriculture and salaried work in the city among Yucatec Maya from the state of Yucatan. Cult Agric 24(1):41–47. https://doi.org/10.1525/cag.2002.24.1.42
Gurri FD, Balam Pereira G, Moran EF (2001) Well-being changes in response to 30 years of regional integration in Maya populations from Yucatan, Mexico. Am J Hum Biol 3(1):590–602
Gurri-García F (2007) Globalización y cambios en la calidad de vida en familias campesinas de Yucatán, México. Revista de la Academia Mexicana de. Ciencias 58(2):60–68
Gurri-García FD, Alayón-Gamboa JA, Molina-Rosales DO (2001) Adaptabilidad en poblaciones mayas y poblaciones migrantes de Calakmul, Campeche. El Colegio de la Frontera Sur, Campeche. Available at: http://bibliotecasibe.ecosur.mx/sibe/book/000026479

Haas JD, Harrison GG (1977) Nutritional anthropology and biological adaptation. Annu Rev Anthropol 6:69–101

Haenn N (2005) Fields of power, forests of discontent. Culture, conservation, and the state in Mexico. University of Arizona Press, Tucson

Hales CN, Barker DJP (2001) The thrifty phenotype hypothesis: Type 2 diabetes. Brit Med Bull 60(1):5–20

Hales CN, Barker DJP (2013) Type 2 (non-insulin-dependent) diabetes mellitus: the thrifty phenotype hypothesis. 1992. Int J Epidemiol 42(5):1215–1222

Hariri AA et al (2013) Adiposity measurements by BMI, skinfolds and dual energy x-ray absorptiometry in relation to risk markers for cardiovascular disease and diabetes in adult males. Dis Markers 35(6):753–764

Hoffman DJ et al (2000) Why are nutritionally stunted children at increased risk of obesity? Studies of metabolic rate and fat oxidation in shantytown children from São Paulo, Brazil. Am J Clin Nutr 72(3):702–707. Available at: http://www.ncbi.nlm.nih.gov/pubmed/10966887. Accessed 24 Sept 2018

Hoorweg J et al (1996) Nutrition in agricultural development: land settlement in Coast Province, Kenya. Ecol Food Nutr 35(1):161–178

Hutchinson DL, Larsen SC (1988) Determination of stress episode duration from linear enamel hypoplasias: a case study from St. Catherines Island. Georgia Hum Biol 60(1):93–110. Available at: https://www.jstor.org/stable/41463980?seq=1#metadata_info_tab_contents. Accessed 24 Sept 2018

INEGI (1991) Yucatán, conteo de población y vivienda 1990: resultados definitivos, tabulados básicos. INEGI. Available at: https://books.google.com.mx/books?id=DaJSDwAAQBAJ&pg=PR11&lpg=PR11&dq=Yucatán,+resultados+definitivos,+XI+censo+general+de+población+y+vivienda&source=bl&ots=uZLRBetd0I&sig=O_Bg8KxjbnMFsFyx_OHv1oRhx9k&hl=es-419&sa=X&ved=2ahUKEwiNrJrfrtTdAhVK44MKHTCwCYUQ6A. Accessed 24 Sept 2018

King KW (1971) The place of vegetables in meeting the food needs in emerging nations. Econ Bot 25:6–11

Kreshover SJ (1960) Metabolic disturbances in tooth formation. Ann N Y Acad Sci 85:161–167

Lanphear KM (1990) Frequency and distribution of enamel hypoplasias in a historic skeletal sample. Am J Phys Anthropol 81(1):35–43

Lazos CE (1995) Del maíz a la naranja en el sur de Yucatán: Auge y dinámica de la huerta. In: Hernández XE, Bello E, Levy S (eds) La milpa en Yucatán. Un sistema de producción agrícola tradicional, Tomo 2. Colegio de Posgraduados, México, pp 527–563

Leatherman TL, Goodman A (2005) Coca-colonization of diets in the Yucatan. Soc Sci Med 61:833–846

Little M (1991) Imperialism, colonialism and the new science of nutrition: the Tanganyika experience, 1925–1945. Soc Sci Med 32(1):11–14

Malina RM et al (2007) Overweight and obesity in a rural Amerindian population in Oaxaca, Southern Mexico, 1968–2000. Am J Hum Biol 19(5):711–721

Manzi G et al (1999) Discontinuity of life conditions at the transition from the Roman imperial age to the Early Middle ages: Example from central Italy evaluated by pathological dento-alveolar lesions. Am J Hum Biol 11(3):327–341. Available at: http://www.ncbi.nlm.nih.gov/pubmed/11533954. Accessed 24 Sept 2018

Mariaca-Méndez R (2015) La milpa maya yucateca en el siglo XVI: evidencias etnohistóricas y conjeturas. Etnobiología 13(1):1–38

Masterson EE et al (2018) Dental enamel defects predict adolescent health indicators: a cohort study among the Tsimane' of Bolivia. Am J Hum Biol 30(3):e23107

Maunders J, Goodman A, Froment A (1992) The ecology of dental enamel hyperplasias among seven Cameroonian groups. In: Lukacs J (ed) Culture, ecology and dental anthropology. Kamla Raj Enterprises, Delhi, pp p109–p116

May RL, Goodman AH, Meindl RS (1993) Response of bone and enamel formation to nutritional supplementation and morbidity among malnourished Guatemalan children. Am J Phys Anthropol 92(1):37–51

Murguía R (1981) Investigación sobre la diferenciación social del crecimiento y desarrollo corporal. Escuela Nacional de Antropología e Historia, México, D.F

Nietschmann B (1972) Hunting and fishing focus among the Miskito Indians, eastern Nicaragua. Hum Ecol 1(1):41–67

de Onis M et al (2007) Development of a WHO growth reference for school-aged children and adolescents. Bull World Health Organ 85(9):660–667. Available at: http://www.ncbi.nlm.nih.gov/pubmed/18026621. Accessed 24 Sept 2018

Otero Prevost DE et al (2017) La incorporación y el aumento de oferta de alimentos industrializados en las dietas de las unidades domésticas y su relación con el abandono del sistema de subsistencia propio en las comunidades rurales mayas de Yucatán, México. Cuadernos de Desarrollo Rural 14(80):1–16

Pastore L et al (2008) Oral manifestations of celiac disease. J Clin Gastroenterol 42(3):224–232

Pindborg J (1970) Pathology of dental hard tissues. WB Saunders Company, Philadelphia. Available at: https://www.sciencedirect.com/science/article/pii/0030422072902873?via%3Dihub. Accessed 24 Sept 2018

Pindborg JJ (1982) Aetiology of developmental enamel defects not related to fluorosis. Int Dental J 32(2):123–134

Popkin BM (1994) The nutrition transition in low-income countries: an emerging crisis. Nutr Rev 52(9):285–298

Press I (1975) Tradition and adaptation: life in a modern Yucatan Maya village. Greenwood Press, Westport, Conn. Available at: https://trove.nla.gov.au/work/10881270?q&versionId=12713775. Accessed 24 Sept 2018

Rappaport RA et al (1971) The flow of energy in an agricultural society. Sci Am 225(3):116–133

Rashid M et al (2011) Oral manifestations of celiac disease: a clinical guide for dentists. J Can Dent Assoc 77:b39

Rawson IG, Valverde V (1976) The etiology of malnutrition among preschool children in rural Costa Rica. J Trop Pediatr Environ Child Health 22(1):12–17. Available at: http://www.ncbi.nlm.nih.gov/pubmed/1048193. Accessed 24 Sept 2018

Re Cruz A (1996) In: State University of New York Press (ed) The two milpas of Chan Kom : scenarios of a Maya village life, Albany

Reategui PE (1976) Nutrition. In: Baker PT, Little MA (eds) Man in the Andes: A multidisciplinary study of high altitude Quechua. Dowden, Hutchinson & Ross, Stroudsburg, PA, pp 208–236

Redfield R (1950) A village that chose progress: Chan Kom revisited University of Chicago Press, Chicago. Available at: https://www.amazon.com/VILLAGE-THAT-CHOSE-PROGRESS-REVISITED/dp/B006RC0YG0. Accessed 24 Sept 2018

Redfield R, Villa Rojas A (1934) Chan Kom: a Maya village. Carnegie Institution of Washington, Washington, DC. Available at: https://books.google.es/books?id=hPkUAAAAIAAJ&hl=es&source=gbs_book_other_versions. Accessed 24 Sept 2018

Reed NA (1964) The caste war of Yucatan. Stanford University Press, California

Rosales González M (1988) Oxkutzcab, Yucatán, 1900–1960: campesinos, cambio agrícola y mercado. Instituto Nacional de Antropología e Historia. Mérida, México. Available at: https://books.google.com.mx/books/about/Oxkutzcab_Yucatán_1900_1960.html?id=qr9EAAAAYAAJ&redir_esc=y. Accessed 24 Sept 2018

Rose JC, Condon KW, Goodman AH (1985) Diet and dentition: developmental Disturbances. In: Gilbert B, Mielke J (eds) The analysis of prehistoric diets. Academic Press, Orlando, pp 281–306

Rugg-Gunn AJ, Al-Mohammadi SM, Butler TJ (1997) Effects of fluoride level in drinking water, nutritional status, and socio-economic status on the prevalence of developmental defects of dental enamel in permanent teeth in Saudi 14-year-old boys. Caries Res 31(4):259–267

Rush EC, Freitas I, Plank LD (2009) Body size, body composition and fat distribution: comparative analysis of European, Maori, Pacific Island and Asian Indian adults. Brit J Nut 102:632–641. Available at: https://www.researchgate.net/publication/23993852. Accessed 24 Sept 2018

Santos RV, Coimbra CEA (1999) Hardships of contact: enamel hypoplasias in Tupí-Mondé Amerindians from the Brazilian Amazonia. Am J Phys Anthropol 109(1):111–127. Available at: http://www.ncbi.nlm.nih.gov/pubmed/10342468. Accessed 24 Sept 2018

Sarnat BG, Schour I (1942) Enamel hypoplasia (chronologic enamel aplasia) in relation to systemic disease: a chronologic, morphologic and etiologic classification. J Am Dent Ass 29(1):67–75. Available at: https://www.sciencedirect.com/science/article/pii/S0002817742910092. Accessed 23 Sept 2018

Schoepf BG, Schoepf C (1987) Food crisis and agrarian change in the eastern highlands of Zaire. Urban Anthropol Stud Cult Syst World Econ Dev 16(1):5–37. Available at: https://www.jstor.org/stable/40553100. Accessed 23 Sept 2018

Sohn YS, Moran E, Gurri F (1999) Deforestation in north-central Yucatan (1985–1995): mapping secondary succession of forest and agricultural land use in Sotuta using the cosine of the angle concept. Photogramm Eng Remote Sensing 65(8):947–958

Steinberger J et al (2005) Comparison of body fatness measurements by BMI and skinfolds vs dual energy x-ray absorptiometry and their relation to cardiovascular risk factors in adolescents. Int J Obes 29(11):1346–1352. Available at: http://www.nature.com/articles/0803026. Accessed 24 Sept 2018

Sun Q et al (2010) Practice of epidemiology comparison of dual-energy x-ray absorptiometric and anthropometric measures of adiposity in relation to adiposity-related biologic factors. Am J Epidemiol 172(12):442–1454. Available at: http://aje.oxfordjournals.org/. Accessed 24 Sept 2018

Sweeney EA et al (1969) Factors associated with linear hypoplasia of human deciduous incisors. J Dent Res 48(6):1275–1279

Sweeney EA, Saffir AJ, de Leon R (1971) Linear hypoplasia of deciduous incisor teeth in malnourished children. Am J Clin Nut 24(1):29–31. Available at: http://www.ncbi.nlm.nih.gov/pubmed/5539052. Accessed 24 Sept 2018

Terán S, Rasmussen CH (1994) La milpa de los mayas: la agricultura de los mayas prehispánicos y actuales en el noreste de Yucatán. Danida, Mérida, México. Available at: http://bibliotecasibe.ecosur.mx/sibe/book/000015072. Accessed 23 Sept 2018

Terán S, Rassmussen C (1992) Estrategia agrícola y religión. In: Zizumbo Villarreal D, Arias Reyes LM, Rassmusen CH (eds) La modernización de la milpa en Yucatán: utopía o realidad. Centro de Investigaciones Científicas de Yucatán. Colegio de Postgraduados y Ministerio de Cultura de Dinamarca, Mérida, México, pp 257–266

Tuz Chi LH (2013) Aj balam yúumtsilo'ob: cosmovisión e identidad en los rituales agrícolas de los mayas peninsulares. Secretaría de Educación del Gobierno del Estado de Yucatán, Mérida, México. Available at: http://bibliotecasibe.ecosur.mx/sibe/book/000057902. Accessed 23 Sept 2018

Ulijaszek SJ, Kerr DA (1999) Anthropometric measurement error and the assessment of nutritional status. Br J Nut 82(3):165–177

Villa Rojas A (1985) Estudios etnológicos: los mayas. Universidad Nacional Autónoma de México, México, D. F

Villanueva Mukul E (1985) Crisis henequenera y movimientos campesinos en Yucatán. Instituto Nacional de Antropología e Historia, México, pp 1966–1983

Villanueva Mukul E (1994) La formación de las regiones en la agricultura: el caso de Yucatán. Maldonado Editores, Mérida, México

Wells JCK (2011) The thrifty phenotype: an adaptation in growth or metabolism? J Hum Biol 23:65–75. Available at: http://citeseerx.ist.psu.edu/viewdoc/download?doi=10.1.1.476.869&rep=rep1&type=pdf. Accessed 24 Sept 2018

Wells A, Joseph GM (1996) Summer of discontent, seasons of upheaval: Elite politics and rural insurgency in Yucatán. Stanford University Press, California, pp 1876–1915

Williams A (1973) Dietary practices in three Mexican villages. In: Smith J (ed) Man and his foods. University of Alabama Press, Alabama, pp 51–74

Wilmsen EN (1978) Seasonal effects of dietary intake on Kalahari San. Fed Proc 37(1):65–72

Zhou L, Corruccini RS (1998) Enamel hypoplasias related to famine stress in living Chinese. Am J Hum Biol 10:723–733

Chapter 9
Hydration, Lactation, and Child Health Outcomes in Yucatec Maya

Amanda Veile, Sunny Asaf, Erik Otárola-Castillo, and Karen L. Kramer

Abstract We examine the ways that water supply shapes lactation strategies and child morbidity patterns in Yucatec Maya subsistence maize farmers in México. We outline the history of Maya water use, with a focus on one subsistence farming community in northern Campeche. Prior to the community's founding in the early twentieth century, the Maya lived in small family units near *sartenejas* (water that collects in natural limestone basins). After the community was established around an abandoned hacienda, women retrieved water by hand from a central well. A water pump was then built in the late 1970s, increasing the efficiency of water collection and decreasing the energetic burden of women. In the early 2000s, water lines were extended from the pump to each house. We demonstrate here that water access and storage still pose challenges to the Maya farmers. We describe these challenges, and then analyze relationships between water access and storage practices on Maya breastfeeding practices and child morbidity outcomes.

9.1 Introduction

This chapter examines the relationships between water access and storage practices, breastfeeding patterns, and child morbidity outcomes in a rural Yucatec Maya community, using perspectives from human biology and ecology. Water is essential to sustain life; however many people worldwide lack access to clean drinking water.

A. Veile (✉) · E. Otárola-Castillo
Department of Anthropology, Purdue University, West Lafayette, IN, USA
e-mail: aveile@purdue.edu; eoc@purdue.edu

S. Asaf
Richland County Health Department, Sidney, MT, USA

K. L. Kramer
Department of Anthropology, University of Utah, Salt Lake City, UT, USA
e-mail: karen.kramer@anthro.utah.edu

H. Azcorra, F. Dickinson (eds.), *Culture, Environment and Health in the Yucatan Peninsula*, https://doi.org/10.1007/978-3-030-27001-8_9

This is a key point in several health initiatives driven by the United Nations International Children's Emergency Fund (UNICEF) and the World Health Organization (WHO), which strive to improve global water access (UNICEF 2016; WHO 2018). Dehydration has many negative effects on human health, especially in young children (Maughan 2012), and may be a limiting factor affecting women's ability to breastfeed (Bentley 1998). At the same time, water can act as a vehicle for pathogens and contamination, introducing risk of microbial infection (Cabral 2010). Ecological, geological, and socioeconomic factors all influence access to potable water worldwide, and also shape localized cultural water management techniques.

In a case study of this, we outline the historical and contemporary use and storage of water in the Yucatán Peninsula in Mexico. We focus on one community of maize farmers, located in the central hilly region of the peninsula (the Puuc region) where aquifer yields, economic development, and population densities are relatively low (Bauer-Gottwein et al. 2011). The local environment lacks standing or running surface-level water. Water is therefore drawn from underground aquifers, though specific patterns of water procurement and use have changed over time. Since ~2000, the community has relied on piped water, but occasional delivery dysfunction renders access unreliable, with possible implications for child health.

In this chapter, we evaluate water sources and water storage systems in the case study community, and determine if they are associated with breastfeeding frequency and duration, as well as child health digestive and respiratory morbidity. We aim to answer the following two questions: (1) Is childhood morbidity linked to water source or methods of water storage? (2) Are breastfeeding durations linked to water source and method of water storage? To answer these, we begin with a literature review of sources obtained from searches for peer-reviewed articles and books using combinations of the following keywords: water, water management, sanitation, drinking, breastfeeding, morbidity, Maya, Yucatán, and motherhood.

We first address geologic characteristics and hydrobiology of the Yucatán Peninsula that have created a need for water storage and management strategies. This is followed by a brief history of the Yucatán, with a focus on historical Maya water collection techniques as a cultural adaptation to meet human water needs. A brief overview of the biology of hydration is then provided, with special attention paid to the hydration of infants, mothers, and children, and the links between water sanitation and child health. The literature review is followed by a detailed description of the ethnographic and historical context of the focus community of Yucatec Maya subsistence farmers in northern Campeche. We then describe the data collection and analysis methods used to ascertain relationships between water access and storage and breastfeeding outcomes and child morbidity. These relationships are outlined and synthesized in the subsequent results and discussion sections.

9.2 Background

9.2.1 Geologic Characteristics and Hydrobiology of the Yucatán Peninsula

Information on the geology of the Yucatán peninsula, which is comprised of Campeche, Yucatán, and Quintana Roo states, is widely available. A working paper for the International Institute for Applied Systems Analysis by Gelting (1995) describes the Yucatán peninsula as a karstic environment. A karst is an area that is characterized by a predominantly limestone geology and a lack of permanent surface water and a relative lack of soil formation. In Yucatán, some groundwater from precipitation moves into karst aquifers and caves; once underground, it flows rapidly towards the ocean (Bauer-Gottwein et al. 2011). Due to the lack of soil to act as a filter, and the high permeability of limestone, precipitation seeps underground quickly and can erode the landscape easily which leads to the formation of sinkholes (Doehring and Butler 1974). During the rainy season, water collects in small depressions and pools called *chaltuns* (Kramer 2005).

The water table underground exists as an easily contaminated and thin layer of freshwater on top of brackish, salinated water. While not the entirety of the Yucatán peninsula is karstic and lacks surface water, the case study community is in a karstic area, and residents do not have access to surface water. The water available underground may be easily contaminated, and in some parts of the Yucatán Peninsula, pollution of this water has led to endemic, waterborne diseases (Back 1995). The lack of filtration as the water seeps underground, and the warm climate, can make underground water sources favorable environments for pathogens (Doehring and Butler 1974).

9.2.2 History of Water in Yucatán

Yucatec Maya culture and groundwater use were intimately linked, and water seems to have influenced the growth of Maya civilization even in ancient times (2000$_{\mathrm{BCE}}$ to 900$_{\mathrm{CE}}$; Back 1995). During the Classic and Post-Classic periods, Ancient Maya looked to the god of water, *Chac*, for assistance in water management (Miller and Taube 1993). The ancient Maya often built their large cities near large sinkholes called *cenotes* (see Chap. 2), and major engineering efforts can be seen in their ancient hydraulic works (Silverstein et al. 2009). Drinking water was obtained from caves and caverns which were accessed by steps and ladders. The water itself was carried out in clay jars or lifted up with rope, and in some areas small aqueducts and canals were used. Outside of urban centers, it is likely that Maya villagers had more flexible settlement patterns and relied on *aguadas* (large, open still-water reservoirs) or *sartenejas* (water that collects in natural limestone basins); these are still used by some Maya families today. Despite evidence for very sophisticated lowland Maya water management techniques (Akpinar-Ferrand 2011; Back 1995; Isendahl 2011), it is possible that repeated drought conditions depleted water reservoirs, contributing ultimately to the rapid Maya population decline during the ninth century (Kuil et al. 2016; Lucero 2002).

In the Colonial/Modern period (from AD 1546 until present date), Spanish settlements were dependent on groundwater sources such as *cenotes* and drilled wells for drinking water and to support livestock populations (Normark 2016). Steam driven mechanical pumps were introduced an in 1880, the first windmill in Merida (capital of Yucatán state) was installed. This quickly became a popular way to get water, and by 1930, there were more than 20,000 windmills. These windmills are now abandoned due to pollution and sewage and Merida now uses a city well with electric pumps. Doehring (1974) also describes rope and bucket methods in more recent memory to get potable water from aquifers along with low and high discharge pumps in more urban settings.

Citing an original survey, González-Gómez et al. (2011) discuss modern challenges of water access in Yucatán state, which is better studied than rural Northern Campeche, where our case study was conducted. At the time of the survey, 75% of Yucatán state's water was licensed for agricultural use. Of 541 water purification plants in Mexico overall, none were in Yucatán. In addition, piped water to homes were not being treated beforehand due to lack of water purification facilities. Water access problems were more evident in inland areas with higher poverty. Household wells were not government managed, and groundwater was considered very contaminated. Results of the survey by González-Gómez et al. also indicate that households drinking untreated water were poorer on average than those that did not, and households with more income drank more bottled water than lower income houses. There was not a significant income-based difference in the quality of pumped water between households. More generally, the availability of potable water is important because it allows people to stay adequately hydrated and provides a source for water required for normal physiological processes.

9.2.3 General Biology of Hydration

Water is necessary to sustain life and composes the majority of the human body. In order to meet the water requirement, the National Research Council recommends water intake of 1 mL per kcal expended (Sawka et al. 2005; Thornton 2010). Water turnover is a constant biological process; adults lose on average from 1–4 liters of water a day from urine, around 50 mL of water from fecal matter, and 800 mL from sweat and respiration (Vargas 2001). Around 1.3 liters of water is regained daily through drinking, about 800 mL through eating, and 250 mL from metabolic water (Vargas 2001). Thus, 5–10% of water turns over daily, and a water intake of 3.7 liters for men and 2.7 liters for women is considered adequate (Sawka et al. 2005). Reduced water intake or increased losses can cause acute or water deficit, though day-to-day hydration generally remains normal with access to food and fluids

(Sawka et al. 2005). We note that the numbers above are derived from a limited population—healthy individuals from the United States and Canada—which is not necessarily representative across human populations. Clearly, water requirements are contingent on a number of factors such as temperature, age, and activity level. In other words, the range of human water requirements is defined by environmental and individual contexts.

9.2.4 Hydration of Infants, Mothers, and Children

Infants and young children have a higher ratio of total body water: body mass compared to adults, making them especially vulnerable to dehydration (Jéquier and Constant 2010). Their water requirements are typically met by either human or formula milk consumption (Sawka et al. 2005). Exclusively breastfed infants obtain all essential nutrients, including water, from breast milk (McManaman and Neville 2003). Prolonged breastfeeding (>1 year) supports better health and growth outcomes for infants than formula feeding and protects against infectious morbidity (Anatolitou 2012; Salam et al., 2015; Sellen 2007).

Lactating women require increased water intake (by ~0.03 L) increase due to increased water production and expulsion (Sawka et al. 2005). Bentley (1998) comprehensively reviewed literature on hydration and lactation, including maternal thirst mediation by oxytocin, which increases with suckling. She concludes that lactation persists despite nutritional and environmental stress, even in settings where technology (for example, air conditioning) is not available to modulate dehydration in lactating women. She notes that overall, little is known about maternal physiological and behavioral responses to heat stress and dehydration, especially outside of industrialized, Western countries. Rosinger's recent work has demonstrated developments in the area of hydration and lactation (Rosinger and Tanner 2014; Rosinger 2014). He showed that lactating Tsimane mothers in the Bolivian Amazon are at greater risk for dehydration than non-lactating women, perhaps due to poor water quality (Rosinger 2014). Given the risk for diarrheal diseases from water contamination, this is a salient concern for lactating mothers who live in areas with poor water quality. In our study, we generally expected to find that purified water used or water storage receptacles might alleviate maternal dehydration, and as such be associated with longer breastfeeding durations.

9.2.5 Water Sanitation and Child Health

Water sanitation also affects health of children, particularly with respect to diarrheal risk and stunting (Fink et al. 2011, Merchant et al. 2003, Wolf et al. 2014). For example, in low-income peri-urban neighborhoods in Peru, Checkley et al. (2004) found that poor household sanitary conditions were associated with slower child growth. Furthermore, children in households with small water receptacles had a higher frequency (but not duration) of diarrheal morbidity than children in houses with large water receptacles. While receptacle size was negatively correlated with socioeconomic status (SES), it is possible that frequent water replacement contributed to the lower risk of digestive morbidity. Water-related interventions to improve child health include improving access to safe water, improving water quality at the source, treating household water, and storing it safely (WHO/UNICEF 2004). A meta-analysis of 21 studies showed that point-of-use chlorine water treatment reduced childhood diarrhea and risk of stored water *E. coli* (Arnold and Colford 2007). Though we have no data on water quality or treatment in the Yucatec Maya case study community, we generally expected that purified water use and water storage systems would be associated with decreased infectious morbidity in children.

9.3 Ethnographic Context

Many contemporary rural Yucatec Maya continue to live as small-scale subsistence farmers (Back 1995). However, some parts of Yucatán peninsula are widely known for archaeological ruins and coastal resorts, and have become increasingly urbanized and globalized. Anthropological research has focused in food commoditization and decreasing Maya diets quality ("coca-colonization") in the context of rising tourism and economic change (Leatherman and Goodman 2005). Understanding food commoditization is important in understanding the context of the Yucatán Peninsula as a whole. Globalization of diets can cause macro- and micronutrient deficiencies, negatively affecting Yucatec Maya population health outcomes (Leatherman and Goodman 2005). The case study community has not seen the same economic changes and growth in tourism as more coastal cities on the Yucatán peninsula. Mainly traditional foods (hand-made tortillas and other maize-based foods, beans, vegetables and poultry) are consumed (Fig. 9.1) and many women still dress in the traditional huipil (Fig. 9.2a, b). Still, globalized food products like soda are now available in community stores, and are used for special occasions or when hosting guests.

9.3.1 Yucatec Maya Case Study Community

We analyze data drawn from a cohort of Maya children who inhabit a subsistence farming community in the Puuc region of the Yucatan Peninsula, in northern Campeche, Mexico. Though increasingly mechanized (Fig. 9.3), villagers use a traditional *milpa* form of agricultural subsistence, in which maize, beans, and squash are planted on plots at the edge of the community; families also maintain household gardens where vegetables and herbs are grown (Birol et al. 2009; De Frece and Poole 2008; Kramer 2005). Despite being small-scale subsistence farmers, the

Fig. 9.1 Traditional hand-made tortillas consumed in the community

Yucatec Maya in the case study community have historically had fairly low child mortality and morbidity rates (Veile and Kramer 2017), and do not experience a pronounced "seasonal hungry time" (Kramer 2005). The social structures and kinship systems that shape and inform households within Yucatec Maya communities is outside the scope of this chapter, but are outlined in detail in the book *Maya Children* by Karen Kramer (2005). Kramer's book describes the case study community using the lens of the economic value of children; additional details on energetic and epidemiologic transitions in the case study community are provided in Veile and Kramer (2018).

The history of water and water management in the case study community is obtained from published and unpublished ethnographic work by Kramer since 1992, and Veile since 2011. The interior of the Yucatan Peninsula, where the case study community is located, was not colonized by the Spanish until the late 1800s. Pre-Hispanic Maya farmers and villages were fairly dispersed and many obtained water from *aguadas* and *sartanejas* (Kramer 2005). The Spanish built haciendas and excavated wells in the region, which were abandoned following the Mexican Revolution. Maya farmers then began to congregate around a 50-meter-deep well at an abandoned hacienda. In ~1910, the case study community was founded around the well under its current name (Kramer 2005). At that time women typically hauled water from the well by hand using ropes and buckets. Water was obtained in this way until the mid-1970s, when the Mexican government installed a gas-powered mill and a water pump, which increased water collection efficiency (Kramer and McMillan 1998, 1999). Water still had to be carried back and forth from the well, but less time and energy were expended in water collection following the pump installation. Using return-rate data from women hand-drawing water from the well over a three-month period when the gas-powered pump was broken, Kramer and

Fig. 9.2 (**a**, **b**) Traditional dress (huipil) of Maya women

McMillan determined that the pump saves women 2.5 h and 325 calories per day (1998, 1999). This was associated with a subsequent fertility transition, in which women had earlier ages at first birth and larger completed family sizes (Kramer and McMillan 1999).

Following installation of the water pump, little else changed with respect to water management in the case study community until the early 2000s, when a piped water system was provided through a government project. Piped water was brought to nearly every house, though it is considered by community members to be costly

Fig. 9.3 Maya men bagging maize

and often unreliable. Furthermore, new houses are constantly built by new young families as the community grows. Due to space constraints, many new homes are built on the outskirts of the community, and many do not have access to piped water. In particular, houses on hills often cannot receive water easily due to inadequate water pressure. The lack of piped water in some homes, and general unreliability of the piped water supply, is mediated by sharing between families, and through the use of water storage units. These units can be kept indoors (typically five-gallon jugs with handles) and outdoors (typically large barrel-shaped receptacles with a lid, from which water is removed by hand using buckets as shown in Fig. 9.4). Some households purchase purified water, which sporadically is available in small community stores, in addition to storing water.

The community is located in a humid, subtropical setting, and children are exposed to domestic animals, their feces and standing water, especially in the rainy season. While a handful of families have a covered bathroom facility with a septic reservoir excavated into the bedrock, others have shallow, hand-dug latrines or practice open defecation in a designated area, which is likely a source of pathogenic contamination. The recent introduction of plastic and cardboard containers has posed a challenge in terms of trash disposal. Some villagers burn or bury their trash, others maintain above-ground trash pits some distance from the house.

Breastfeeding patterns have been extensively described for the case study community (Veile and Kramer 2015, 2017). Widespread breastfeeding initiation and prolonged breastfeeding durations (>2.5 years) are the norm, even in the face of a number of modernizing influences (Veile and Kramer 2017). This is likely the case because breastfeeding women do not engage in wage labor, and because there are no social taboos against prolonged or public breastfeeding in the community.

Fig. 9.4 Large barrel-shaped receptacles typically used to storage water outdoor

However, formula-feeding is also practiced by mothers who birth in hospitals, especially following a cesarean birth or upon encouragement by physicians (Veile and Kramer 2017). Though formula-feeding can be a source of infant morbidity where water quality is poor (Kramer and Kakuma 2012), we have noted that mothers in the case study community generally formula feed in conjunction with, not instead of, breastfeeding (Veile et al., 2019).

9.4 Methods

9.4.1 Data Collection and Analysis Methods

All data were collected by a government-trained local health promoter and non-local government physician. Child morbidity, breastfeeding and formula feeding status (as reported by mothers) were recorded at monthly health clinic visits from 2007 to 2016, as part of a government program. We combined child morbidity data into two binary outcomes for "any gastrointestinal illness" (GI) and "any respiratory illness" (RI). GI symptoms included stomach pain, diarrhea, parasites, and vomiting, while RI symptoms included cold, cough, sore throat, respiratory problems, and ear infection. Household water storage method and water source were obtained from a separate, community-wide household sanitation survey conducted in 2104. Maternal and child ages were obtained from demographic surveys collected by KK since 1992, and cross-checked with clinic records. Children with only one observation were excluded from analysis. The final sample consists of 49 mothers and 91 children aged 0–5 years (2350 observations total).

9.4.2 Data Analysis

Prior to building longitudinal prediction models, we computed descriptive statistics for the following predictor variables: maternal and child age (continuous variables), child sex (categorical variable), water-related variables (piped water, purified, storage indoors, storage outdoors, all categorical 0/1), and *Oportunidades* enrollment (categorical 0/1). *Oportunidades* is a health and cash transfer program (now called *Prospera*) which was fully implemented in Mexico in 2004 that is known to be associated with child development outcomes (Fernald et al. 2008). We also computed descriptive statistics for the following longitudinal outcome variables: child morbidity (digestive and respiratory, categorical 0/1), and breastfeeding frequency, meaning the child was still being breastfed at the time of the observation (categorical 0/1) and duration (continuous, modeled using Kaplan-Meier survival analysis).

Repeated-measures, binary logistic generalized estimating equations (GEEs) were used to longitudinally model child respiratory and digestive illness with respect to predictor variables (indoor/outdoor water storage and purified water use) while accounting for child age, sex, maternal age, and *Oportunidades* status. Repeated-measures, binary logistic GEEs were also used to determine the probability of a child being breastfeed at any observation while accounting for child age, sex, maternal age, and *Oportunidades* status. All GEEs account for maternal and child ID as random effects. Breastfeeding durations (medians and 95% confidence intervals) were computed using the Kaplan-Meier survival analysis procedure for two factors (purified water versus no purified water, internal water storage versus no internal water storage, external water storage versus no external water storage), respectively. All descriptive statistics and longitudinal models were conducted using SPSS version 24 (IBM 2016).

9.5 Results

9.5.1 Descriptive Statistics

All 49 families in the sample had piped water access in their homes. None relied on surface water sources, 6 (12%) used purified water (in addition to piped), eight families (16%) used internal water storage only, 10 (20%) used external water storage only, and one family (2%) used both internal and external water storage methods. Thirty-three families (67%) were enrolled in *Oportunidades*. Childhood digestive and diarrheal morbidity rates were low (3.7% and 2.6%, respectively) and did not differ by child age ($B = -0.005$, $p = 0.625$ and $B = -0.009$, $p = 0.282$, respectively) or sex ($B = -0.268$, $p = 0.416$ and $B = -0.450$, $p = 0.173$, respectively). Children were on average 31.2 months old (±17.2, range = 0–60 months). Mothers were on average 35.0 years old (± 6.4, range = 15.5–50.5 years) and had on average 2.2 children (±0.9, range = 1–4) in the current sample (e.g., 2.2 children aged 0–5 from 2007 to 2016). Median breastfeeding duration was 32.0 months (95% CI: 30.9–33.1) and 58/91 children (64%) were formula-fed at some point in the first 5 years of life. We note that the latter statistic denotes "ever" formula-fed, not the duration or magnitude of formula feeding.

9.5.2 Analytical Results

Results of the longitudinal models are provided in Table 9.1. Purified water use did not reduce risk of childhood digestive or respiratory illnesses, and purified water use was also an insignificant predictor of breastfeeding frequency. Outside water storage was not associated with respiratory morbidity or breastfeeding outcomes, but was protective against digestive morbidity. Inside water storage was not associated with digestive morbidity or breastfeeding outcomes, but was protective against

Table 9.1 Results of longitudinal modeling

	Respiratory morbidity	Digestive morbidity	Breastfeeding frequency
Predictor	OR (95% CI)	OR (95% CI)	OR (95% CI)
Purified water[a]	1.712 (0.681–4.305)	0.879 (0.320–2.419)	0.520 (0.206–1.316)
Outside storage[b]	0.425 (0.151–1.199)	0.172 (0.061–0.484)*	1.746 (0.594–5.134)
Inside storage[b]	0.458 (0.212–0.990)*	0.643 (0.279–1.485)	1.373 (0.490–3.846)
No storage	N/A	N/A	N/A

All analyses accounted for maternal and child age, child sex, and *Oportunidades* status. Significant results are noted with an asterisk

[a]Purified water is modeled relative to no purified water, regardless of water storage method

[b]Outside and inside water storage are modeled relative to having no water storage mechanism

Table 9.2 Median breastfeeding duration based on purified water use and water storage mechanisms

Predictor	Median	Std. error	95% CI
Purified water[a]	29.949	1.281	27.439–32.459
Outside storage[b]	24.068*	1.294	21.531–26.605
Inside storage[b]	33.634	1.200	31.281–35.986
Total	31.048	0.545	30.943–33.080

Significant results are noted with an asterisk
[a]Purified water is modeled relative to no purified water, regardless of water storage method
[b]Outside and inside water storage are modeled relative to having no water storage mechanism

respiratory morbidity. Breastfeeding durations, as a function of purified water use and water storage method, are provided in Table 9.2. Breastfeeding durations did not differ significantly between children who lived in homes where purified water was or was not used, nor did they differ significantly in houses where inside water storage methods were or were not used. Breastfeeding durations were, however, significantly shorter (by nearly seven months) in children who lived in houses with external water storage methods, compared to the sample median for all children.

9.6 Conclusion

In this chapter, we evaluate water sources and water storage systems in a Yucatec Maya farming community and determine if they are associated with breastfeeding frequency and duration and childhood digestive and respiratory morbidity. We found that all houses had access to piped water in their homes, that 38% of families rely on some form of water storage, and that 12% of families at least occasionally purchase purified water from village stores. These practices are probably in place because there are sporadic interruptions in water delivery service, and villagers need to maintain a water supply for washing and drinking during those periods. We also found that overall childhood digestive and respiratory morbidity rates were quite low (<7% of observations). This is likely because of the Maya tradition of prolonged breastfeeding (Veile and Kramer 2017) and also because of recent interventions by government programs like *Oportunidades*, that aim to improve child health outcomes (Fernald et al. 2008). Finally, purified water use was not associated with childhood morbidity nor either of the breastfeeding outcomes. As we expected, water storage systems were associated with lower childhood morbidity rates. Large, outdoor water receptacles were associated with decreased digestive morbidity, whereas smaller indoor water storage receptacles were associated with decreased respiratory morbidity. Outdoor water storage receptacles were also associated with decreased breastfeeding durations.

9.6.1 Childhood Morbidity Outcomes

While water purification may protect against childhood morbidity (particularly diarrheal) in many settings (Conroy et al. 1996; Sima et al. 2012), we find that use of purified water had no effect on childhood morbidity in the Yucatec Maya case study community. This is probably because very few (if any) families rely exclusively on purchased purified water, which is viewed as costly and inconvenient. As such, purified water is likely consumed by children in conjunction with piped and stored water. We also expect that this finding is linked to more general aspects of children's water consumption that we have observed. Even if children drink only purified water at home, they roam about the village on foot frequently from an early age, carried by alloparents in infancy, and in mixed-age groups from early childhood on. It is not uncommon for thirsty children to stop by a friend or relative's house to scoop water out from their external water storage receptacles.

Water storage systems were associated with childhood morbidity patterns. Inside water storage receptacles were associated with ~50% reduction in childhood respiratory morbidity. This finding may be related to phenomena such as more frequent handwashing, personal cleanliness, and/or better hydration in the presence of indoor water storage receptacles. The same may be true with outdoor water storage receptacles, which are associated with a major decrease in the frequency of childhood digestive morbidity. It is also possible that families without water storage receptacles may differ from families with them, in ways that affect disease transmission. For example, families with water storage systems may simply be able to maintain cleaner homes, or they may live in houses that are better-ventilated. We are unable to assess these possibilities with the data available.

Outside water storage systems facilitate more frequent handwashing and drinking, especially for small children who may struggle to independently pour water from a 5-gallon indoor water receptacle. With a small bucket, children can easily extract from the large water barrels an appropriate amount of water for handwashing, which is linked to reduced illness in a number of studies (Rabie and Curtis 2006). However, some studies from Bangladesh and Guatemala found no effect of handwashing on child morbidity in rural and poor communities (Arnold et al. 2009; Huda et al. 2012). While some authors (e.g., Arnold et al. 2009) attribute this finding to poor compliance, other research indicates that socioeconomic status is more important than specific behaviors in determining handwashing, other sanitary behaviors, and their relationships to health outcomes. For example, in urban Bangladesh, the only handwashing-related variable that was associated with decreased child respiratory morbidity was the presence of an indoor water source for which to wash hands, which is also an indicator of high socioeconomic status (Luby and Halder 2008). We do not expect socioeconomic status to be as strongly associated with child health outcomes in the Maya case study community. As we have previously noted, wealth disparities in the community are minimal, and because of government programs and subsidized health insurance, all families have equal access to medical care (Veile and Kramer 2018).

9.6.2 Breastfeeding Outcomes

Because lactation is associated with increased water intake requirements, we examined the relationship between water source and storage methods and breastfeeding outcomes in the Yucatec Maya. We expected that purified water or indoor/outdoor storage receptacles would alleviate maternal dehydration, and be associated with more frequent breastfeeding and longer breastfeeding durations. However, we found that breastfeeding frequency (likelihood of breastfeeding at any observation, over time) was not associated with any of the water-related predictor variables. Purified water use and outdoor water storage receptacles were also not associated with differences in breastfeeding duration. The only significantly different breastfeeding outcome we found was shorter breastfeeding durations (by nearly seven months) in houses with indoor water storage receptacles. This is interesting in light of our finding that children have decreased digestive morbidity rates in houses with outdoor water storage receptacles. A possible explanation is that Maya mothers may breastfeed children with digestive illness for longer durations as a protective measure. As such, children without digestive illness would be breastfed for shorter durations. Indeed, our previous research shows that Maya mothers are aware of the benefits of breastfeeding (Veile and Kramer). Therefore, they may adjust their breastfeeding practices in response to perceived infant need (Veile and Kramer 2017). It is also possible that mothers in houses with outdoor water receptacles may differ in other ways, such as having a high workload that drives them to cease breastfeeding infants at younger ages. Future studies are underway that will allow us to pursue these questions further.

9.6.3 Future Directions

The research and data here reported could be expanded by collecting data on dehydration rates, systematically monitoring the reliability of piped water access throughout the year, and collecting detailed observations of handwashing and water use patterns. Further data collection efforts that include monitoring water storage methods on a monthly basis, determining how people choose their water storage methods, testing the quality and contamination of water based on storage method, and analyzing breast milk composition with respect to water storage method, would greatly improve and expand on the results reported here. Understanding the associations between water source, storage method, breastfeeding practices and childhood morbidity outcomes, can be useful in informing broader policy in terms of public health, programming and the development of water management infrastructure.

References

Akpinar E (2011) Aguadas: a significant aspect of the southern Maya lowlands water management systems. Doctoral dissertation, U. of Cincinnati

Anatolitou F (2012) Human milk benefits and breastfeeding. J Pediatr Neonat Individual Med 1(1):11–18

Arnold B, Arana B, Mäusesahl D et al (2009) Evaluation of a pre-existing, 3-year household water treatment and handwashing intervention in rural Guatemala. Int J Epidemiol 38(6):1651–1661

Arnold B, Colford JM Jr (2007) Treating water with chlorine at point-of-use to improve water quality and reduce child diarrhea in developing countries: a systematic review and meta-analysis. Am J Trop Med Hyg 76(2):354–364

Back W (1995) Water management by early people in the Yucatán, Mexico. Environ Geol 25(4):239–242

Bauer-Gottwein P, Gondwe BRN, Charvet G et al (2011) The Yucatan Peninsula karst aquifer, Mexico. Hydrogeol J 19(3):507–524

Bentley GR (1998) Hydration as a limiting factor in lactation. Am J Hum Biol 10:151–161

Birol E, Villalba ER, Smale M (2009) Farmer preferences for "milpa" diversity and genetically modified maize in Mexico: a latent class approach. Environ Dev Econ 14(4):521–540

Cabral, J. P. (2010). Water microbiology. Bacterial pathogens and water. International journal of environmental research and public health, 7(10), 3657–3703.

Checkley W, Gilam RH, Black RE et al (2004) Effect of water and sanitation on childhood health in a poor Peruvian peri-urban community. Lancet 363(9403):112–118

Conroy RM, Elmore-Meegan M, Joyce T et al (1996) Solar disinfection of drinking water and diarrhoea in Maasai children: a controlled field trial. Lancet 348(9043):1695–1697

De Frece A, Poole N (2008) Constructing livelihoods in rural Mexico: milpa in Mayan culture. J Peasant Stud 35(2):335–352

Doehring DO, Butler JH (1974) Hydrogeologic constraints on Yucatán's development: the peninsula's water resource is threatened by man-induced contamination. Science 186(4164):591–595

Fernald LC, Gertler PJ, Neufeld LM (2008) Role of cash in conditional cash transfer programmes for child health, growth, and development: an analysis of Mexico's Oportunidades. Lancet 371(9615):828–837

Fink G, Günther I, Hill K (2011) The effect of water and sanitation on child health: evidence from the demographic and health surveys 1986–2007. Int J Epidemiol 40(5):1196–1204

Gelting RJ (1995) Water and population in the Yucatán peninsula. International Institute for Applied Systems Analysis, Laxenburg

González-Gómez F, Guardiola J, Grajales AL (2011) The challenges of water access in Yucatán, Mexico. P I Civil Eng-Munic 64(1):45–53

Huda TMN, Unicomb L, Johnston RB et al (2012) Interim evaluation of a large scale sanitation, hygiene and water improvement programme on childhood diarrhea and respiratory disease in rural Bangladesh. Soc Sci Med 75(4):604–611

IBM Corp (2016) IBM SPSS Statistics for Windows, Version 240. IBM Corp, Armonk, NY

Isendahl C (2011) The weight of water: a new look at pre-hispanic Puuc Maya water reservoirs. Anc Mesoam 22(1):185–197

Jéquier E, Constant F (2010) Water as an essential nutrient: the physiological basis of hydration. Eur J Clin Nutr 64:115–112

Kramer, K. L., & McMillan, G. P. (1998). How maya women respond to changing technology. Human Nature, 9(2), 205.

Kramer, K. L., & McMillan, G. P. (1999). Women's labor, fertility, and the introduction of modern technology in a rural Maya village. Journal of Anthropological Research, 55(4), 499–520.

Kramer MS, Kakuma R (2012) Optimal duration of exclusive breastfeeding. Cochrane Database Syst Rev 8

Kramer K (2005) Maya children: helpers at the farm. Harvard University Press, Cambridge, MA

Kuil L, Carr G, Viglione A et al (2016) Conceptualizing socio-hydrological drought processes: the case of the Maya collapse. Water Resour Res 52(8):6222–6242
Leatherman TL, Goodman A (2005) Coca-colonization of diets in the Yucatán. Soc Sci Med 61:833–846
Luby SP, Halder AK (2008) Associations among handwashing indicators, wealth, and symptoms of childhood respiratory illness in urban Bangladesh. Trop Med Int Health 13(6):835–844
Lucero LJ (2002) The collapse of the Classic Maya: a case for the role of water control. Am Anthropol 104(3):814–826
Maughan RJ (2012) Hydration, morbidity, and mortality in vulnerable populations. Nutr Rev 70(2):S152–S155
McManaman JL, Neville MC (2003) Mammary physiology and milk secretion. Adv Drug Deliv Rev 55(1):629–641
Merchant AT, Jones C, Kiure A et al (2003) Water and sanitation associated with improved child growth. Eur J Clin Nutr 57(12):1562
Miller M, Taube K (1993) The gods and symbols of ancient Mexico and the Maya. Thames and Hudson, London
Normark, J. (2016). The Chicxulub impact and its different hydrogeological effects on Prehispanic and Colonial settlement in the Yucatan peninsula. Wiley Interdisciplinary Reviews: Water, 3(6), 871–884.
Rabie T, Curtis V (2006) Handwashing and risk of respiratory infections: a quantitative systematic review. Tropical Med Int Health 11(3):258–267
Rosinger A (2014) Dehydration among lactating mothers in the Amazon: a neglected problem. Am J Hum Biol 27(4):576–578
Rosinger A, Tanner S (2014) Water from fruit or the river? Examining hydration strategies and gastrointestinal illness among Tsimane' adults in the Bolivian. Public Health Nutr 18(6):1086–1108
Salam RA, Das JK, Bhutta ZA (2015) Current issues and priorities in childhood nutrition, growth, and infections. J Nutr 145(5):1116S–1122S
Sawka MN, Cheuvront SN, Carter R (2005) Human water needs. Nutr Rev 63(6):S30–S39
Sellen DW (2007) Evolution of infant and young child feeding: implications for contemporary public health. Annu Rev Nutr 27:123–148
Silverstein JE, Webster D, Martinez H et al (2009) Rethinking the great earthwork of Tikal: a hydraulic hypothesis for the Classic Maya polity. Anc Mesoam 20(1):45–58
Sima L, Desai MM, McCarty KM et al (2012) Relationship between use of water from community-scale water treatment refill kiosks and childhood diarrhea in Jakarta. Am J Trop Med Hyg 87(6):979–984
Thornton S (2010) Thirst and hydration: physiology and consequences of dysfunction. Physiol Behav 100:15–21
UNICEF (2016) Water, sanitation and hygiene www.unicef.org/wash/3942_4456.html Accessed 11 Jan 2018
Vargas LA (2001) Thirst and drinking as a biocultural process. In: de Garine I (ed) Drinking: anthropological approaches. Berghahn Books, New York, NY, pp 11–21
Veile A, Kramer K (2015) Birth and breastfeeding dynamics in a modernizing indigenous community. J Hum Lact 31(1):145–155
Veile A, Kramer K (2017) Shifting weanling's optimum: breastfeeding ecology and infant health in Yucatán. In: Tomori C, Palmquist AEL, Quinn EA (eds) Breastfeeding: new anthropological approaches. Routledge, Abingdon
Veile, A., & Kramer, K. L. (2018). Pregnancy, birth, and babies: Motherhood and modernization in a Yucatec Village. In Maternal Death and Pregnancy-Related Morbidity among Indigenous Women of Mexico and Central America (pp. 205–223). Springer, Cham
Veile, A., Faria, A. A., Rivera, S., Tuller, S. M., & Kramer, K. L. (2019). Birth mode, breastfeeding and childhood infectious morbidity in the Yucatec Maya. American Journal of Human Biology, 31(2), e23218.

WHO (2018) Water sanitation hygiene. World Health Organization. Available at https://www.who.int/water_sanitation_health/water-quality/en/. Accessed 11 Jan 2018

WHO/UNICEF Joint Monitoring Programme for Water Supply and Sanitation (2004) Meeting the MDG drinking water and sanitation target: a mid-term assessment of progress. WHO, Geneva

Wolf J, Prüss-Ustün A, Cumming O et al (2014) Systematic review: assessing the impact of drinking water and sanitation on diarrhoeal disease in low-and middle-income settings: systematic review and meta-regression. Tropical Med Int Health 19(8):928–942

Chapter 10
Patterns of Activity and Somatic Symptoms Among Urban and Rural Women at Midlife in the State of Campeche, Mexico

Lynnette Leidy Sievert, Laura Huicochea-Gómez, Diana Cahuich-Campos, and Daniel E. Brown

Abstract Explanations for variation in symptom frequencies at midlife generally focus on hormonal changes associated with the menopausal transition. This chapter emphasizes culture-specific patterns of activity among women aged 40–60 in the hot and humid environment of the state of Campeche, Mexico, and the relationship between activities and somatic symptoms. A total of 543 Maya and non-Maya women participated in this study—305 from the city of Campeche and 238 from 12 rural communities. Semi-structured interviews collected demographic, biobehavioral, reproductive, and activity information. The nine activities and seven symptoms varied across the communities, and there was heterogeneity in explanatory factors. For example, increasing age, decreasing socioeconomic status (SES), and more time spent sweeping/mopping increased the likelihood of muscle/joint pain. Washing clothes by hand, the number of people living in the household, menopausal status, and SES increased the likelihood of headaches. Working outside in the garden or agricultural fields was associated with palpitations and dizziness, but not after controlling for SES. Only shortness of breath was consistently associated with high ambient temperatures. Women in rural communities reported more somatic symptoms, and also more hours of physical activity. However, SES, not physical activity, was the variable most often associated with somatic symptoms. These results reflect the great complexity of human ecology, where physical, biological, and sociocultural factors are enmeshed.

L. L. Sievert (✉)
Department of Anthropology, University of Massachusetts Amherst, Amherst, MA, USA
e-mail: leidy@anthro.umass.edu

L. Huicochea-Gómez · D. Cahuich-Campos
Sociedad y Cultura, El Colegio de la Frontera Sur, Unidad Campeche, Campeche, México
e-mail: lhuicochea@ecosur.mx; dcahuich@ecosur.mx

D. E. Brown
Department of Anthropology, University of Hawai'I at Hilo, Hilo, HI, USA
e-mail: dbrown@hawaii.edu

H. Azcorra, F. Dickinson (eds.), *Culture, Environment and Health in the Yucatan Peninsula*, https://doi.org/10.1007/978-3-030-27001-8_10

10.1 Introduction

Women at midlife are a critical, but often an invisible, part of every community. They work behind the scenes to maintain homes, support children, care for grandchildren, tend to the needs of their parents and parents-in-law, earn extra money for households, sustain churches, libraries, and other community institutions, and help husbands in the fields, yard, and gardens. The ecology of women at midlife is largely sociocultural in that they are embedded in a web of relationships. In the Yucatan Peninsula, these sociocultural systems are situated in the context of a hot, tropical environment, often in difficult economic circumstances. In some of these communities, women experience hard physical labor in a difficult geography along with a lack of convenient potable water. The question of interest in this chapter is: can we read the effects of sociocultural contexts and physical activity in the **symptoms** reported by women at midlife? In particular, this chapter will emphasize socioeconomic status and the number of people in the household as potential measures of social stress, religious participation as a potential measure of stress or support, ambient temperature, and patterns of physical activity among women aged 40–60 in the state of Campeche, Mexico, and the relationship between these variables and somatic symptoms.

There is a constant interaction between people and their milieu. Humans respond to all aspects—physical, biological, and sociocultural—of the surrounding environment, and in turn the environment is influenced by human activities. Ecological studies of any species are notoriously complex; this complexity is magnified when dealing with the intricacies of the human social environment (Kormondy and Brown 1998). Reactions to one's surroundings often depend more upon one's **perception** of these surroundings than some objective measure of the environment (Lazarus and Folkman 1984; Brown 2016). To make things more daunting, ecological effects can be viewed at both the individual and population level. Responses to the environment include both physiological and behavioral/cultural components (Dressler 1991; Dressler and Bindon 2000), and these components, in turn, interact in complicated ways. For instance, the behavior of hard physical labor in a hot climate leads to physiological changes associated with heat **acclimatization** and physical fitness (Frisancho 1993). These adaptations, in turn, permit increased physical activity that enhances the adaptations (Bouchard et al. 1994). The biological response to exercise is also affected by the individual's perception of the physical activity within a sociocultural context of what activities are appropriate for a given status (Brown et al. 1998). The need for physical activity is determined by broad requirements for resource acquisition set by the larger ecological context.

Characteristics and perceptions of the environment have an effect on health and the experience of specific symptoms associated with health. Fatigue, for example, can result from physical labor, iron deficiency, or respiratory diseases—among many physical factors—but also from psychological factors such as depression (Stadje et al. 2016) or stress (e.g., Pawlikowska et al. 1994). An underlying assumption of this work is that stress is positively associated with somatic symptoms.

Everyday stress is part of the ecological context in which women live, love, and work. We must be careful, however, to note that one person's stress can be another's joy. For example, multiple grandchildren living in the household may signify a burden or a happy family (Chen et al. 2017). Religion can play a supportive or stressful role in the context of women's lives (Zimmer et al. 2016). In addition, physical activity can reduce stress in most instances, but exhaust an individual and increase stress at other times. In an earlier study of an urban subgroup of women from this same population, the Perceived Stress Scale (PSS) was significantly and positively associated with the likelihood and severity of fatigue, muscle/joint pain, difficulty sleeping, and headaches (Sievert et al. 2018).

Research on symptom frequencies at midlife generally focuses on hormonal changes associated with the **menopausal transition** (Freeman et al. 2007). This focus works well for symptoms that have a clear relationship with declining levels of estrogen, such as hot flashes and vaginal dryness (Dhanoya et al. 2016; Freedman 2005; Gold et al. 2004; Randolph et al. 2005). However, in many populations, women aged 40–60 are more likely to complain of fatigue, stiffness, or aches and pains than hot flashes. For example, in the study of symptoms and stress in the city of Campeche mentioned above (Sievert et al. 2018), tiredness (reported by 78% of participants), muscle/joint pain (76%), difficulty sleeping (64%), and headaches (58%) were all reported with more frequency than hot flashes (49%).

There is mixed evidence suggesting that the decline in levels of estrogen may be related to joint pain (Chlebowski et al. 2013; de Klerk et al. 2009); however, other variables are likely contributors, such as body mass and levels of physical activity (Gay et al. 2014; Liu et al. 2016). Difficulty sleeping may result from hot flashes (Smith et al. 2018; Thurston et al. 2008; Woods and Mitchell 2010)—directly related to fluctuating levels of estrogen, but other variables are likely contributors, such as psychological stress (Staner 2003). Migraine headaches may be less likely after menopause, but there is evidence that the frequency of tension-type headaches increases (Oh et al. 2012). In addition to these four symptoms, we are also interested in three other general somatic symptoms: **palpitations**, **dizziness**, and shortness of breath. It is likely that all of these symptoms have a psychosomatic component, in the sense that psychosocial distress may be expressed through the body (De Gucht and Fischler 2002; Gureje et al. 1997; Hange et al. 2013; Keyes and Ryff 2003; Simon et al. 1996). For example, in some cultural contexts, it may be more appropriate to complain of back pain than anger or depression (Kleinman and Kleinman 1985). That women may somaticize stress is suggested by our findings that the PSS was significantly and positively associated with somatic symptoms (Sievert et al. 2018).

There may be somatization of emotional distress, but the fact remains that for many of the women in our study, the ecological context includes hard physical labor in a hot, tropical environment. Some of these women work with their husbands in the fields (*milpa*) where the geography is rough. In addition, there are social/gender roles and responsibilities. Many of these women are the head of their households, taking care of children, and caring for elder family members and sometimes grandchildren.

The purpose of this study was to examine nine physical activities to understand how much work and what kind of work women were doing in each of four sites in the state of Campeche. We hypothesized that more physical activity would result in a higher likelihood of somatic symptoms after controlling for factors intrinsic to the woman (age, menopausal status, and body mass index) and factors external to the woman (number of household members, socioeconomic status (SES), religious attendance, and, in a subset of women, ambient temperature). With regard to ambient temperature, we hypothesized that heat would increase the likelihood of somatic symptoms.

10.2 Communities

The study was carried out in the Mexican state of Campeche from 2013 to 2015. The state is part of the Yucatán Peninsula, bordering the Gulf of Mexico. The climate is hot and humid with the coolest 24-hour average temperatures from December (23.7 °C) through February (24.3 °C), and the hottest 24-hour average temperatures (29.0 °C) lasting from May through October. The mean annual rainfall is 112 cm (www.worldclimate.com). Twelve percent of the population speaks an indigenous language, most often Maya (INEGI 2014). There is a gradient of marginalization across the three municipalities sampled in this study (Cahuich-Campos et al. 2018). A total of 543 Maya and non-Maya women participated in this study—305 from the city of Campeche and 238 from 12 rural communities. The communities are grouped into three municipalities: Campeche, Hopelchén, and Calakmul (Cahuich-Campos et al. 2018). Table 10.1 illustrates variation in the basic services available to women in the different municipalities of Campeche.

The city of Campeche is the state capital and is home to approximately 220,000 residents (2010 Census). Almost all urban households have access to running water, electricity, and sewage (Table 10.1) (INEGI 2015). There is a diversity of health

Table 10.1 Characteristics of the municipalities

	Campeche Municipality	Hopelchén	Calakmul
Population, 2015 census	283,025	40,100	28,424
Characteristics (in percentages)			
Households with running water	77.5	51.6	12.4
Households with electricity	99.2	96.9	95.4
Households with sewage	97.6	72.3	73.5
Population affiliated with health services	89.9	86.9	92.1
Population without schooling	4.3	22.5	14.6

Panorama sociodemográfico de Campeche, 2015.Encuesta Intercensal 2015. Instituto Nacional de Estadística y Geografía. México
http://internet.contenidos.inegi.org.mx/contenidos/productos//prod_serv/contenidos/espanol/bvinegi/productos/nueva_estruc/inter_censal/panorama/702825082116.pdf

services available, both public and private. For this study, 305 Maya and non-Maya participants were drawn from offices, businesses, schools, in front of the city market, and by giving presentations to groups of women in private homes in eight city neighborhoods (*colonias*). In the city and surrounding areas, women work in stores, government offices, and public schools (Huicochea et al. 2017).

About 40 min by bus from the capital city are the rural Maya communities of Suc Tuk and Ich Ek. In these communities, 78 Maya women were recruited following presentations to community leaders. Local medical providers facilitated community presentations through introductions. Being close to the city of Campeche, these communities enjoy access to urban amenities. Women in both communities work in the home, take care of kitchen gardens, embroider, sell their products, and help their husbands in the fields. The economic activities of Ich Ek rotate around the production of agriculture, breeding of small livestock and backyard animals, and the cultivation of bees. Honey from native species, *Melipona beecheii*, is produced by a group of women who belong to an organization called *koolel kab* (women who work with bees). Thanks to the help of community elders, the women rescued a prehispanic tradition on the verge of extinction. Nests of bees were started in *hobones* or hollow logs, and women learned the basics of bee cultivation. This activity has contributed to the strengthening of other cultural and medicinal traditions and knowledge associated with the use of honey (Carvajal Correa and Huicochea-Gómez 2010).

In the rural, isolated communities of the municipality of Calakmul, 84 women participated in the study following presentations to community leaders and other public informational meetings. These communities are ethnically non-Maya because this land was given to migrants from other Mexican states, e.g., Veracruz, Tabasco, and Chiapas, in the 1960s through 1980s (García Gil and Pat Fernández 2000). In the rural, isolated Maya communities of the municipality of Hopelchén, 76 women participated in the study after presentations were given to community leaders and to the general public. In Hopelchén, we were helped with translation by community assistants who spoke Maya. In Calakmul, we were helped by translators who spoke Chol.

In the rural municipalities of both Calakmul and Hopelchén, many homes lack basic infrastructure. Schooling was not universally accessible to these women who are now 40–60 years of age (Table 10.1). Currently, primary health services in these rural communities are provided by PROSPERA, a federal health program that addresses both health care needs and sociopolitical action (SEDESOL 2016). There are also rural clinics and a mobile health "caravan." In both Calakmul and Hopelchén, women work in the home and also carry out agricultural activities. In Calakmul, women help with the harvest and drying of jalapeño peppers. They also attend to small stores, clean houses, or wash and iron clothes. At times, women in both Calakmul and Hopelchén participate in the cleaning of schools, clinics, churches, or municipal buildings. In Hopelchén, women attend to small stores, embroider traditional dresses, and weave hammocks.

Ethical approval was obtained from the Institutional Review Boards of UMass Amherst, the University of Hawaii at Hilo, and the *Comité de Ética de la Secretaría de Salud del Estado de Campeche*.

10.3 Methods

Semi-structured interviews collected demographic, biobehavioral, reproductive, and dietary information. The symptom list was based on an "everyday symptom list" that has been used in many studies (Avis et al. 1993; Dennerstein et al. 1993; Obermeyer et al. 2007), including in Mexico (Sievert and Espinosa-Hernandez 2003). Symptoms were assessed by asking, "During the past two weeks, have you felt any of the following?" (*Durante las dos semanas pasadas ha sentido alguna de las siguientes molestias?*) Symptom intensities were reported as: 0 = none (*nada*); 1 = a little (*un poco*); 2 = a lot (*mucho*); and 3 = extreme (*muchísimo*). The specific somatic symptoms examined in this chapter are dizziness (*vértigos o mareos*), fatigue (*cansancio o falta de energía*), heart palpitations (*palpitaciones cardiacas*), headaches (*dolor de cabeza*), pain in muscles or joints (*dolores musculares o articulares*), shortness of breath (*dificultad para respirar*), and difficulty sleeping (*dificultad para dormir*).

A measure was developed for nine culture-specific physical activities. Women were asked how many times, and for how many hours, they carried out a specific activity. Specifically, we queried walking while carrying weight (*caminar cargando peso*), walking outside (*caminar afuera*), sweeping or mopping (*barrer o fregar el hogar*), washing clothes by hand (*lavar ropa a mano*), going out and carrying home water (*buscar agua*), grinding by hand or making tortillas (*moler a mano o tortear*), working in the garden or patio (*trabajar en el solar*), working in the fields (*trabajar en la milpa*), and participating in structured activities (e.g., running, dancing, Zumba, sports, Pilates, yoga, relaxation, or meditation). The number of times per week was multiplied by the number of hours for a measure of hours of activity per week for each woman.

Three variables were considered to be intrinsic, i.e., mostly independent of the environment: age, menopausal status, and body mass index (BMI). Chronological age was determined by date of interview minus date of birth. Menopausal status was defined by STRAW + 10 stages: (1) menstruation was regular, (2) there were changes in the number of days or quantity of blood, (3) menstrual cycles were more or less frequent than before, (4) there was a change of more than 6 days in the frequency of periods, (5) two months or more months passed without a period, and (6) more than 12 months passed without a period (Harlow et al. 2012). Stages 1 and 2 were categorized as pre-menopausal, stages 3–5 as peri-menopausal, and stage 6 as postmenopausal. Stature was measured with a Seca 213 stadiometer to the nearest 0.1 cm. Weight was measured to the nearest 0.1 kg with a digital scale. BMI was computed as kg/m^2.

Five variables were considered to be measures of the participant's environment: study site, household composition, socioeconomic status (SES), religion, and ambient temperature. For the purposes of this study, household composition was treated as the number of individuals living in the household. For a subset of women in the city of Campeche ($n = 305$), we have an additional variable of the number of people living in the household who need extra help (e.g., a disabled husband, a child with developmental delays, or a parent with infirmities). An SES index was created from 10 dimensions related to housing construction and infrastructure, such as type of architecture, number of bedrooms, access to drinking water, type of cooking fuel, and type of bathroom. Most women in the study were Catholic (70%), but a sizable number were Presbyterian, Evangelical, Pentecostal, Jehovah's Witness, or members of other Protestant groups. For the purposes of this study, external religiosity was measured by asking, "How often do you go to church or religious meetings?" Responses were divided into infrequent (never to a few times a year) or frequent (a few times a month to more than once per week) attendance. Ambient temperature was measured in the subgroup of women who wore a hot flash monitor ($n = 155$) (Sievert et al. 2018).

The nine activities (hours per week) and seven somatic symptoms (yes/no) were examined across the four sites (one urban and three rural) by ANOVA and chi-square as appropriate. Each of the nine activities was examined in relation to symptom frequencies (yes/no) by t-tests. Symptoms (yes/no) were examined in relation to age, menopausal status, BMI, the number of people living in the household, the number of people in the household requiring extra help (in the city, only), SES, frequent/infrequent church attendance, and ambient temperature (in a subgroup) by chi-square and *t*-test as appropriate. Finally, variables significant at $p < 0.05$ in bivariate analyses were examined in logistic regression analyses in relation to each symptom.

10.4 Results

Sample characteristic are shown in Table 10.2. There was no difference across the four sites in age at interview or menopausal status, but women in Calakmul had a smaller BMI than women at other sites (range for entire sample, 18.27–57.73 kg/m^2). With regard to environmental variables, women in the city of Campeche had a significantly higher mean SES index (range for entire sample, 17–39) and fewer people lived in the household compared to rural sites (range for entire sample 1–11). In the city of Campeche, the number of dependents ranged from 0 to 4. Women in Hopelchén demonstrated the highest level of external religiosity, with 92% of participants claiming to frequently attend religious services (from a few times a month to more than once per week). Among the subset of women who wore a hot flash monitor, those living in the city of Campeche were exposed to a cooler mean ambient temperature compared to the other sites (for entire sample, temperature ranged from 21 °C to 38 °C).

The most common symptoms in the total sample (Table 10.3) were fatigue (80%) and muscle/joint pain (80%), followed by headaches (63%), difficulty sleeping (61%), palpitations (44%), dizziness (44%), and shortness of breath (33%). There were no differences across the sites in frequency of fatigue or difficulty sleeping. With the exception of palpitations, all symptoms were most frequent in Hopelchén. Except for difficulty sleeping, all symptoms were least frequent in the city of

Table 10.2 Sample characteristics

	Total ($n = 543$)	City of Campeche ($n = 305$)	Ich Ek Suc Tuc ($n = 78$)	Calakmul ($n = 84$)	Hopelchén ($n = 76$)	*p* value
Intrinsic variables						
Age at interview, years, $n = 543$ Mean (s.d.)	47.4 (4.9)	47.5 (5.0)	47.4 (4.6)	47.3 (5.0)	47.1 (4.4)	0.961
Menopausal status, $n = 543$, % Pre Peri Post	40.5 24.1 35.4	43.6 22.3 34.1	39.7 24.4 35.9	41.7 21.4 36.9	27.6 34.2 38.2	0.229
BMI, kg/m^2 $n = 536$ Mean (s.d.)	30.9 (5.3)	30.6 (5.5)	31.8 (4.6)	30.0 (5.0)	31.8 (5.0)	0.040
Environmental variables						
SES $n = 543$ Mean (s.d.)	29.7 (4.8)	33.1 (2.4)	25.9 (3.4)	25.0 (3.0)	25.3 (3.4)	<0.001
# individuals in the household, $n = 518$ Mean (s.d.)	4.3 (1.9)	3.8 (1.4)	5.2 (1.9)	4.7 (2.2)	5.6 (2.4)	<0.001
# in the household needing extra help, $n = 304$ Mean (s.d.)	–	0.5 (0.8)	–	–	–	–
Church attendance, % Infrequent Frequent	29.2 70.8	36.8 63.2	28.2 71.8	21.4 78.6	7.992.1	<0.001
Ambient temperature, °C $n = 150$ Mean (s.d.)	29.3 (3.1)	27.6 (2.1)	32.6 (3.1)	31.5 (2.6)	29.7 (1.4)	<0.001

Chi-square used to compare percentages across study sites
ANOVA used to compare means across study sites

Table 10.3 Frequency of somatic symptoms across study sites (in percentages)

	N	Total	City of Campeche	Ich Ek Suc Tuc	Calakmul	Hopelchén	*p* value
Fatigue	543	80.3	78	83	80	88	0.192
Muscle/joint pain	541	80.0	76	82	86	88	0.044
Headaches	541	63.0	58	61	70	79	0.003
Difficulty sleeping	543	61.1	64	53	55	67	0.123
Palpitations	543	44.4	36	55	58	50	<0.001
Dizziness	542	44.3	34	50	58	63	<0.001
Shortness of breath	538	32.5	26	29	45	48	<0.001

Chi-square used to compare frequencies of symptoms across study sites

Campeche. Women in the most isolated communities in the municipalities of Calakmul and Hopelchén reported significantly more shortness of breath compared to women in the communities of Ich Ek and Suc Tuc which are closer to the city ($p = 0.035$). In addition, women in Calakmul and Hopelchén reported significantly more muscle/joint pain ($p = 0.020$), headaches ($p = 0.001$), dizziness ($p < 0.001$), palpitations ($p < 0.001$), and shortness of breath ($p < 0.001$) compared to women in the city of Campeche.

The most bothersome symptoms, measured by the responses *mucho* or *muchisimo* were muscle/joint pain (35%), fatigue (29%), difficulty sleeping (25%), and headaches (21%) (Table 10.4). Across the four communities, the only significant differences for bothersomeness were palpitations (more bothersome in Calakmul) and dizziness (more bothersome in Calakmul and Hopelchén).

Each symptom (yes/no) was examined in relation to age at interview, menopausal status, and BMI. Only muscle/joint pain was associated with age at interview so that women with pain were older (47.6 (s.d. 4.6) years vs. 46.5 (s.d. 4.9) years, $p = 0.023$). Peri-menopausal women had the highest frequency of fatigue (87%) compared to pre- (76%) and post-menopausal (81%) women ($p = 0.028$). Post-menopausal women had the highest frequency of muscle/joint pain (85%) compared to pre- (75%) and peri-menopausal (80%) women ($p=0.039$). Peri-menopausal women tended to have more headaches ($p = 0.06$). There were no associations between menopausal status and difficulty sleeping, palpitations, dizziness, or shortness of breath. Women with fatigue and muscle/joint pain had higher BMIs (31.1 (s.d. 5.4) vs. 29.9 (s.d. 4.7), $p = 0.041$, and 31.1 (s.d. 5.4) vs. 28.9 (s.d. 4.9), $p = 0.038$, respectively). Therefore, the only intrinsic variables included in logistic regression models were age, menopausal status, and BMI for muscle/joint pain; menopausal status and BMI for fatigue; and menopausal status for headaches.

Each symptom was also examined in relation to number of people in the household, number of people needing extra help in the household (in the city only), SES, and religious attendance. Headaches and shortness of breath were associated with the number of people in the household so that women with these symptoms lived with more people (4.5 (s.d. 2.0) vs. 4.1 (s.d. 1.7), $p = 0.040$, and 4.6 (s.d. 2.1) vs. 4.2 (s.d. 1.7), $p = 0.019$, respectively). There were no relationships between any symptom

Table 10.4 Frequency of bothersome symptoms ("*Mucho*" or "*Muchisimo*") (%)

	N	Total	City of Campeche	Ich Ek Suc Tuc	Calakmul	Hopelchén	*p* value[a]
Fatigue	543	28.5	26	32	27	37	0.247
Muscle/joint pain	541	34.9	34	29	46	34	0.089
Headaches	541	21.4	18	20	28	30	0.052
Difficulty sleeping	543	25.0	26	21	25	28	0.760
Palpitations	543	12.2	10	12	21	12	0.040
Dizziness	542	10.0	7	6	20	16	0.002
Shortness of breath	538	8.0	7	7	11	11	0.523

[a]A lot/extreme (*Mucho/Muchisimo*) vs. None/a little (*Nada/Un poco*)
Chi-square used to compare frequencies of bothersome symptoms across study sites

frequency and the number of people needing extra help in the household in the city of Campeche. In the total sample, women with muscle/joint pain, headaches, palpitations, dizziness, and shortness of breath had significantly lower SES scores (all, $p < 0.01$). Finally, there were no associations between church attendance (infrequent/frequent) and any somatic symptom (yes/no). Therefore, the environmental variables included in logistic regression models were number of people in the household and SES for headaches and shortness of breath, and SES for muscle/joint pain, palpitations, and dizziness. Regarding the mean interior temperature at the study site for a subset of women ($n = 150$) who wore hot flash monitors, women who reported fatigue, dizziness, and shortness of breath experienced significantly hotter temperatures in their homes (Table 10.5).

Apart from the time spent sweeping and mopping, time spent in various physical activities varied across the communities (Table 10.6). With the exception of structured activity, which was practiced more often in the city of Campeche, women in the rural communities reported carrying out more hours of each physical activity.

Table 10.5 Mean interior temperature (°C) at the study site in relation to symptoms ($n = 150$)

	No	Yes	*p* value
Fatigue	28.1	29.6	0.031
Muscle/joint pain	29.5	29.3	0.827
Headaches	29.5	29.3	0.771
Difficulty sleeping	29.4	29.3	0.837
Palpitations	28.9	29.8	0.088
Dizziness	28.6	30.1	0.005
Shortness of breath	28.6	30.5	<0.001

T-test used to compare mean temperatures by symptom (yes/no)

Table 10.6 Mean number of hours per week spent in various activities

	N	Total	City of Campeche	Ich Ek Suc Tuc	Calakmul	Hopelchen	*p* value
Sweep or mop the house	542	4.8	3.0	4.9	4.6	3.6	0.360
Walk outside	542	3.6	2.3	5.4	5.4	4.9	0.001
Work in the yard, garden, or patio	542	2.7	0.9	3.8	5.0	5.9	<0.001
Wash clothes by hand	540	2.3	1.4	2.7	3.6	4.0	<0.001
Walking carrying weight	538	1.6	1.1	3.6	1.5	1.5	0.006
Grind corn by hand or make tortillas	542	1.5	0.3	3.4	3.4	3.7	<0.001
Work in the fields	542	1.1	0.1	1.9	3.0	1.9	<0.001
Structured activity	542	0.9	1.4	0.5	0.2	0.2	<0.001
Go outside and carry home water	541	0.4	0.2	0.7	1.2	0.4	<0.001
All activities	537	18.7	12.4	27.5	27.6	26.0	<0.001

ANOVA used to compare mean hours of activity across study sites

For all activities added together, women in the city of Campeche reported 12 h of physical activity per week compared to 26 or more hours per week in the rural communities ($p < 0.001$).

Women who reported fatigue spent more time sweeping and mopping floors, and participated in structured activities less often, than women who did not report fatigue (Table 10.7). Women with muscle/joint pain spent more time sweeping/mopping floors, washing clothes by hand, and going out for water than women without muscle/joint pain. Women with headaches spent more time washing clothes by hand. No activities were significantly associated with difficulty sleeping. Palpitations and dizziness were associated with more time spent going out for water, making tortillas by hand, working in the yard or garden (*solar*), and working in the fields (*milpa*). Women who reported shortness of breath spent more hours going for water and making tortillas. Women with dizziness and shortness of breath spent less time in structured activities during the week.

The logistic regression model for fatigue included sweeping/mopping, structured activity, menopausal status, and BMI. Women who participated in structured activities were less likely to report fatigue, and peri-menopausal women were twice as likely to report fatigue. In the subgroup analysis, ambient temperature was not significant, and did not alter the results. The logistic regression model for muscle/joint pain included sweeping/mopping, washing clothes by hand, going out for water, age at interview, menopausal status, BMI, and SES. Sweeping/mopping increased the likelihood of muscle/joint pain, as did age at interview. A higher SES decreased the likelihood of muscle/joint pain (Table 10.8).

The logistic regression model for headaches included washing clothes by hand, menopausal status, number of people living in the household, and SES. Washing clothes by hand and being peri-menopausal increased the likelihood of headaches. The logistic regression model for palpitations included bringing home water, making tortillas, working in the garden/yard, working in the fields, and SES. Women with higher SES had a lower risk of palpitations, and bringing home water approached significance ($p = 0.052$). The logistic regression model for dizziness included bringing home water, making tortillas, working in the garden/yard, working in the fields, structured activity, and SES. As SES increased, the likelihood of having dizziness decreased. When ambient temperature was added to the model with a subset of participants, ambient temperature was not associated with dizziness, but bringing home water became significant (OR 1.910, 95% CI 1.037–3.515), so that women who spent more time bringing home water were at a higher risk of dizziness. Finally, the logistic regression model for shortness of breath included bringing home water, making tortillas, structured activity, SES, and number of people in the household. Women with a higher SES were at less risk of shortness of breath. When ambient temperature was added to the model, with a subset of the women, increasing temperature increased the likelihood of difficulty breathing (OR 1.216, 95% CI 1.042–1.418) and bringing home water also became significantly associated shortness of breath (OR 1.373, 95% CI 1.027–1.834).

Table 10.7 Somatic symptoms in relation to mean number of hours (s.d.) spent in physical activities (only significant relationships shown)

		Fatigue	Muscle/joint pain	Headaches	Difficulty sleeping	Palpitations	Dizziness	Shortness of breath
Walking carrying weight	No Yes							
Walk outside	No Yes							
Sweep or mop	No Yes	3.73 (4.1) 5.00 (6.6) $p = 0.013$	3.66 (3.9) 5.03 (6.7) $p = 0.006$					
Wash clothes by hand	No Yes		1.6 (2.6) 2.5 (3.8) $p = 0.007$	1.8 (3.2) 2.6 (3.8) $p = 0.018$				
Go out and carry home water	No Yes		0.18 (0.7) 0.50 (1.7) $p = 0.002$			0.28 (1.1) 0.63 (2.0) $p = 0.016$	0.30 (0.3) 0.24 (0.6) $p = 0.032$	0.29 (1.2) 0.70 (2.1) $p = 0.016$
Grind corn by hand or make tortillas	No Yes					1.28 (3.0) 1.86 (3.5) $p = 0.040$	1.20 (2.9) 2.0 (3.6) $p = 0.008$	1.27 (2.9) 2.04 (3.8) $p = 0.017$
Work in the yard, garden, or patio	No Yes					2.15 (4.2) 3.33 (7.2) $p = 0.025$	1.96 (3.7) 3.58 (7.5) $p = 0.002$	
Work in the fields	No Yes					0.62 (3.2) 1.59 (4.9) $p = 0.009$	0.56 (2.5) 1.67 (5.4) $p = 0.004$	
Structured activity	No Yes	1.51 (2.8) 0.76 (2.8) $p = 0.010$					1.07 (2.4) 0.71 (1.7) $p = 0.045$	1.07 (2.4) 0.59 (1.6) $p = 0.005$

T-test used to compare mean number of hours spent in each physical activity by symptom (yes/no)

Table 10.8 Logistic regression results for symptoms

	Fatigue	Muscle/ joint pain	Headaches	Palpitations	Dizziness	Shortness of breath
Age		**1.063** **1.000–1.129**				
Menopausal status Pre- (ref) Peri- Post-	**2.104** **1.151–3.846** 1.273 0.779–2.081	1.177 0.668–2.075 1.378 0.716–2.653	**1.663** **1.032–2.680** 1.102 0.728–1.667			
BMI	1.044 0.999–1.091	1.040 0.995–1.087				
# people in the household			1.033 0.922–1.158			1.024 0.915–1.145
SES		**0.925** **0.877–0.975**	0.961 0.917–1.007	**0.946** **0.906–0.987**	**0.932** **0.892–0.974**	**0.944** **0.898–0.992**
Sweeping/ mopping	1.045 0.998–1.095	**1.050** **1.003–1.100**				
Washing clothes by hand		1.032 0.952–1119	**1.070** **1.002–1.143**			
Going outside and carrying home water		1.189 0.933–1.514		1.132 0.999–1.283	1.091 0.969–1.229	1.123 0.996–1.268
Making tortillas by hand				0.989 0.929–1.053	0.992 0.931–1.057	1.018 0.950–1.090
Working in garden				1.016 0.992–1.095	1.029 0.990–1.070	
Working in fields				1.042 0.992–1.095	1.045 0.993–1.100	
Structured activity	**0.896** **0.820–0.979**				0.960 0.879–1.048	0.906 0.811–1.012
Subgroup analysis, $n = 155$ Adding ambient temperature	Temperature n.s. and did not change results				Temperature n.s.; searching for water **significant**	Temperature **significant** searching for water **significant;** SES n.s.

Odds ratio and 95% confidence intervals. Only variables significantly associated with a symptom at $p = 0.05$ in bivariate analyses were included in the logistic regression for that symptom. Thus, variables in each regression model differed. Significant variables are bolded

10.5 Discussion

Our aim was to understand reported symptom experience within an individual's ecological context. Specifically, analyses focused on the associations between the reported frequency of seven somatic symptoms at midlife (fatigue, muscle/joint pain, headaches, difficulty sleeping, palpitations, dizziness, and shortness of breath) and physical activities in light of various intrinsic (age, menopausal status, BMI) and environmental characteristics (types of activity, socioeconomic status, household composition, and environmental temperature). There is heterogeneity in these associations based on the community in which the women lived and the specific symptom under study.

Women were drawn from very different sites across the state of Campeche, from the capital city to rural isolated Maya and non-Maya communities in the municipalities of Hopelchén and Calakmul (Huicochea Gómez et al. 2017). Across these sites, there were no differences in age at interview or menopausal status, but BMI differed so that women in non-Maya Calakmul were leaner than women at other sites. If BMI is a factor in muscle/joint pain, then less pain might be expected among women in Calakmul. Across the entire sample, women with muscle/joint pain did have higher BMIs; however, as Table 10.3 shows, the lean women in Calakmul reported more, not less, muscle/joint pain compared to women in Ich Ek/Suc Tuc and the city of Campeche. Something beyond BMI is influencing the experience of muscle/joint pain in the rural community. For example, there may be a difference in ethnic-related concepts of whether it is appropriate to report pain (Sievert and Brown 2016), although in this study there were no Maya/non-Maya differences in the likelihood of reporting the somatic symptoms of interest (data not shown). As the logistic regression results show, increasing age, decreasing SES, and more time spent sweeping/mopping increased the likelihood of muscle/joint pain.

With regard to environmental variables, women in the city of Campeche had a significantly higher mean SES index and fewer people lived in the household compared to the rural sites. This suggests that if economic or social stressors are factors in determining symptom frequencies, then there might be fewer somatic symptoms in the city. Across the entire sample, women with muscle/joint pain, headaches, palpitations, dizziness, and shortness of breath had significantly lower SES scores. SES remained associated with all but headaches in the logistic regression models. Also, across the entire sample, women with headaches and shortness of breath lived with more people in the household. As expected, compared to women in the most rural municipalities, women in the city were significantly less likely to report muscle/joint pain, headaches, dizziness, palpitations, and shortness of breath. Reduced stress may play a role in the lower symptom frequency.

Women in Hopelchén demonstrated the highest level of external religiosity, but religiosity can be associated with either lower or higher levels of stress (Zimmer et al. 2016). In fact, as Table 10.3 demonstrates, women in Hopelchén reported the highest frequency of every symptom except for palpitations. Church attendance is not dampening symptom frequencies in Hopelchén, and may be a contributor.

However, in the entire sample, there were no associations between infrequent/frequent church attendance and any somatic symptom.

Finally, the city of Campeche registered a cooler mean temperature compared to the other sites. If heat exacerbates discomfort, then it may be that women in rural communities register more somatic symptoms, and this is what we see in Table 10.3. As seen in Table 10.5, women who reported dizziness and shortness of breath experienced significantly hotter temperatures in their homes. In logistic regression models, only shortness of breath was associated with higher temperatures.

We were most interested in whether or not activities particular to each community were associated with somatic symptoms. As Table 10.6 shows, all women frequently sweep or mop their home. Not surprisingly, women in rural communities spend significantly more time in all activities, particularly walking outdoors or working in the yard, garden, or patio. Women in Calakmul were more likely to work in the agricultural fields, women in Ich Ek/Suc Tuc were more likely to carry weight, and women in the city were more likely to carry out structured activities, e.g., in a gym. Women in the most rural communities spent more time washing clothes by hand and making tortillas by hand. How do these activities correlate with somatic symptoms? Mopping and sweeping were associated with fatigue—but not after controlling for menopausal status and BMI. Mopping and sweeping were also associated with muscle/joint pain which was more frequent in the rural communities. This remained significant even after controlling for age, SES, and other variables. Walking outdoors was not associated with any symptom, but working in the yard, garden, or patio, and working in agricultural fields were associated with palpitations and dizziness which were more frequent in rural communities. Neither remained significant after controlling for SES. Structured activity was associated with lower reported frequencies of fatigue, dizziness, and shortness of breath. Women with these symptoms may be less likely to participate in gym programs, dancing, or other activities. Finally, washing clothes by hand was associated with muscle/joint pain and headaches, while making tortillas was associated with palpitations, dizziness, and shortness of breath—all characteristic of the most rural communities. Headaches remained significantly associated with washing clothes by hand after controlling for menopausal status, SES, and the number of people living in the household.

Women have extensive social networks as part of their ecological context, but our quantitative data were not as fine-tuned as they should have been to assess the relationship between household composition and somatic symptoms. In future analyses, qualitative data will be helpful in better understanding the source of social stressors and social support.

We expected physical activity to be more clearly associated with somatic symptoms, but the evidence from this study is mixed. In bivariate analyses, bringing home water was the activity most often associated with symptoms, i.e., with muscle/joint pain, palpitations, dizziness, and shortness of breath. In the subset of women with ambient temperature measures, bringing home water was significantly associated with dizziness and shortness of breath after controlling for temperature. Bringing home water was also almost significantly associated with palpitations ($p = 0.05$). The difficulty of going outside to carry home water may explain why

Calakmul has some of the highest symptom frequencies. However, the other rural communities in Hopelchén, also have high symptom frequencies.

We expected high temperatures to exacerbate somatic difficulties, but only shortness of breath was consistently associated with the high temperatures characteristic of the state of Campeche.

In summary, it appears that women across all sites are equally tired and have equal difficulty sleeping. Also, with the exception of palpitations and dizziness, the intensity of symptoms (e.g., "*nada*," "*un poco*," "*mucho*," or "*muchisimo*") did not differ across communities. On average, rural communities reported more somatic symptoms, and also more hours of physical activity. However, SES, not physical activity, was the variable most often associated with somatic symptoms. This suggests that the reporting of somatic symptoms may have a psychosomatic etiology as well as a component of physical wear and tear.

These results reflect the great complexity of human ecology, where physical, biological, and sociocultural factors are not only interconnected but also enmeshed in making up the biocultural whole of humans. This complexity is quite obvious when one attempts to understand the causes of symptom variability in diverse communities.

References

Avis N, Kaufert P, Lock M et al (1993) The evolution of menopausal symptoms. Bailliere Clin Endocrinol Metab 7:17–32

Bouchard C, Shephard RJ, Stephens T (1994) Physical activity, fitness, and health: international proceedings and consensus statement. Human Kinetics Publishers, Champaign, IL

Brown DE (2016) Stress biomarkers as an objective window on experience. In: Sievert LL, Brown DE (eds) Biological measures of human experience across the lifespan: making visible the invisible. Springer International Publishing, Cham, pp 117–141

Brown DE, James GD, Nordloh L et al (1998) Comparison of factors affecting daily variation of blood pressure in Filipino-American and Caucasian nurses in Hawaii. Am J Phys Anthropol 106(3):373–383

Cahuich-Campos D, Huicochea-Gómez L, Sievert LL et al (2018) Factores socio-ambientales determinantes del uso de la herbolaria durante el climaterio en Campeche, México. Rev Etnobiol 16(2):98–113

Carvajal Correa M, Huicochea-Gómez L (2010) Ceremonia del wahil kol en la comunidad de Ich Ek, Campeche; identidad y patrimonio cultural de los mayas peninsulares. In: Huicochea Gómez L, Cahuich Campos MB (eds) Patrimonio biocultural de Campeche. Experiencias, saberes y prácticas desde la Antropología y la Historia. EL Colegio de la Frontera Sur, San Cristóbal de las Casas, Chiapas, pp 83–103

Chen F, Bao L, Shattuck RM et al (2017) Implications of changes in family structure and composition for the psychological well-being of Filipina women in middle and later years. Res Aging 39(2):275–299

Chlebowski R, Cirillo DJ, Eaton CB et al (2013) Estrogen alone and joint symptoms in the Women's Health Initiative Randomized Trial. Menopause 20(6):600–608

de Klerk BM, Schiphof D, Groeneveld FP et al (2009) No clear association between female hormonal aspects and osteoarthritis of the hand, hip and knee: a systematic review. Rheumatology 48(9):1160–1165

De Gucht V, Fischler B (2002) Somatization: a critical review of conceptual and methodological issues. Psychosomatics 43:1–9
Dennerstein L, Smith AMA, Morse C et al (1993) Menopausal symptoms in Australian women. Med J Australia 159:232–236
Dhanoya T, Sievert LL, Muttukrishna S et al (2016) Hot flushes and reproductive hormone levels during the menopausal transition. Maturitas 89:43–51
Dressler WW (1991) Stress and adaptation in the context of culture: depression in a southern black community. State University of New York Press, Albany, NY
Dressler WW, Bindon JR (2000) The health consequences of cultural consonance: cultural dimensions of lifestyle, social support and arterial blood pressure in an African American community. Am Anthropol 102(2):244–260
Freedman RR (2005) Pathophysiology and treatment of menopausal hot flashes. Semin Reprod Med 23:117–125
Freeman EW, Sammel MD, Lin H et al (2007) Symptoms associated with menopausal transition and reproductive hormones in midlife women. Obstet Gynecol 111:230–240
Frisancho AR (1993) Human adaptation and accommodation. University of Michigan Press, Ann Arbor, MI
García Gil G, Pat Fernández JM (2000) Apropiación del espacio y colonización en la Reserva de la Biosfera Calakmul, Campeche, México. Rev Mexicana del Caribe 10:212–231
Gay A, Culliford D, Leyland K et al (2014) Associations between body mass index and foot joint pain in middle-aged and older women: a longitudinal population-based cohort study. Arthritis Care Res 66(12):1873–1879
Gold EB, Block G, Crawford S et al (2004) Lifestyle and demographic factors in relation to vasomotor symptoms: baseline results from the Study of Women's Health Across the Nation. Am J Epidemiol 159:1189–1199
Gureje O, Simon GE, Ustun TB et al (1997) Somatization in cross-cultural perspective: a World Health Organization study in primary care. Am J Psychiatry 154:989–995
Hange D, Mehlig K, Lissner L et al (2013) Perceived mental stress in women associated with psychosomatic symptoms, but not mortality: observations from the Population Study of Women in Gothenburg, Sweden. Int J Gen Med 6:307–315
Huicochea Gómez L, Sievert LL, Cahuich Campos D et al (2017) An investigation of life circumstances associated with the experience of hot flashes in Campeche, Mexico. Menopause 24(1):52–63
INEGI (2014) Anuario estadístico y geográfico de Campeche. Instituto Nacional de Estadística y Geografía, Aguascalientes. Available via DIALOG www.inegi.org.mx. Accessed 15 January 2019
INEGI (2015) Panorama sociodemográfico de Campeche, 2015. Encuesta Intercensal 2015. Instituto Nacional de Estadística y Geografía, Aguascalientes. Available via DIALOG http://internet.contenidos.inegi.org.mx/ Accessed 15 January 2019
Keyes CL, Ryff CD (2003) Somatization and mental health: a comparative study of the idiom of distress hypothesis. Soc Sci Med 57(10):1833–1845
Kleinman A, Kleinman J (1985) Somatization: the interconnections in Chinese society among culture, depressive experiences, and the meanings of pain. In: Kleinman A, Good B (eds) Culture and depression. University of California Press, Berkeley, CA, pp 429–490
Kormondy EJ, Brown DE (1998) Fundamentals of human ecology. Prentice Hall, Upper Saddle River, NJ
Lazarus RS, Folkman S (1984) Stress, appraisal and coping. Springer, New York
Liu SH, Driban JB, Eaton CB et al (2016) Objectively measured physical activity and symptoms change in knee osteoarthritis. Am J Med 129(5):497–505
Obermeyer CM, Reher D, Saliba M (2007) Symptoms, menopause status, and country differences: a comparative analysis from DAMES. Menopause 14:788–797
Oh K, Jung KY, Choi JY et al (2012) Headaches in middle-aged women during menopausal transition: a headache clinic-based study. Eur Neurol 68(2):79–83

Pawlikowska T, Chalder T, Hirsch SR et al (1994) Population based study of fatigue and psychological distress. Br Med J 308:763–766
Randolph JF Jr, Sowers M, Bondarenko I et al (2005) The relationship of longitudinal change in reproductive hormones and vasomotor symptoms during the menopausal transition. J Clin Endocrinol Metab 90:6106–6112
SEDESOL (2016) Conoce todo sobre Prospera. Secretaria de Desarrollo Social, Mexico City. Available via DIALOG https://www.gob.mx/sedesol/articulos/conoce-todo-sobre-prospera. Accessed 25 July 2018
Sievert LL, Brown DE (eds) (2016) Biological measures of human experience across the lifespan: making visible the invisible. Springer International Publishing, Cham
Sievert LL, Espinosa-Hernandez G (2003) Attitudes toward menopause in relation to symptom experience in Puebla, Mexico. Women Health 38(2):93–106
Sievert LL, Huicochea-Gómez L, Cahuich-Campos D et al (2018) Stress and the menopausal transition in Campeche, Mexico. Women's Midlife Health 4:9. https://doi.org/10.1186/s40695-018-0038-x
Simon G, Gater R, Kisely S et al (1996) Somatic symptoms of distress: an international primary care study. Psychosom Med 58:481–488
Smith RL, Flaws JA, Mahoney MM et al (2018) Factors associated with poor sleep during menopause: results from the Midlife Women's Health Study. Sleep Med 45:98–105
Stadje R, Dornieden K, Baum E et al (2016) The differential diagnosis of tiredness: a systematic review. BMC Fam Pract 17:147
Staner L (2003) Sleep and anxiety disorders. Dialogues Clin Neurosci 5(3):249–258
Thurston RC, Bromberger JT, Joffe H et al (2008) Beyond frequency: who is most bothered by vasomotor symptoms? Menopause 15(5):841–847
Woods NF, Mitchell ES (2010) Sleep symptoms during the menopausal transition and early postmenopause: observations from the Seattle Midlife Women's Health Study. Sleep 33(4):539–549
Zimmer Z, Jagger C, Chiu C-T et al (2016) Spirituality, religiosity, aging and health in global perspective: a review. SSM Popul Health 2:373–381

Part II
Human Ecology from a Bioarchaeological Perspective

Chapter 11
Environmental and Cultural Stressors in the Coastal Northern Maya Lowlands in Pre-Hispanic Times

Andrea Cucina

Abstract Tropical coastal environments expose individuals to a wide array of stressful conditions, but at the same time are supposed to grant settlers a qualitatively richer and more varied diet. The present study analyzes caries and linear enamel hypoplasia (LEH) in two Classic Period (CE 250–750) coastal sites (Jaina and Xcambó), located in the Yucatan peninsula's northern coastline, in comparison with inland sites. The analysis reveals the absence of a clear pattern, showing high levels of caries and LEH both in Xcambó and in more inland sites, in contrast to lower levels of such stressful indicators at Jaina and in other sites located in the interior of the peninsula. Archaeological and bioarchaeological evidence rules out that social status and wealth, as well as sample composition by sex, or by age at death, can be at the base of such differences. Results, instead, highlight the concept that well-being is the result of the entangled and intertwined interaction between the environmental, biological, and cultural factors that shaped each and every human community.

11.1 Introduction

Human adaptation to tropical and subtropical environments can be challenging as it exposes single individuals and whole communities to sometimes-hostile environs. This is particularly true if we consider the biodiversity, and the facility for water-borne or mosquito-borne diseases to spread close to bodies of water (Kormondy and Brown 1998; Sattenspiel 2000). However, human well-being is not hampered exclusively by the harshness of the surrounding environment; it is the complex result of biological, environmental, and cultural constrains, including diet and access to dietary resources, among the others.

A. Cucina (✉)
Facultad de Ciencias Antropológicas, Universidad Autónoma de Yucatán,
Mérida, Yucatán, México
e-mail: cucina@correo.uady.mx

H. Azcorra, F. Dickinson (eds.), *Culture, Environment and Health in the Yucatan Peninsula*, https://doi.org/10.1007/978-3-030-27001-8_11

According to the model developed by Goodman et al. (2013), environmental constrains (like stressors and limiting resources) can be buffered by the cultural system, though the latter can also produce stressors (what the authors called "culturally induced stressors") (Goodman and Martin 2002). In bioarchaeology, well-being and health (or the lack of) are assessed through the analysis of past people's mortal remains; they provide information on pathological conditions, developmental stress, morbidity, growth, and eventually death (Goodman et al. 2013).

Stressors encompass a wide range of conditions, from climatic to environmental, nutritional, or infectious. Some stressors, which put at risk the individual's well-being and even life, are well defined and their etiology can be recognized in the skeletal and dental system. Others, on the other hand, are general and nonspecific (Goodman et al. 2013), and can be the result of multiple factors that interact with one another leaving observable marks in the human remains. **Developmental stress**, a nonspecific indicator of physiological insult, can be detected through the analysis of some skeletal and dental markers. Linear enamel hypoplasia (LEH) is one of the most common and reliable markers of physiological stress during development. It is a deficiency in enamel thickness produced by a temporary growth disruption caused by physiologically stressful events during the amelogenetic phase of **enamel** formation (Goodman and Rose 1990). It appears as horizontal lines or grooves of reduced enamel thickness, which may lead to the exposure of dentine in the most severe cases (Hillson 1996). Goodman and Rose (1990) associated LEH to a large number of potential physiological conditions that altered the individual's **homeostasis**; among the more than one-hundred possible etiological causes (Kreshover 1960) stand quantitative or qualitative nutritional deficiencies often related to differential access to resources and infectious and metabolic disease (Hillson 2000). Differently from other indicators of systemic disruption from bony segments, like Harris lines (Goodman and Martin 2002), dental enamel does not remodel once it is laid down; therefore, LEH represents a permanent, indelible record that persists into adulthood of stressful events that occurred during infancy and childhood (Hillson 2000).

As regards the "limiting resources," Goodman and Martin (2002) state that the most important ones are likely the most basic ones: food, water, and shelter. Limited access to any of them may hinder the individuals' resistance to diseases. Yet, not only is the quantity of the basic resources that is important, but also the quality and variety, in particular of food, may be detrimental to health and well-being. There can be little doubt that a diet in which some particular types of starchy food predominate, like maize in Mesoamerica, may expose the individuals to health problems and to the risk of suffering from nutritional deprivation and diseases (Larsen 1995). In the specific case of maize, for example, it is deficient in several essential **amino acids** (lysine, isoleucine, and tryptophan), while vitamin B3 (niacin), although being present, has limited bioavailability (though nixtamalization process allows it to be released more easily—Ellwood et al. 2013; Sahai et al. 2001).

In bioarchaeological research, stressors linked to the quality of diet have often been inferred based on the frequency of infectious diseases, being carious lesions the most commonly used indicator (Lukacs 1989, 1992; Larsen et al. 1991). Caries is an infectious disease that became endemic with the introduction of agriculture (i.e., when subsistence shifted from hunting and gathering to one based on domesticated, carbohydrate-rich plants) almost everywhere among the human population (Larsen 1995; Turner 1979). It represents a long-standing health problem affecting human well-being, health, and quality of life both at individual and population level. After the industrial revolution, refined sugars added into the diet have worsened even more oral health by enhancing the oral bacteria's acidification process, with an eventual mineral loss in the enamel and dentine, leading to initiation/progression of dental caries (Hillson 2008; Takahashi and Nyvad 2011).

However, its etiology is multifactorial and complex. In fact, caries prevalence is related to a wide array of extrinsic factors, like diet and the kind of food ingested, daily habits, lifestyle, and socioeconomic level, as well as to intrinsic ones, such as hormonal level, dental morphology, and oral pH (Aufderheide and Rodríguez-Martín 1998; Cucina et al. 2011; Hillson 2008; Larsen et al. 1991; Lukacs 2008, 2011a, b). Therefore, we have to be cautious in linking frequency of caries with a diet based on carbohydrates (i.e., domesticated plants) and to the social status related to diet (access to limited or more ample and varied dietary resources) because, although carious lesions increased in frequency after the adoption of agriculture, they may be abundant in preagricultural or nonagricultural groups (Nelson et al. 1999; Watson 2008) or may be augmented by the intake of other kinds of cariogenic foods (Cucina et al. 2011).

Under theoretical background, coastal populations of the peninsula of Yucatan should have benefitted from healthier living conditions than their inland counterparts as a consequence of a varied and more protein-rich (and less cariogenic) diet from marine resources in comparison to the milpa-based subsistence economy that characterized the inland environments (Ortega Muñoz 2015; Steckel and Rose 2002).

In this chapter, I focus on the two abovementioned stressors (LEH and caries) among coastal Maya populations in northern Yucatan relative to inland sites, to discuss lifestyle and living conditions in relation to environmental and cultural constraints during the Classic period (CE 250–900). Some of the data comes from own published research, alongside a set of unpublished database.

11.2 Materials

The analysis and discussion center on four dental collections dated to the Classic Maya period (CE 250–900) from northern coastal (Xcambó and Jaina) and inland (Yaxuná and Noh Bec) Yucatán (Fig. 11.1). Xcambó, Yaxuná, and Noh Bec dental collections are currently housed in the Laboratory of Bioarchaeology and Histomorphology, School of Anthropological Sciences, Autonomous University of Yucatán (Mérida). Jaina, in turn, is housed in the Osteology Laboratory, Direction of Physical Anthropology, Museum of Anthropology (Mexico City). Due to access to materials, only a small sample of the dental collection of Jaina could be studied for LEH.

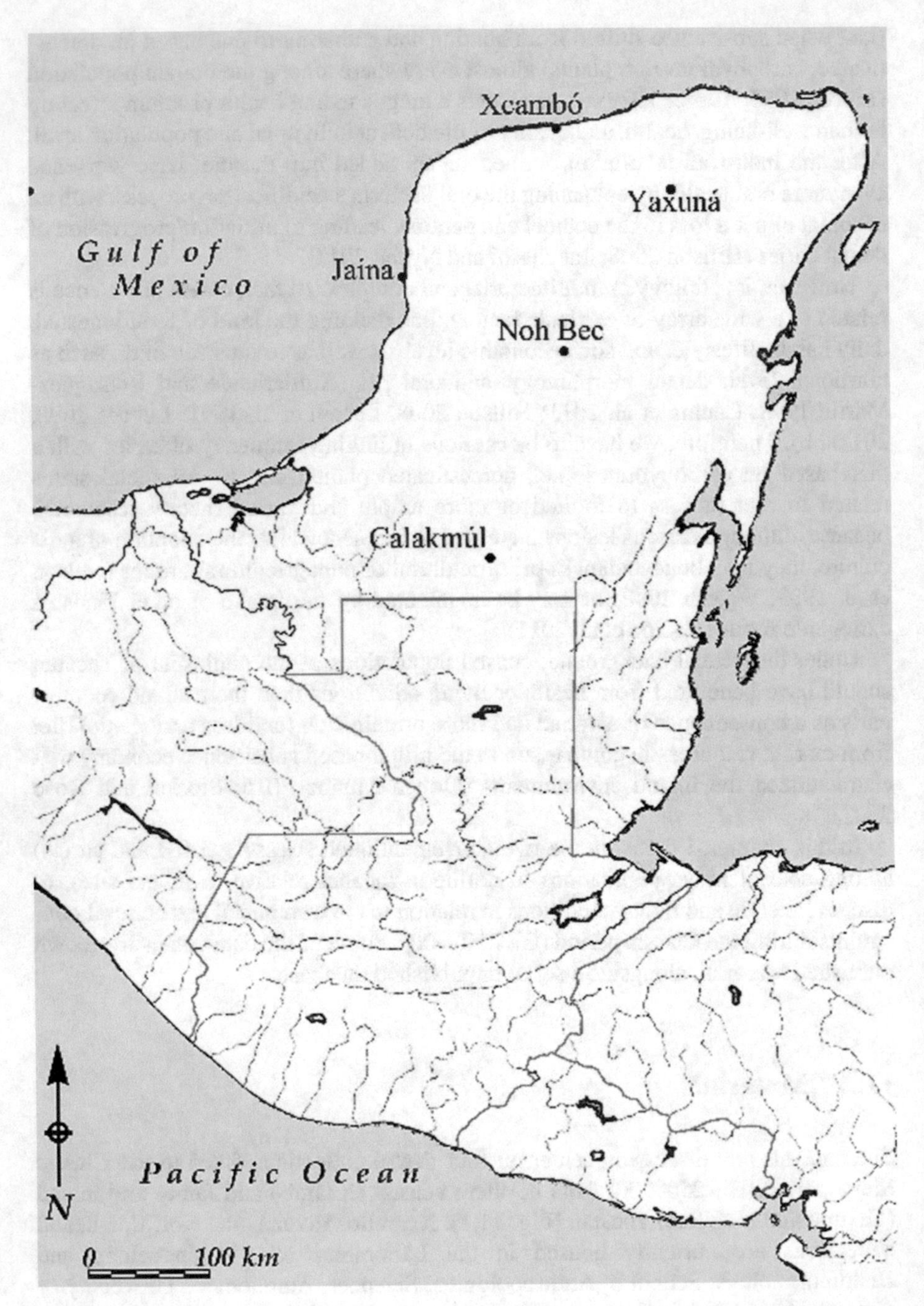

Fig. 11.1 Map of the sites analyzed in the study, including the location of the site of Calakmul for comparative purposes

Skeletal individuals were selected following different procedures for caries and LEH. In the case of caries, the samples consist of adult individuals only, since the presence of subadult specimens (with both deciduous and recently erupted permanent teeth) introduces a bias, and underestimates frequencies of affection (Cucina et al. 2011). On the contrary, sample for the analysis of linear enamel hypoplasia includes every individual available (regardless of age at death) showing at least one maxillary central incisor and/or mandibular canine, which are the most susceptible teeth to developmental stress (Goodman and Armelagos 1985; Cucina 2011) with the crown completely formed. Differently from caries, LEH is the marker of developmental stress, so age at death does not introduce any bias.

11.2.1 Xcambó

The site of Xcambó was excavated by archaeologist Thelma Sierra Sosa (INAH Yucatán) between 1996 and 2000. It is located on a small natural mound (700 m long by 150 m wide), surrounded by natural marshland, situated approximately one mile from the northern coast of the Yucatan peninsula (Fig. 11.1). The archaeological intervention allowed to recover round 600 skeletal individuals, chronologically dated to the Early Classic (CE 250–550) and the Late Classic (CE 550–700/750). The site was abandoned by CE 750 as a consequence of a shift in the regional political sphere of influence (Sierra Sosa et al. 2014). During the Early Classic, the site was mainly a small center responsible for the production of the Xtampu salt mines as well as other smaller centers for salt production and marine products (Sierra Sosa 2004). Xcambó experienced a complete transformation during the Late Classic period, when it became an autonomous port of trade related to the production and administration of marine salt. Population density increased, and life revolved around two main ceremonial and administrative plazas (Sierra Sosa et al. 2014). Archaeological evidence indicates that during the Late Classic the inhabitants of Xcambó enjoyed a relatively high social status (though no political elite lived at the site). In this period, skeletal evidence indicates that occupational activities decreased in a significant manner in the male segment of the society (Maggiano et al. 2008). This suggests, in other words, that the male segment was less subject to strenuous physical activities in comparison with the earlier period of occupation. In turn, women's lifestyle and activities remain unaltered during the entire time span of occupation of the site. The skeletal evidence is in line with the shift in the site's socioeconomic function during the Late Classic, when Xcambó became a politically autonomous port of salt production, trade, and administration.

Materials analyzed for LEH correspond to 275 individuals, 67 of which are subadults and 208 adults. Among the adult cohort, 108 are males and 93 are females (seven individuals could not be sexed). As regards chronology, 38 individuals (25 adults and 13 subadults) belong to the Early Classic period, and 208 to the Late Classic.

Caries were scored in 167 adult individuals, equally distributed between sexes (83 females and 81 males). Of the total, 23 individuals were dated to the Early Classic, and 140 to the Late Classic. Only four individuals could not be assigned to any specific chronological period.

11.2.2 Noh Bec

The site of Noh Bec is located in the southern corner of the present-day state of Yucatán (Fig. 11.1). The skeletal collection comes from a salvage excavation by the Yucatán INAH Center (Instituto Nacional de Antropología e Historia—National Institute of Anthropology and History) related to road construction in 2005–2006 (Peña Castillo 2006). The road construction affected the peripheral part of the site, which seemingly formed a concentric pattern, with the main core located in the middle of the sampled area. Among the ceremonial building complex stood a ball court, similar and comparable to those of the Late Classic period in Northern Yucatán. Between the site's several plazas, many residential basements were erected (Peña Castillo 2006). The excavation recovered a total of 51 burials (in a relatively poor state of preservation) from these basements, 45 of which had been interred in formal cists (Peña Castillo 2006), and all were accompanied by a large number of funerary objects. Although material record indicates that the site had been in use since the Preclassic times, the period that is mostly represented at the site and nearby minor centers is the Late Classic (CE 550–800) with the ceramic variety shared mainly with the Chenes region in the western side of the peninsula (Peña Castillo 2006).

Skeletal material analyzed for LEH corresponds to 24 individuals, 20 of whom are adults and 4 are subadults. Carious lesions, instead, were scored on 43 adult individuals. Unfortunately, the poor state of preservation limits the analyses by sex, for it could not be determined in 79% of the adult sample.

11.2.3 Yaxuná

The site of Yaxuná is located in the northern Maya lowlands in the modern state of Yucatan, Mexico, some 20 km southwest of the site of Chichén Itzá (Fig. 11.1). The archaeological excavations carried out by the Selz Foundation and by the PIPCY projects between 1980s and 2015 recovered 48 individuals dated to the Classic period (CE 250–900). Being located in between the borders of the eastern and western cultural spheres of the northern Maya lowlands, its continued cultural ties to the southern lowland indicate that the site performed strategic economic and political roles during its time of occupation (Tiesler et al. 2017). The archaeological excavations have brought back to light two of the few known royal tombs from northern Maya lowlands dated to the Classic period (Tiesler et al. 2017), both belonging to the Early Classic. Yaxuná was eventually abandoned (though there is evidence of minor occupation in later times) by the rise of Chichén Itzá as the region's main center of power by the Terminal Classic.

Twenty-eight individuals were analyzed for carious lesions, while the presence of LEH could be assessed in 22 individual specimens.

11.2.4 *Jaina*

Jaina is a small island located along the western shores of the Yucatan peninsula, in the modern state of Campeche, close to the border with the state of Yucatan (Fig. 11.1). It is very close to the peninsula's shores, so that in its closest part, it is separated from the mainland by about 70 m of seawater. It is thought that the "island" was actually built intentionally at the end of the Preclassic, for its soil is in sharp contrast with the one commonly present along the coast. Overall, it is about 900 m long (north–south) and 600 m wide (east–west). According to Benavides Castillo (2011), the island's period of occupation ranges between CE 300–900. Its economic and political development is closely tied to the maritime trade route that put the island in contact with many sites in the peninsula, as well as in Central Mexico. The presence of material objects associated to the elite suggests that the latter administered a political and economic power of a site receiving and distributing goods and services (Benavides Castillo 2011).

The dental collection corresponds to the materials recovered during different field works between 1957 and 1974. It consists of 103 individuals from different age groups and sexes. Of these, 67 are adults (above age 20) and could be scored for carious lesions: 27 are males, 19 are females, and 21 are of undetermined sex. Most of the specimens forming this collection were recovered in 1973–1974. According to López Alonso and Serrano Sánchez (1984), this skeletal material was recovered in the southern part of the island, where no architectural structures existed. It can be dated to the Late and Terminal Classic (CE 550–750, and 750–900) based on associated ceramic fragments. It must the stressed that this skeletal collection, despite its enormous archaeological importance, benefits from very little **taphonomic** and contextual documentation, which hampers detailed analyses and interpretations, like those carried out in other northern lowland Maya sites. The often-deteriorated preservation further limits bioarchaeological interpretations.

11.3 Methods

Carious lesions result from the chemical demineralization of the dental tissue (enamel and dentine), which is caused by the oral bacteria's acidic by-products (Fig. 11.2). Caries are recorded macroscopically (with the help of a 4× magnifying glass) in every permanent tooth in the adult segment of the population (20 years and

Fig. 11.2 Grade 3 carious lesion in Late Classic period Xcambó

Fig. 11.3 Severe linear hypoplastic defect affecting the enamel of a mandibular canine from Late Classic Xcambó

above). Five degrees of severity are recognized, in a 0–4 scale, with zero indicating absence of lesions; grade 1 indicates a small pit that affects only the enamel; in grade 2, the pit has penetrated into the **dentine**; grade 3 represents a cavitation that has opened up the pulp chamber, and grade 4 represents a cavitation that has destroyed at least half of the dental crown (Cucina et al. 2011). However, only caries grade 2 and above are considered as present (Cucina et al. 2011), in order not to count as carious lesions also minor hypoplastic pits from different etiologies (Cucina et al. 2011).

Linear enamel hypoplasia (LEH) is a deficiency in enamel thickness caused as a response to a physiological developmental disruptive event during the phase of deposition of the enamel (Goodman and Rose 1990) (Fig. 11.3). It is recorded in every permanent tooth using a 4× magnifier under tangential light. For the purpose of this study, alongside the individual presence, it was quantitatively recorded in central maxillary incisors and mandibular canines (i.e., the number of defects affecting the tooth crown was specifically recorded) (Goodman and Armelagos 1985).

Due to relatively small sample size of some of the collections, the chronology of LEH was not taken into consideration.

11.4 Results

11.4.1 Hypoplasia

Table 11.1 shows the average number of hypoplastic lesions affecting teeth (maxillary central incisors and mandibular canines), alongside the general frequency of affection.

At Yaxuná only 11 individuals for the maxillary central incisor and 12 for the mandibular canine could be scored. On average, 2.08 defects affect the incisor, and 0.82 the canine. Frequency of affection per type of tooth is, respectively, 58.3% and 36.4%. At Noh Bec, 24 individuals presented at least one permanent tooth that permitted to be scored for LEH. The whole sample (100%) showed markers of developmental stress. When analyzed more in detail, ten maxillary central incisors showed at least one defect with 2.2 number of defects on average (s.d. 1.14); the mandibular canine was equally affected in 100% of the case and showed 2.5 defects per tooth on average (N = 14, s.d. 1.02). On the contrary, at Jaina LEH could be scored only on the mandibular canine. Only seven individuals provided information on the presence of developmental defects, and all of them (100%) show hypoplastic lesions in their mandibular canine. The mean number of defects is 3.1 (s.d. 1.6).

Last, developmental stress manifested at Xcambó in more than 90% of the tooth types considered, both for the early phase of occupation (91.3% and 96.6%, respectively, for incisor and canine) and for the later one (95.3% and 96.1%, respectively). In terms of the average number of defects, the earlier phase shows 4.1 and 4.9 defects on average in the incisor and canine, while in the later phase I found 3.5 and 4 defects on average, respectively.

Table 11.1 Mean number of LEH and frequency of affection in the maxillary central incisor (I1') and in the mandibular canine (C') in the dental collections analyzed

	I1'				C'			
	N[a]	Mean	SD	%[b]	N	Mean	SD	%
Yaxuná	12	2.08	2.5	58.3	11	0.82	1.25	36.4
Early Xcambó	23	4.1	2.3	91.3	29	4.9	1.8	96.6
Late Xcambó	107	3.5	2	95.3	144	4	2.2	96.1
Noh Bec	10	2.2	1.14	100	14	2.5	1.02	100
Jaina	–	–	–	–	7	3.1	1.6	100
Calakmul[c]	15	2.1	1.06	95.5	21	1.9	1.48	91.7

[a]Sample size (N) refers to the teeth for which the number of defects could be assessed
[b]Frequency of affected teeth (%) also considers those specimens for which LEH could be detected on the buccal side of the crown, but the number of defects could not be assessed
[c]Personal, unpublished data

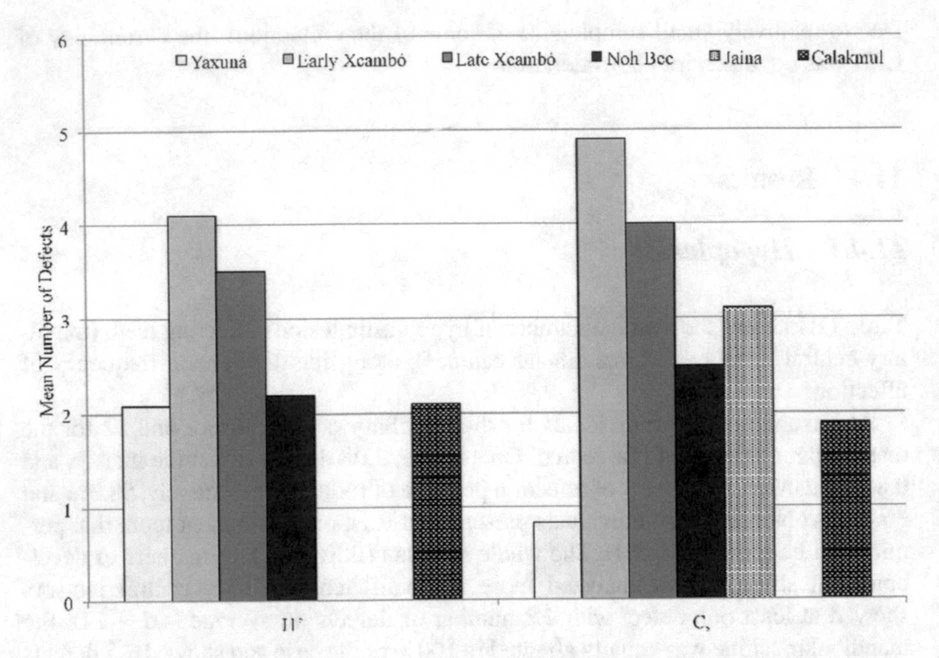

Fig. 11.4 Mean number of defects in the maxillary central incisor (I1') and in the mandibular canine (C') in the five collections analyzed. The comparative collection of Calakmul is separated from the rest of the samples

Figure 11.4 shows the comparison of mean number of defects among the five collections analyzed, while Fig. 11.5 shows the same comparison for the frequency of affected teeth.

As Table 11.1 and Figs. 11.4 and 11.5 show, Yaxuná is the site with the lowest environmental impact among all the collections analyzed, both for the mean number of defects and for the frequency of teeth affected by LEH. Although not reported in detail, such low values at Yaxuná are due to the presence of several teeth (5 incisors out of 12, and 7 canines out of 11) that were free from markers of developmental stress. On the other hand, however, the high values of standard deviation in relation to the mean indicate that those teeth with LEH presented several defects each. Noh Bec and Jaina (the latter only for the canine) follow Yaxuná. In these two cases, all the teeth presented marks of LEH, though defects were distributed more evenly than at Yaxuná. Last, Xcambó presents the highest average values, with means ranging from 3.5 to 4.9 per tooth. The frequency of affected teeth indicates that at Xcambó (both periods) very few individual teeth were free from defects, while at Noh Bec and Jaina all the teeth presented linear defects.

Comparative information from Calakmul, located farther south, close to the modern borders between Mexico and Guatemala, shows a level of developmental stress that is in line with the evidence from Noh Bec, which is located at the southern fringes of the modern state of Yucatan. Nonetheless, the two regions are characterized by a different ecosystem, with Calakmul belonging to the high canopy rainy forest and Noh Bec to the low canopy semiarid biotome.

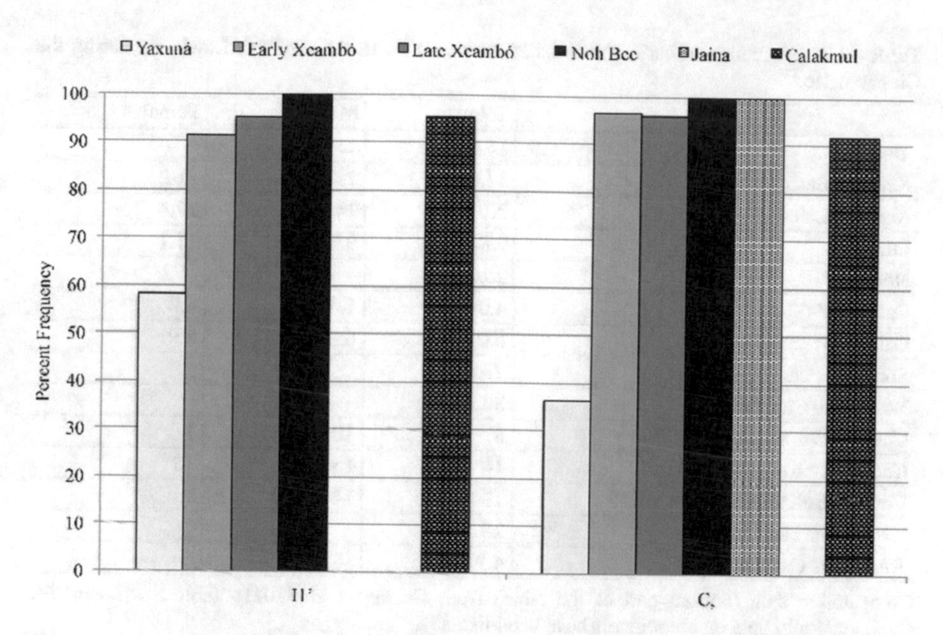

Fig. 11.5 Frequency of defects in the maxillary central incisor (I1') and in the mandibular canine (C') in the five collections analyzed. The comparative collection of Calakmul is separated from the rest of the samples

11.4.2 Caries

Carious lesions have been scored on the adult segment of the population. As mentioned above, in the case of Yaxuná and Noh Bec the information considers the sample as a whole, regardless of sex, due to the poor state of preservation of the skeletal material and small sample size, which do not permit a more detailed analysis.

Based on the tooth-count method, in which the unit of analysis is the tooth and not the individual, Table 11.2 shows the overall frequency of affected teeth. For Early and Late Classic Xcambó, and for Jaina, frequency is presented also by sex.

As we can note (see also Fig. 11.6), overall frequencies of carious lesions range from 9.8% to 28%. No geographic pattern can be appreciated, for Jaina and Xcambó are both coastal sites, while Noh Bec and Yaxuná are inland. Instead, a striking difference between males and females appears between Jaina and both periods at Xcambó. At Jaina, in fact, though females present higher frequency than males, such difference is minimal (9.6% versus 11.1%) and difference is not statistically significant ($X^2 = 0.8$, 1 d.f., $p = 0.36$); on the contrary, at Xcambó differences by sex are highly significant ($X^2 = 14.99$, 1 d.f., $p = 0.000$; 63.73, 1 d.f., $p = 0.000$, respectively, for the Early and Late Classic periods).

In the comparative analysis with collections from the literature (Table 11.2 and Fig. 11.6), the northern Lowland's samples fall well within the range of variability. Similar to LEH, also in this case no clear geographical pattern emerges, with both

Table 11.2 Frequency of carious lesions in various sites in the northern Lowlands during the Classic period

	Total	Males	Females
Yaxuná	14.9		
Xcambó Early	14.6	7.4	21.2
Xcambó Late	20.2	14	27.4
Jaina	9.8	9.6	11.1
Noh Bec	28.0		
Calakmul[a] (elite)	4.0	1.4	8.8
Calakmul[a] (commoners)	6.0	6.3	6.9
Southern Peten[b]	23.4		
Wild Cane Cay[c] (Belize)	36.2		
Cuello[d] (Belize) 1200–650 BC	8	10	3
Cuello[d] (Belize) 650–400 BC	14	15	15
Cuello[d] (Belize) 400 BC–CE 250	12	14	7
Kichpanha[e] (Belize)	28.5		
Ambergris Caye[f] (Belize)	4.9		

Comparative data (bottom part of the table) from Cucina et al. (2011: Table 5). Except for Calakmul, only data on commoners have been listed
[a]Cucina and Tiesler (2003)
[b]Personal unpublished data
[c]Seidemann and McKillop (2008)
[d]Saul and Saul (1997)
[e]Magennis (1999)
[f]Glassman and Gerber (1999)

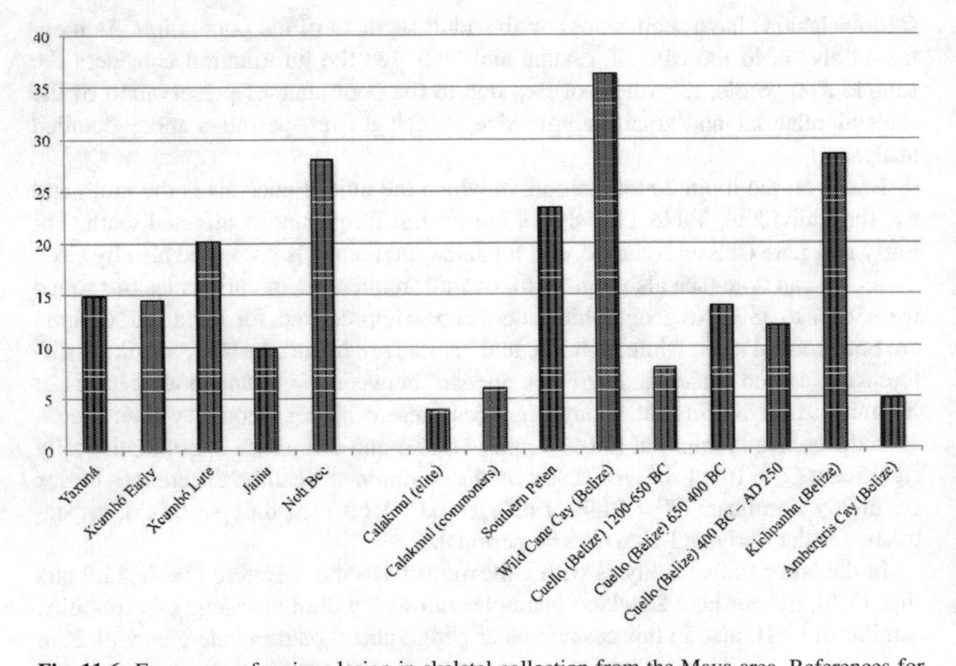

Fig. 11.6 Frequency of carious lesion in skeletal collection from the Maya area. References for the comparative sites are in Table 11.2

coastal and inland sites distributed both among the lowest and highest ends of the range. The comparison by sex shows that females show higher frequencies than males in both elite and commoners at Calakmul, but not at Cuello.

11.5 Discussion

The effects of environmental and cultural stressors on human beings depend on a wide variety of factors, from individual host resistance to subsistence economy, from population density and transmission of pathogens to socially determined access to resources and more (Goodman et al. 2013; Sattenspiel 2000). Food and nutrition are no doubt among the main factors that have an impact on the individuals' well-being, both positive and negative, and both direct and indirect. To the same extent, the environment in which an individual is born and spends his/her life is a powerful factor in exposing to, or buffering from, stressful impact.

Given that stress is multifactorial in nature, a simple and linear model to explain well-being (or the lack of)—like coastal communities enjoying better living conditions—may be valid only to a limited extent. In fact, if from one side coastal settlements are supposed to grant access to a varied and protein-richer diet (Glassman and Garber 1999; Ortega Muñoz 2015), it is also true that in specific (i.e., tropical) geographical contexts living along the coast may expose people to a pathocenotic burden in the form of water-borne and air-borne pathological vectors (see Sattenspiel 2000).

The present analysis reveals contradictory results between coastal and inland sites, in terms of markers of both cultural stressors (carious lesions) and developmental stress related to environmental pressure (linear enamel hypoplasia). This confirms that relationships between health and well-being are intertwined and entangled, that environmental impact and diet are multifactorial in nature, and that there is seemingly not a specific (geographical) pattern upon which one can base hypotheses. In this perspective Ortega Muñoz (2015), in an analysis of several dental and skeletal indicators of health (among which are caries and LEH), found that LEH was lower in inland sites than in coastal ones, though the opposite was true in the comparison between Classic versus Postclassic sites; at the same time, and contrarily to the theoretical expectations, he also reported that coastal sites presented a higher impact of carious lesions than inland centers. Unfortunately, the author followed the individual-count methodological approach instead of the tooth-count method employed in this study, so his results cannot be used for more detailed comparative purposes.

As described above, linear enamel hypoplasia is a marker of physiological stress during growth and development (Goodman and Rose 1990; Cucina 2011). Although initially described as an indicator of nutritional stress, more than one hundred factors have been associated to its development, which makes it a nonspecific indicator of stressful impact (Kreshover 1960; Goodman and Rose 1990). So its presence on the enamel surface of permanent teeth must not be interpreted as a "direct indicator of harsh living conditions coupled with limited access to resources" (Méndez Colli et al. 2009); rather, it should be interpreted as a permanent record of events of metabolic disruptions that were strong enough to go over the physiological thresholds of individual's resistance and trigger a catabolic response by the ameloblasts during enamel formation (Cucina 2011).

The analysis of LEH on permanent maxillary central incisors and mandibular canines, based on the frequency of affection, but regardless of the number of defects per tooth, shows that all the sites (except for Yaxuná) go past the 90% thresholds. This indicates that virtually every individual did suffer from some kind of physiological disruption at least once in his/her early years of life. This should not come as a surprise, given that all these people lived in pre-antibiotic eras, and their only mean to cope with environmental pressure was their own physiological resistance. It must be remembered, however, that all the individuals who could be scored for LEH had been able to survive to the stressful events that produced the hypoplastic grooves on the enamel surface; those who succumbed directly to the environmental insult could not complete the formation of their permanent teeth, and therefore do not form part of the sample.

A different perspective on environmental impact arises when the mean number of defects is considered. Xcambó (Early and Late) clearly manifests the higher level of stressful impact per person, with an average number of defects well above the other samples, for both the maxillary central incisor and the mandibular canine. Jaina follows it, though this collection could only be scored for the canine. The inland sites of Yaxuná and Noh Bec (and Calakmul for comparison) fall below the coastal ones.

This pattern, with all the limitations imposed by sample size, suggests that coastal samples were more intensely exposed to environmental impact than inland ones. Interestingly, alongside being coastal sites, both Jaina and Xcambó are peculiar also for their geographical position. The former is located on an island on the open sea, while the latter, though technically not an island, is located on a mound surrounded by marshland very close to the coastal line; both sites developed within well-defined but limited boundaries dominated by mangroves (Benavides Castillo 2011; Sierra Sosa 2004; Sierra Sosa et al. 2014), and in both cases population is considered to have been fairly dense for the limited extension of the areas (Benavides Castillo 2011; Ortega Muñoz et al. 2018). Despite the relatively high amount of marine resources (Benavides Castillo 2011; Sierra Sosa et al. 2014), which should have granted good nutritional intake, high population density can be detrimental for the population, particularly in circumscribed areas, for it facilitates the transmission of infectious and gastrointestinal diseases, which can severely affect the health of the settlers. Mortality tables both at Jaina and Xcambó (Hernandez Espinosa and Marquez Morfin 2007; Tiesler et al. 2005) indicate the presence of many infants among the dead, reinforcing the idea of a heavy pathogenic burden.

Differences in the frequency of LEH among samples might be due, to some extent, also to the different demographic composition of the skeletal collections, i.e., the larger presence of subadults that are supposedly more heavily affected by LEH. Goodman and Armelagos (1988) and Storey and McAnany (2006) first proposed such idea. The reason behind this difference can be tracked to the fact that developmental stress may increase the individual's physiological frailty, exposing the child to subsequent morbidity, and eventually to early mortality; on the other hand, as Duray (1996) suggested, physiologically weaker individuals represent an easier target and are thus more prone to suffer from environmental stressors. Cucina (2011) reported that at Xcambó individuals dying in their subadult age classes are significantly more affected by LEH than adults. Given that the number of subadults is demographically representative at the site, this may be a potentially causative factor for the higher mean number of defects at Xcmbó, while the same cannot be said for all the other collections.

However, as Cucina (2011) noted, the real difference between subadults and adults was in the extremes (i.e., the few individuals with ten or more defects were all subadults, while those with zero or just one defect were all adults). All the other ones, sporting from two to eight/nine defects, were equally distributed among all the age classes. To this, we have to consider also that mean values of LEH in the adult segment of the population at Xcambó correspond to 3.27 and 3.93, respectively, for the maxillary central incisor and mandibular canine (against 4.42 and 5.26 for subadults) (Méndez Colli et al. 2009). Such a figure is still higher than the ones estimated for all the other dental collections analyzed in this study, confirming the evidence that LEH at Xcambó was indeed much higher than in the rest of the samples. Unfortunately, the limited information available for LEH in subadults at Jaina restricts this discussion to Xcambó only.

Inland sites, on the other hand, seem to have experienced slightly better developmental conditions, in particular Yaxuná, whose mean value for the lower canine (<1) is the lowest in the area. However, the mean number of defects in its maxillary central incisor is in line with the values calculated at Noh Bec and Calakmul.

Differently from islands, inland sites are not limited by natural boundaries, which to some extent should limit the close contact among people and with potentially hazardous wastes. Considering the pathocenotic load of subtropical environments, with the presence of air-borne and water-borne vectors of diseases, two defects per tooth per individuals (on average) during infancy still denote a relatively low environmental impact for inland sites.

Leatherman and Goodman (1997) stated that socioeconomic disparities are at the base of differences in health and living conditions. Yet, in the present study socioeconomic level does not seem to be a factor in the onset of developmental stress. Xcambó and Jaina are characterized by a relatively high socioeconomic level, though no true political elite lived there (Benavides Castillo 2011; Sierra Sosa et al. 2014); at Yaxuná despite its sample includes a couple of royal people, most of the individuals recovered from the site and analyzed in this study are not members of the elite (Tiesler et al. 2017). Last, the skeletal collection of Noh Bec was recovered as a salvage excavation during the construction of a road; construction carefully avoided large structures and palaces, and all the human remains were unearthed from commoners' residential compounds. Therefore, socioeconomic disparities are not supposed to be at the roots of the differential impact between coastal and inland sites, and reasons should be more likely searched for in environmental rather than cultural factors.

A different scenario arises when carious lesions are considered. Caries are the product of chemical demineralization of enamel and dentine due to oral bacteria's acidic by-products (Hillson 1996), and their presence is closely related to a diet rich in carbohydrates (though other intrinsic and extrinsic factors are involved in their development).

The Maya population is notorious to have heavily relied on a maize-based subsistence (White 1999), so that this population has always been exposed to carious lesions. Nevertheless, access to varied and cariostatic resources in coastal and island sites should theoretically protect local people from excessive insurgence of caries. Still, as Seidemann and McKillop (2008) found out, no geographical or chronological pattern can be detected in the region. The expected low frequency of caries in coastal or island sites was reported at Ambergris Caye (Glassman and Garber 1999), but not at Wild Cane Cay (Seidemann and McKillop 2008), which shows one of the highest frequencies. In a similar fashion, low rates of caries are reported at the inland site of Calakmul (Cucina and Tiesler 2003), while high frequency characterizes the people at Cuello (Saul and Saul 1997). Similarly, Ortega Muñoz (2015) reports higher frequencies in the Peninsula's east coast in comparison to inland sites.

Sample composition (sex, age, social status) is an important variable that must be taken into consideration at the moment of making inferences and drawing conclusions about the high or low frequency of oral lesions. As regards social status, Cucina and Tiesler (2003) noted that male elite at Calakmul presented very low rates of caries, but female elite did manifest frequency as high as that observed in commoners. Only five individuals formed the elite female group, so rate of carious lesions may have been biased by the small sample size and (likely) by the age at death of the royal women. Notwithstanding this, the authors consider the differences they found in comparison with the male elite segment of the population as the result of differential access to resources, which seemingly put the royal females in line with their commoners' counterpart.

Besides, females are known to be more prone than males to suffer from carious lesions for a wide array of factors, both extrinsic (access to resources, daily habits) and intrinsic (hormone levels, pregnancies and lactation) (Larsen et al. 1991; Lukacs 2008, 2011a, b). In the present study, in those contexts in which caries could be assessed by sex (i.e., Xcambó and Jaina), females do manifest more lesions than males. In some cases, differences are statistically not significant (as in the case of Jaina), while in others they reach the 0.05 statistical thresholds of significance (at Xcambó).

In line with the discrepancies described above, also the present study falls into the unpredictability model that seemingly characterized the Maya realm. Jaina, with an overall frequency of 9.8%, is in sharp contrast with Xcambó, in particular with the sample from the Late Classic (overall 20.2%). The same can be said for the two inland sites, with Yaxuná reaching 14.9% in contrast with 28% at Noh Bec.

If we visualize a diet rich in marine resources at Jaina, in general it can be thought that the limited number of carious lesions is mirroring the resources provided by the environmental context. Unfortunately, however, the archaeologically decontextualized origin of the Jaina sample limits a thorough understanding on the reasons behind such figures. Instead, the same association to the environmental resources should not be true at Xcambó, which also during the early phase of (continuous) occupation reached a frequency similar to that described at the inland site of Yaxuná. Yet, Sierra Sosa (2004) and Sierra Sosa et al. (2014) reported a very large amount of marine resources at the site, which makes it difficult to think that diet was heavily based on maize.

Differently from Jaina, the accurate contextualization of the human skeletal remains from Xcambó allows a better understanding of the general context. During the Early Classic, Xcambó was a small settlement of fishermen/agriculturalists dedicated (also) to salt production. According to the model by Larsen et al. (1991), frequency of caries suggests that males engaged in physical activities outside the settlement, and females in household activities. During the Late Classic (Sierra Sosa 2004), Xcambó turned into an autonomous port of trade and administration of salt, with commercial ties with places as far south as Veracruz. Living conditions improved, as well as the general wealth. In this period, females continued to be engaged in the same physical activities as in the Early Classic, being more independent from the port's administrative activities, while males were relieved from harsh physical labors, being actively involved in administrative tasks (Maggiano et al. 2008). Nonetheless, contrarily to the expectations, caries rates increased in both sexes by about 7 points percent each. Given the scenario, it is difficult to think that subsistence drastically switched to maize in the later period of occupation.

As Cucina et al. (2011) pointed out, the ameliorated lifestyle exposed the Late Classic settlers to a stronger impact of oral infectious conditions, instead of acting as a cultural buffer against it. A more sedentary lifestyle for males presumably increased the number of times they were snacking food during the day; such a behavior increases the cariogenic effects of oral bacteria (Larsen et al. 1991). However, as mentioned above, it is not likely that such an impact originated from a higher intake of maize, in particular because caries increased also in females, who had not become more sedentary. Rather, it is more reasonable to think that the wealthier status permitted the population of both sexes to access new (cariogenic) foods previously not ingested or ingested in lower quantities, like honey as a sweetener for cacao (Cucina et al. 2011). In this perspective, Vega Lizama and Cucina (2014) have shown in two modern samples from the peninsula that cultural stressors introduced by globalized, ameliorated living conditions led to a significant increase of carious lesions both in males and females.

11.6 Conclusions

In conclusion, we must be very cautious to try make inferences on well-being based on the geographical location (inland versus coastal) of the settlements. Living in coastal/island settlements surely granted access to protein-richer diet, but at the same time exposed people to a set of environmental/cultural stressors not experienced in other contexts. Health and well-being are the result of a complex biological and sociocultural interaction. A close and holistic analysis is necessary to have a clearer picture of what living conditions used to be in past populations. Unfortunately, extrinsic limitations, like preservation and samples' representativeness, often prevent from such accurate and wide-angled point of view.

References

Aufderheide AC, Rodríguez-Martín C (1998) Dental caries. In: Aufderheide AC, Rodríguez-Martín C (eds) The Cambridge Encyclopedia of human paleopathology. Cambridge University Press, Cambridge, pp 402–404

Benavides Castillo A (2011) Jaina: ciudad, puerto y mercado. Dissertation, Universidad Nacional Autónoma de México, Mexico City

Cucina A (2011) Maya subadult mortality and individual physiological frailty: an analysis of infant stress by means of linear enamel hypoplasia. Chld Past 4:105–116

Cucina A, Tiesler V (2003) Dental caries and ante mortem tooth loss in the northern Petén area, Mexico: a biocultural perspective on social status differences among the Classic Maya. Am J Phys Anthropol 122:1–10

Cucina A, Perera Cantillo C, Sierra Sosa T et al (2011) Carious lesions and maize consumption among the Prehispanic Mayas: an analysis of a coastal community in northern Yucatán. Am J Phys Anthropol 154:560–567

Duray SM (1996) Dental indicators of stress and reduced age-at-death in prehistoric native Americans. Am J Phys Anthropol 99:275–286

Ellwood EC, Scott MP, Lipe WD et al (2013) Stone-boiling maize with limestone: experimental results and implications for nutrition among SE Utah preceramic groups. J Archaeol Sci 40:35–44

Glassman DM, Garber JF (1999) Land use, diet, and their effects on the biology of prehistoric Maya of northern ambergris cay, Belize. Part II: Paleopathology. In: White CD (ed) Reconstructing ancient Maya diet. The University of Utah Press, Salt Lake City, pp 119–132

Goodman AH, Armelagos GJ (1985) Factors affecting the distribution of enamel hypoplasias within the human permanent dentition. Am J Phys Anthropol 68:479–493

Goodman AH, Armelagos GJ (1988) Childhood stress and decreased longevity in a prehistoric population. Am Anthropol 90:936–944

Goodman AH, Martin DL (2002) Reconstructing health profiles from skeletal remains. In: Steckel RH, Rose JC (eds) The backbone of history. Cambridge University Press, Cambridge, pp 11–60

Goodman AH, Rose JC (1990) Assessment of systemic physiological perturbations from dental enamel hypoplasias and associated histological structures. Yearb Phys Anthropol 33:59–110

Goodman AH, Martin DL, Armelagos GJ et al (2013) Indicators of stress from bone and teeth. In: Cohen MN, Armelagos GJ (eds) Paleopathology at the origins of agriculture, 2nd edn. University Press of Florida, Gainesville, pp 13–50

Hernandez Espinosa PO, Marquez Morfin L (2007) El escenario demográfico de Jaina prehispánica durante el Clásico. In: Hernandez Espinosa PO, Marquez Morfin L (eds) La población prehispánica de Jaina. INAH, Mexico City, pp 33–62

Hillson SW (1996) Dental anthropology. Cambridge University Press, Cambridge

Hillson SW (2000) Dental anthropology. In: Katzenberg MA, Saunders SR (eds) Biological anthropology of the human skeleton. Wiley-Liss, New York, pp 249–286

Hillson S (2008) The current state of dental decay. In: Irish JD, Nelson GC (eds) Technique and application in dental anthropology. Cambridge University Press, Cambridge, pp 111–135

Kormondy EJ, Brown DE (1998) Fundamentals of human ecology. Prentice Hall, Upper Saddle River, New Jersey

Kreshover SJ (1960) Metabolic disturbance in tooth formation. Ann N Y Acad Sci 85:161–167
Larsen CS (1995) Biological changes in human populations with agriculture. Annu Rev Anthropol 24:185–213
Larsen CS, Shavit R, Griffin MC (1991) Dental caries evidence for dietary change: an archaeological context. In: Kelley MA, Larsen CS (eds) Advances in dental anthropology. Willey-Liss, Inc., New York, pp 179–202
Leatherman TL, Goodman AH (1997) Expanding the biocultural synthesis toward a biology of poverty. Am J Phys Anthropol 102:1–3
López Alonso S, Serrano Sánchez C (1984) Practicas funerarias prehispánicas en la isla de Jaina Campeche. Investigadores Recientes en el Área Maya 2:441–452. Sociedad Mexicana de Antropología, Mexico City
Lukacs JR (1989) Dental paleopathology: methods for reconstructing dietary patterns. In: Iscan MY, Kennedy KAR (eds) Reconstruction of life from the skeleton. Alan R. Liss, New York, pp 261–286
Lukacs JR (1992) Dental paleopathology and agricultural intensification in South Asia: new evidence from the bronze age Harappa. Am J Phys Anthropol 87:133–150
Lukacs JR (2008) Fertility and agriculture accentuate sex differences in dental caries rates. Curr Anthropol 49:901–914
Lukacs JR (2011a) Gender differences in oral health in South Asia: metadata imply multifactorial biological and cultural causes. Am J Hum Biol 23:398–411
Lukacs JR (2011b) Sex differences in dental caries experience: clinical evidence, complex etiology. Clin Oral Investig 15:649–656
Magennis AL (1999) Dietary change at the lowland Maya site of Kichpanha, Belize. In: White CD (ed) Reconstructing ancient Maya diet. The University of Utah Press, Salt Lake City, pp 133–150
Maggiano IS, Schultz M, Kierdorf H (2008) Cross-sectional analysis of long bones, occupational activities and long distance trade of the Classic Maya from Xcambó. Archaeological and osteological evidence. Am J Phys Anthropol 136:470–477
Méndez Colli C, Sierra Sosa T, Tiesler V (2009) Enamel hypoplasia at Xcambó, Yucatán, during the classic Maya period: an evaluation of the impact of coastal marshland on ancient human populations. HOMO J Comp Hum Biol 60(4):343–358
Nelson GC, Lukacs JR, Yule P (1999) Dates, caries and early tooth loss during the Iron age of Oman. Am J Phys Anthropol 108:333–343
Ortega Muñoz A (2015) Dental health and alimentation among the Quintana Roo Mayas: coastal and inland sites of the classic-Postclassic periods. Am J Hum Biol 27:779–791
Ortega Muñoz A, Sierra Sosa T, Tiesler V et al (2018) Demographic growth simulation of Xcambo, Yucatan, Mexico. A coastal settlement of Early and Late Classic period. Lat Am Antiq 29:591. https://doi.org/10.1017/laq.2018.11
Peña Castillo A (2006) Archaeological report of salvage excavation, Noh Bec-El Escondido, Yucatán. Report on File, INAH Yucatán Center, Mérida
Sahai D, Mua JP, Surjewan I et al (2001) Alkaline processing (nixtamalization) of white Mexican corn hybrids for tortilla production: significance of corn physiological characteristics and process conditions. Cereal Chem 78:116–120
Sattenspiel L (2000) Tropical environments, human activities, and the transmission of infectious diseases. Yearb Phys Anthropol 43:3–31
Saul J, Saul F (1997) The Preclassic skeletons from Cuello. In: Whittington SL, Reed D (eds) Bones of the Maya: studies of ancient skeletons. Smithsonian Institution Press, Washington, DC, pp 28–50
Seidemann RM, McKillop H (2008) Dental indicators of diet and health for the Postclassic coastal Maya on wild cane cay, Belize. Anc Mesoam 18:1–11
Sierra Sosa T (2004) La arqueología de Xcambó, Yucatán, centro administrativo salinero y puerto comercial de importancia regional durante el Clásico. Dissertation, Universidad Nacional Autónoma de México

Sierra Sosa T, Cucina A, Price TD et al (2014) Maya coastal production, exchange, and population mobility. A view from the Classic period port of Xcambó, Yucatan, Mexico. Anc Mesoam 25:221–238

Steckel RH, Rose JC (eds) (2002) The backbone of history. Health and nutrition in the western hemisphere. Cambridge University Press, Cambridge

Storey R, McAnany P (2006) Children of K'axob. Premature death in a formative Maya village. In: Ardren T, Hutson SR (eds) The social experience of childhood in ancient Mesoamerica. The University Press of Colorado, Boulder, pp 53–72

Takahashi N, Nyvad B (2011) The role of bacteria in the caries process: ecological perspectives. J Dent Res 90:294–303

Tiesler V, Cucina A, Sierra Sosa T et al (2005) Comercio, dinámicas biosociales y estructura poblacional del asentamiento costero de Xcambó, Yucatán. Inv Cult Maya 13:365–372

Tiesler V, Cucina A, Stanton TW et al (2017) Before Kukulkán. Bioarchaeology of Maya life, death, and identity at Classic period Yaxuná. University of Arizona Press, Tucson

Turner CGII (1979) Dental anthropological implications of agriculture among the Jomon people of Central Japan. Am J Phys Anthropol 51:619–636

Vega Lizama E, Cucina A (2014) Maize dependence or market integration? Caries prevalence among indigenous Maya communities with maize-based versus globalized economies. Am J Phys Anthropol 153:190–202

Watson JT (2008) Prehistoric dental disease and the dietary shift from cactus to cultigens in Northwest Mexico. Int J Osteoarchaeol 18:202–212

White CD (ed) (1999) Reconstructing ancient Maya diet. The University of Utah Press, Salt Lake City

Chapter 12
History of Health and Life of Pre-Hispanic Maya Through Their Skeletal Remains

Lourdes Márquez Morfín and Patricia Olga Hernández Espinoza

Abstract Health studies performed in pre-Hispanic osteological series show a close link between health, the natural physical environment and the degree of social complexity. From a standardized methodology, it is possible to know some aspects of past societies such as the structure by age and sex of the population, as well as the population density of the settlement. In the case of Terminal Classic Maya societies, we have obtained interesting results in making this same association that should be explained according to the degree of health of the individuals, their quality of life, their social organization and the natural physical environment. Some of these results show high frequencies in infectious diseases, which impacted infant mortality rates and life expectancies at birth. Nutritional deficiencies also show high figures, which correspond to a densely populated and urban areas living under a high-pressure demographic regime, which means high fertility and mortality rates.

12.1 Introduction

Research on some pre-Hispanic Maya populations of the Yucatan Peninsula and other regions focused on assessing demographic profiles, height and changes or recording bone indicators to identify health and nutritional problems have allowed to learn specific aspects about their diet, living conditions, diseases, and their relation with socioeconomic, cultural and ecological aspects in the environment of each population. Specific works on the pre-Hispanic Maya population in Yucatan are found in Cetina Bastida and Sierra Sosa (2005), Cucina (2011), Cucina et al. (2011a, b), Cucina and Tiesler Blos (2007), Márquez Morfin and Del Angel (1997), Márquez

L. Márquez Morfín (✉)
Posgrado en Antropología Física, Instituto Nacional de Antropología e Historia/Escuela Nacional de Antropología e Historia, Ciudad de México, México

P. O. Hernández Espinoza
Centro INAH Sonora, Secc. Antropología, Instituto Nacional de Antropologia e Historia, Hermosillo, Sonora, México
e-mail: olga_hernandez@inah.gob.mx

H. Azcorra, F. Dickinson (eds.), *Culture, Environment and Health in the Yucatan Peninsula*, https://doi.org/10.1007/978-3-030-27001-8_12

Morfín (1981, 1984, 1987, 1989, 1991, 1996), Márquez Morfín et al. (2002a), Márquez Morfín and Jaén (1997), Márquez Morfín et al. (1982), Ortega Muñoz (2007), Storey (1999), Tiesler Blos (2001) and Tiesler Blos and Cucina (2004, 2007). The purpose of this chapter is to identify some general health and nutrition trends in two Maya groups (Jaina and Xcaret) from the Yucatan Peninsula coasts and two from the Maya lowlands (Palenque and Copán) in order to compare different physical environments. Results obtained for the four selected groups highlight nutritional issues related to the physical environment, marine diet in coastal sites, and the diet based on maize and several animals in the case of inhabitants of inland cities (Berriel 2002; Rodríguez Suárez 2004). Another aspect commonly analysed are the infectious and parasitic diseases, which were constant in all pre-Hispanic populations. Hence, we shall analyse the results obtained from three different Maya groups from the Classic period (Jaina, Palenque, Copán) and Xcaret as of the contact with Europeans (sixteenth century), taking into consideration the different ecological environment and their possible impact on health. Some explanatory alternatives are shown regarding the causality of the health/disease process in each historical context and its determinant factors related to this process.

The material used for this study address skeletal series from the Classic period, two of them located on the coasts of the Yucatan peninsula, which are Jaina (n = 105), on the current Campeche coast (Hernández Espinoza and Márquez Morfín 2007), and Xcaret (n = 118), on the coast of Quintana Roo (Márquez Morfín et al. 2002b). Two more series located inland in regions other than the Yucatan Peninsula, such as Palenque (n = 126), in the current state of Chiapas (Márquez Morfín et al. 2002a: 13–31; Márquez Morfín and Hernández Espinoza 2006c: 265–289), and Copán (n = 238), in Honduras (studied by Storey et al. 2002), were included. The series were integrated into a database with a total of 587 skeletal remains (Márquez Morfín and Hernández Espinoza 2002, 2003).

The model used for the analysis of interactions between the material conditions of existence and lifestyle was the health determinant adapted for pre-Hispanic populations (Márquez Morfín and Hernández Espinoza 2006b, 2006c). The variables that intervene as determinants of the health, nutrition and demographic processes are the ecological environment, composition, population size and density, its genetic structure, social organization, source of livelihood, and political and cultural aspects (Frenk et al. 1991). We applied a standardized methodology for the registration of indicators, or traces that can be seen in the skeleton, related to several diseases and deficiencies (Goodman and Martin 2002; Steckel and Rose 2002). Demographic variables were obtained for each site using the palaeodemographic methodology (Meindl 2003; Meindl et al. 2008; Meindl and Russell 1998). Results of height calculation from skeletal remains of the four series were used as an indicator of quality of life and stress accumulated throughout life (Márquez Morfín and Del Ángel 1997; Márquez Morfín 1982). The growth process, particularly height according to the age, is one of the relevant aspects in the health analysis. Available data are from the Jaina population (Peña Reyes et al. 2007; Storey et al. 2002). For the evaluation of stress indicators, a percentage of hyperostosis, cribra orbitalia, enamel hypoplasia and periosteal reactions were obtained for each. The statistical method based on the prevalence model proposed by Waldron (2016) and applicable to each stress indicator was also used, which allows the data comparison of different size series.

The chosen strategy consisted in contrasting the results obtained in the different series selected to identify general health and nutrition trends such as the percentage of infections, according to the population density in some of the sites, or to the lack of appropriate sanitary conditions and the differences between them, as in Palenque and Copán. Furthermore, the purpose was to find nutrition aspects related to the physical environment from the site natural resources. Results of the Paleo diet obtained by the analysis of trace elements conclude that the Maya group diet in Xcaret was based on marine resources, as occurred in the nearby site of Chac Mool (Berriel 2002; Rodríguez Suárez 2004). The food preparation process from archaeological and ethnohistorical information is another valuable data. Two physical environments can be contrasted: the cities inland (Palenque and Copán) and those in the Yucatan Peninsula coastal region (Jaina and Xcaret). Preliminary studies of each of these groups highlighted the influence of environmental conditions on the health and nutrition of individuals (Hernández Espinoza and Márquez Morfín 2007; Márquez Morfín et al. 2002b; Storey et al. 2002).

12.2 Background

Living conditions, health and nutrition, as well as different demographic aspects of pre-Hispanic Maya populations, have been the subject of several studies in the Maya lowland region since the end of the twentieth century (Cohen 2007: 78–88; Danforth 1994: 206–211, 1997: 127–137; 1999: 103–118; Glassman and Garber 1999: 119–132; Herrera Flores and Götz 2013:69–98; Magennis 1999: 133–150; Storey 1999: 169; Whittington and Reed 1997: 157–170; Wright 1997: 181–195; 1999: 197–220: Wrigth and White 1996: 147–198). These studies highlighted the main health problems specially related to nutritional deficiencies, infectious and parasitic diseases and population changes. Our work on these topics, based on the skeletal series of Maya sites located in Mexican territory, has revealed results similar to those reported for other Maya groups (Storey et al. 2002; Márquez Morfín 1982, 1991; Márquez Morfín et al. 1982; Márquez Morfín et al. 2002a, b: 13–31; Márquez Morfín and Hernández Espinoza 2003, 2004; 2006a: 113–160; 2006b: 73–102; 2006c: 265–290; Márquez Morfín et al. 2001: 291–313).

12.3 Methodology

The methodology used adapted from the Frenk et al. (1991) model takes into account different factors that have been posed as co-determinants and constraints of the quality of life of individuals, which include the natural physical environment of each settlement; the environment built by human beings such as type of urbanization and settlement services; size and constructive density; demographic structure and density with profiles such as number of inhabitants, their composition by gender and age; and social organization, stratification and social role of individuals, all represented in the skeletal samples. The influence of the physical environment, climate, geographical location and natural resources, among other natural elements, on the health and epidemiological and demographic profiles of a population is undeniable. As in the case of all cities, housing of pre-Hispanic cities was a socially constructed and modified space where aspects such as size of settlement; density of buildings; type of housing; utilities, i.e. drainage; water; waste disposal and organic waste systems; population size; and population composition by age and gender, among other factors, had an impact on the living conditions and health of individuals. The structure of society and cultural aspects such as marriage patterns, migration, income distribution and dieting also play an important role in the quality of life of individuals and society.

12.3.1 General Characteristics of the Sites and Skeletal Series

Urban Copan is located in a valley 600 metres above sea level southeast of the Maya area, with a tropical climate with well-identified wet and dry seasons. Copán is one of the largest and most important Maya centres of the Classic period in the lowlands; its population at its highpoint reached 27,000 inhabitants in the ninth century CE (Storey et al. 2002: 287–289). Skeletons located in this large city are mostly from the Late Classic and Terminal periods (800 to 1000 CE). Copan burials were explored in 9 N-8 set, "the House of Bacabs", the largest outside the Royal Acropolis, as well as in 11 courtyards with over 50 structures that could house 200 individuals at the same time. According to archaeological information, families from noble ancestry lived there, so the proposal is that "the *sample will be treated as an example of privileged individuals within their society*" (Storey et al. 2002: 288). Skeletal remains were explored both in elaborate tombs and in simple graves. As in other Maya centres of the Classic period, the variety in burial types and forms is wide since tombs and cists were found in ceremonial buildings and burials directly under the room floors, or in the patios of housing units, which are indicative of social status at heterarchical level.

Palenque is in the current state of Chiapas, in a jungle region crossed by a river that provided water and food such as fish, turtles and other species, while white-tailed deer (*Odocoileus virginianus*) and other mammals were environmentally common. The city had its peak during the Classic period when most of the monumental buildings were built. The social organization was complex, with a marked stratification and great inequalities (Liendo and Vega 2000; Liendo Stuardo 2000; López Bravo 1995, 2000). The skeletal remains explored from the main site structures belonged to the leading and intermediate groups living in high-class housing units (Márquez Morfín et al. 2002a). The explored burials are of different collective, individual, secondary, primary, direct and indirect types, and skeletal remains were found in different anatomical positions. Funerary customs are an excellent source of information on the social identity since Palenque inhabitants, as in other pre-Hispanic sites, were buried according to their social status. Low social class individuals were buried under the room floors or courtyards of housing units in areas far from the administrative and political centre, while those of the middle class group, i.e. specialized craftsmen, warriors, merchants, architects and artisans, who lived in the architectural complexes peripheral to the monumental buildings were buried in housing units close by these constructions and, finally, those belonging to the highest social hierarchy that inhabited palaces and elite units were deposited in places according to their class, for example, in elaborate tombs or in simple graves. [In the case of Palenque, the tomb of Pakal stands out in the Temple of the Inscriptions] (Dávalos and Romano, 1973; Ruz Luihllier 1958; Ruz Luihllier 1973; Tiesler Blos and Cucina 2004).

Jaina is a small artificial island built by the Maya that is located off the Campeche coast. The site is made up of several monumental structures and about 1400 housing units, 700 of which should have been occupied with at least four inhabitants each (Benavides Castillo 2007: 13–32). The island was located in a mangrove area where molluscs, crustaceans, insects, birds, a great variety of fish, crocodiles, turtles and few mammals abound. This environment provided marine resources, agricultural products and an important protein contribution (López Alonso and Serrano Sánchez 1997; Piña Chan 1968). Dozens of burials were explored in Jaina with valuable offerings, highlighting human figurines that represent different characters and trades. According to the offering quality and quantity criteria, the series is considered to be from several social groups of different classes (López Alonso and Serrano Sánchez 2007, 63–96). Most of the skeletal remains are from the Late Classic period, with a peak period between 600 and 1000 CE. The nature of buildings and the settlement itself is not clearly defined; this was a contemporary Xcambó coastal settlement as other sites in the east of the Peninsula such as Xcaret (Benavides 2007: 13).

Xcaret—The Xcaret cove is located in the municipality of Cozumel, Q. R., on the eastern coast of the Peninsula. The extension of Xcaret site is difficult to define given that there seems to be a continuous settlement along the entire coast. Xcaret was a port of commercial relevance since the Classic period that continued working after the Spanish colonization (Con 1991: 74–123; Márquez Morfín et al. 2002b). The housing area was made up of a series of drystone walls, small platforms and rooms with a central and nuclear zone having groups of structures associated with each group. Some of the buildings rest on platforms that make up squares (Con 1991: 74–123). Water supply was obtained from the *aguadas* and *cenotes*. Its climate is warm, subhumid with a rainy season in summer. This region's flora corresponds to the plant community of subperennifolia medium forest, among which the best-known and very useful species are *chicozapote*, cedar, mahogany and *ramón*, which produces a seed rich in carbohydrates with more protein than maize. Among the most commonly used palms for house roofing are the *chit* and *huano*. Fauna was characterized by the pheasant, quail, turkey, partridge and some wild ducks, as well as turtles, jaguar, tapir and manatee (Con 1991: 71–73), which would be part of its inhabitant diet.

12.3.2 Bone Material

Four skeletal series of different sizes were selected: (1) Urban Copan, 238 individuals, considered of noble ancestry; (2) Palenque, 126 individuals of all ages and both genders; most of them from burials recovered from Groups I, II, IV, B, C, Murciélagos, Northern Group areas and temples XV, de la Cruz, de la Cruz Foliada and de la Calavera, all of the Late Classic period (650–900 CE); (3) Jaina, 105 skeletons of all ages and both genders from different social class housing units; and (4) Xcaret, colonial skeletal series consisting of 118 individuals of all ages, both genders explored under the nave floor of one of the first sixteenth century chapels in the area (Márquez et al. 2002b).[1]

12.4 Results

Age distribution—The Buikstra and Ubelaker techniques (1994) were used to estimate age and gender (adults only) by classifying the skeletons into four age groups: 0–14, 15–29, 30–49 and ≥ 50 (Table 12.1). The age distribution to death in the skeletal series studied reveals great differences, since actually these are rarely broad enough to represent all age groups in order to have a balance between genders, which is one of the great limitations for bioarchaeological studies in the Maya area. There are several reasons, such as partial excavations, poor state of conservation of remains and different funerary practices by age, gender and social class. The age distribution in Jaina (56%) and Copán (46.9%) shows the highest figures in the first group, which is considered to be in line with the expected mortality for subadults. In contrast, this age group is less represented in Palenque (18%) and Xcaret (26.3%). As we have already explained in other palaeodemography works on these series, there are multiple reasons, so the archaeological data of origin, as well as the indicators of social stratification, should be taken into account.[2] These distributions do not comply with the mortality of these populations because they are biased by external factors. In Palenque and Xcaret, the subregistration of subadult individuals is evident since the death percentage for this age group in any ancient population is around 40–50%, and infant mortality between birth and the first year of life is expected to be 30% (Livi-Bacci 2002: 107). In Palenque as well as in other sites of the Classic period Maya area, the predominance of adult men burials in tombs and the lack of children in them are identified, so the age distribution also depends on the excavated sites. For this series, most data come from tombs explored in buildings and elite housing areas from partial excavations of a small area of the site

[1] Skeletal material of pre-Hispanic and colonial Xcaret were excavated by archaeologist María José Con and sent by her to Lourdes Márquez for study.

[2] For more information, see Hernández and Márquez (2007), Márquez and Hernández (2008) and (2013).

Table 12.1 Distribution by age groups of skeletons analysed in Copan, Jaina, Palenque and Xcaret

Skeletal Series	Total	0–14	15–29			30–49			≥50		
			F	M	I	F	M	I	F	M	I
Urban Copan[a]	238	112	10	10	0	40	24	0	24	18	0
		46.9%	4.2%	4.2%	0%	16.7%	10.0%	0%	10%	7.5%	0%
Jaina	105	59	4	6	1	18	9	1	4	2	1
		56%	3.8%	5.7%	1%	17.1%	8.6%	1%	3.8%	1.9%	1%
Palenque	126	23	19	13	5	25	32	8	0	1	0
		18%	14.8%	10.2%	3.9%	19.5%	25%	6.3%	0%	0.8%	0%
Xcaret	118	31	17	6	4	32	21	0	2	3	2
		26.3%	14.4%	5.1%	3.4%	27.1%	17.8%	0%	1.7%	2.5%	1.7%
Totales	587	225	50	35	10	115	86	9	30	24	3
	100%	38.3%	8.5%	6%	1.7%	19.6%	14.7%	1.5%	5.1%	4.1%	0.5%

Source: From database prepared by Márquez Morfín and Hernández Espinoza (2002 and 2003); *F* female, *M* male, *I* Unidentified

[a]Data from Storey et al. (2002)

(Márquez Morfín and Hernández Espinoza 2006c). Hence, the cultural aspects related to funeral Maya practices should be taken into account. On the other hand, appropriate sample composition in terms of age, as in the case of the Jaina and Copán series, is an excellent example or reference of infant and children under 5 years old mortality for ancient Maya populations. High mortality in this case was associated with fragile and precarious life and health conditions. Only 50% of children born survived until 15 years of age. For the 15–29 years group, Copán and Jaina have similar figures, while in the other series reports are insufficient. The highest percentage of adults is observed in the 20–49 years group. Copán has the only series with the highest percentage of the ≥50 years group that, due to its social origin, could indicate a greater survival of individuals at ages corresponding to old age.[3] The series size impacts the results due to its large size. In addition to taking into consideration the high social position of individuals represented in the Copan series, the Palenque series shows a higher index of masculinity that can be explained because skeletal remains are from tombs of the main buildings of this ancient Maya city. The other series indicate an unequal distribution between men and women due to the excavation sites that favoured certain types of buildings and housing or ceremonial units.

Demographic results. The palaeodemographic methodology starts from the distribution by age at death in order to generate mortality profiles by age group, life expectancy at birth, fertility and mortality rates and other indicators. Each series was evaluated in terms of its composition by age groups. There is an underrepresentation of individuals younger than one year due in part to the fact that the series are not from extensive excavations of the site under study with the exception of the Jaina series. Therefore, three series were adjusted, adding the individuals that were needed until reaching a 25–30% representation of the total of series. This simulation allows us to obtain the infant mortality rate that these populations should have due to infectious respiratory and digestive diseases. Mortality studies conducted in Mexican historical series with cause of death (Camacho Martínez 2018; Márquez Morfín and Hernández Espinoza 2016: 7–44) have confirmed the statements made by historians about the European population. Nearly 30% of newborns do not survive the first year due to this type of diseases that usually leave no trace in skeletons. From the new distributions by age groups, a growth rate of 1.5% was applied according to the functional and mathematical formulas of each of the indicators selected for this analysis (Table 12.2).

1. Proportions by age group (conventionally expressed with four decimals): <15 years (<15); productive and childbearing age (15–50) and over 50 years (> 50). Dependency index: the proportion of individuals that depend on the productive population is obtained by dividing the number of individuals of productive and reproductive ages by those under 15 years of age and those over 50 years old. The adjusted sample composition in the four series is approximately 60% of

[3] It should be noted that from the Acropolis and important buildings are represented in this series, whose quality of life could have been better.

Table 12.2 Palaeodemographic indicators

Indicators	Age	Osteological series			
		Jaina	Copán	Palenque	Xcaret
Series composition (age)	<15	0.5441	0.5574	0.5959	0.5710
	15–50	0.4340	0.4211	0.4000	0.4280
	>50	0.0222	0.0221	0.0036	0.0007
ID		1.3	1.4	1.5	1.3
Survival	Ex	26.7	25.6	23.1	27.1
	$q_{(x)}$	206.8 ‰	147.5 ‰	147.7 ‰	126.2 ‰
	TBM	54.4 ‰	39.4 ‰	55.5 ‰	47.2 ‰
	A	16.5	16.1	14.0	14.8
	AA	28.4	28.9	25.2	25.9
	S	1.4%	4.2%	3.9%	3.4%
Fertility	TBN ‰	69.4 ‰	54.5 ‰	70.5 ‰	62.2 ‰
	$\overline{b}$	0.0866	0.0969	0.0874	0.0801
	TGF	5.6	6.8	5.6	5.7
	TBR	2.8	3.4	2.8	2.8
	$R_{(0)}$	1.4	1.5	1.4	1.5
	$\overline{T}$	22.4	27.3	23.1	26.6
	F	4.5	3.7	4.3	4.1

Source: calculations based on the database of Márquez Morfín and Hernández Espinoza (2002, 2003). (E_x) average life expectancy at birth; ($q_{(0)}$) infant mortality (deaths per thousand births); (TBM) gross mortality rate (deaths per thousand inhabitants); (A) average age of the entire living population; (AA) average age of adults (over 15 years); (S) percentage of survivors at age 50; ($\overline{b}$) annualized fertility rate; (TGF) global fertility rate; (TBR) gross reproduction rate; (TNR) net reproduction rate; (TBN) gross birth rate (births per thousand inhabitants); (F) average family size; ($\overline{T}$) reproductive length

children under 15 years of age, against 40% of the adult population, which was expected for this type of population. The Weiss model (1973) assumes that dependents are less than 15 years and over 50 years of age, which are conventional ages to standardize the indicator. Individuals of pre-Hispanic population started working in the fields at an early age, although this does not mean that they were economically independent. In the case of the elderly, there is no information available that would let us know the age at which they were considered as adult-dependent.

2. Mortality and survival indicators. The average life expectancy at birth ($E_{(x)}$), in the four series, shows values of 23–27 years, the lowest for Palenque and the highest in Xcaret. These numbers are considered moderate for this type of population, which have been discussed in previous papers (Hernández Espinoza and Márquez Morfín 2015; Márquez Morfín and Hernández Espinoza 2013). These figures are consistent with those obtained by Ortega (2007) for El Meco, 26.83 years. Tiesler Bloss et al. (2005) proposed through a different methodology a life expectancy for Xcambó of 37.6 years, almost ten more years. This figure seems very high compared even to those on life expectancy in Mexico

City during the Colonial period, whose calculated values range from 15 to 30 years (Márquez Morfín and Hernández Espinoza 2016). Infant mortality ($q_{(0)}$) refers to deaths of children less than 1 year of age per thousand births. Jaina shows the highest figures. The gross mortality rate (as per the acronym in Spanish, TBM), which represents deaths per thousand inhabitants, was similar in Jaina and Palenque. The average age of the entire living population (A) ranges from 14 to 16 years, while the average of adults (older than 15 years) of the living population (AA) ranges between 22 and 28, which indicates a structure of young population in all series. The percentage of survivors aged over 50 years (S) varies from 1.4% in Jaina to 4.2% in Copán, indicating that few people survived after this age.

The calculation of palaeodemographic indicators depends, as already stated, on the structure by age at death of an osteological series, and the better the representation in the different age groups, the more consistent the information obtained shall be. If the series has a good representation of young and median adults, the life expectancy at birth shall be close to the average values of the age groups considered, so it is important to review the composition before starting any calculation. Infant mortality rate is another important indicator of this block, which is inversely proportional to life expectancy: a high infant mortality decreases the life expectancy at birth and a high life expectancy at birth implies a low probability of dying during the first year of life. The values obtained correspond to a moderate mortality since it is considered that there is a good percentage of underrepresentation, even with the adjustment.

Daily life in tropical climates, such as the Maya region, involves exposure to high zoonoses that few children would have survived. This discussion shall be seen later in the description of results of health indicators.

Another demographic indicator is the average age of the living population, which was built with the intervention of mortality levels; hence, the population in all four cases as a whole did not exceed 20 years on average and the mean age of those that survived adolescence (15 years) does not exceed the fourth decade of life probably due to the risks of daily life, both in productive work and in war and, in the case of women, the inherent risks to reproduction.

3. Probable levels of fertility. The specific fertility rate ($\overline{b}$) has no meaning, but is necessary to calculate the other indicators. It is considered that a population adapted successfully to their environment when they have children, and these in turn survive to reproduce themselves. The net reproduction rate ($R_{(0)}$) is the number of daughters a newborn girl can have taking into consideration the risk of dying before finishing her childbearing period. This rate indicates the future capacity for potential growth; the reproductive period expands from 22 to 27 years. The gross birth rate (as per the acronym in Spanish, TBN) is the number of births per thousand inhabitants, which has the highest values in Palenque, while Copán shows the lowest figures. The mean surviving children at the beginning of the reproductive period (conventionally, 15 years old) is identified with (F). The gross reproduction rate (as per the acronym in Spanish, TBR) refers to

the number of daughters per woman who survived at the end of their childbearing period, which was higher in Copán; the other populations had similar values. ($\bar{T}$) is the duration of female childbearing period, in years (Table 12.2). Assuming a moderate growth, the number of children per woman should have been between six and seven children (TFR), and four and five of them (F) survived to replace their generation. Results show 5.6 children in Jaina, Palenque and Xcaret, and 6.8 in Copán, which can be explained due to a population with greater growth. Fertility levels calculated for other Maya population, such as Xcambó, have the same status, with 6.0 children per woman. Tiesler et al. (2005) consider that the Maya society was a pronatalist society, which required a large population to defend its territories and to build its cities. In fact, all pre-Hispanic populations were pronatalist because human capital was necessary for all work. Furthermore, infant mortality was very high and only a proportion would survive to reproduce and replace their generation.

12.4.1 Height

The height of Maya groups has been one of the subjects of greater interest since the last century when some researchers raised the possibility of the decrease in size in individuals that inhabited the lowlands at the end of the Classic period. Based on the results recorded about Tikal, the discussion revealed that the size decreased due to malnutrition and infectious diseases caused by the so-called Maya Collapse. Differences in social status were also considered during the discussion (Haviland 1967). Saul in turn (1972) supported the proposal on biological decay according to analysis of skeletal remains of Altar de Sacrificios. Regarding the possibility that this same process had occurred among the Maya in Yucatan during the Classic period, Márquez (1982) and Márquez and Del Ángel (1997) analysed 14 Maya skeletal series, and integrated data published by Saul on the Altar de Sacrificios series. The relevant aspect in this study is the methodology used based on the absolute measurements of six long bones. Data were disaggregated by gender and by side. A decrease during the Classic period and an increase towards the Postclassic period in the dimensions of femurs in men were registered. For tibias, the reduction is seen from the Classic to the Postclassic periods. The decrease in femur measurement in women was identified from Classic to Postclassic periods, while there is an increase of the tibia towards the Classic period and a further decrease in the Postclassic period (Márquez Morfín 1984, 1987: 51–61). Results were controversial since changes in bone dimensions were observed from one period to another that could be interpreted as the possible decline in size in the Classic period. Even though several explanations are provided in the study about such a decrease in height, it is difficult to draw definitive conclusions given the small number of individuals in the series. There is a wide variability in the regional and temporary Maya height currently considered, which decline causality in some groups should be interpreted in its specific context.

On the other hand, Marie E. Danforth (1994) published a study on the association of the height decrease and the Maya Collapse based on the materials from seven sites of the southern lowlands in which she found the same problem of the small series size and raised the need to have more data to document whether a decrease in height actually occurred. Several skeletal series were analysed from new excavations in Yucatan in order to obtain more information about the height of pre-Hispanic Maya published by Márquez Morfín (1982), (Tiesler 2001). Size limitations of collections and the variability of results remained once again on the discussion table. We consider that a relevant and essential factor in interpretations lies in the duration of site occupation. Most of them had a peak moment, and skeletons, as a whole, correspond to a single period. There are often no skeletons of different periods that allow the analysis of changes in body dimensions per site. Hence, the size values recorded in the four collections analysed in this study (Copan, Palenque, Jaina and Colonial Xcaret) are only a reference in which no temporary changes can be detected except for the Xcaret series, which has some pre-Hispanic skeletal remains of the Postclassic period ($n = 35$) and the wide series of the sixteenth century ($n = 118$). The height calculated for men in the Late Classic period using femurs and applying the Genovés formulas show similar values in Jaina and Palenque, while values are higher in Copan (Table 12.3). For skeletal remains of Colonial Xcaret, a height of 164.2 cm was calculated, which is the highest. In Jaina and Palenque, women size ranges between 148.3 cm and 151.0 cm, respectively. In Copan, the size of women reaches 155.5 cm, similar to the 154 cm obtained for pre-Hispanic Xcaret. The substantial change was observed in Colonial Xcaret, where sizes were smaller, 145.5 cm. If we review the values reported by Tiesler for height from the femur (2001: 262) obtained from the Maya series of the Early Classic to the Terminal periods in Yucatan, the height for men and women were 160.1 cm (19 cases) and 146.6 cm (14 cases), respectively, figures similar to those shown in this study.

The discussion and debate about the height decrease in Maya lowland population led to consider that:

> *by reviewing the recent literature, it can be suggested that not only was there large variability within regions as to who had access to certain resources, but there was also large interregional variability. The high degree of intra and interregional variability suggests that*

Table 12.3 Height calculated for Maya series

Site	Period	Men (*n*)	Women (*n*)	Reference
Jaina	Late classic period	160.6 (7)	151.0 (12)	Hernández Espinoza and Márquez Morfín (2007)
Palenque	Late classic period	160.0	148.3	Márquez Morfín and Hernández Espinoza (2004)
Copan	Late classic period	162.8 (36)	155.5 (46)	Storey et al. (2002: 304)
Xcaret	Postclassic period	164.5 (2)	154.0 (3)	Márquez Morfín et al. (2002a, b)
Xcaret	Colonial period	157.3 (15)	145.5 (31)	Márquez Morfín et al. (2002a, b)

the trend in decreasing stature was not necessarily widespread, and therefore does not reflect a plastic adaptation among the whole Maya population (Masur 2011: 6–24).

Subadult growth. A qualitative/quantitative methodology was designed for Jaina that would allow growth evaluation by age during childhood in order to obtain their height. The data was related to the presence of biological disruption indicators (see Peña et al. 2007: 153–178). We concluded that the estimated height of children in Jaina series were lower than those reported by Faulhaber (1976) in Yucatan at an early age. Two-thirds of the cases of children under 1 year of age were located at the growth deficit level of <−2 standard deviations (−2 s), which represents a chronic process that causes deterioration. It is suggested that there is a close link between the environment where these individuals lived and the type of lesions found in skeletal remains. Jaina population revealed high infant mortality related to multiple factors, which is evidence that the prevailing living conditions and diseases affected the growth potential (Hernández Espinoza and Márquez Morfín 2007: 33–76).

12.4.2 *Methodology for Health/Disease Evaluation*

The methodology used was based on the registration of biological disruption indicators (stress), such as hyperostosis spongiosum, cribra orbitalia, periostitis and enamel hypoplasia that were selected and standardized for their quality to evaluate the health of ancient groups (Goodman and Martin 2002: 11–60). This methodology has been used in several Mesoamerican pre-Hispanic series in order to generate a broad database that allows describing and interpreting demographic and epidemiological trends (Márquez Morfín and Hernández Espinoza 2002 and 2003; Márquez Morfín and Jaén 1997; Storey et al. 2002, Márquez Morfín et al. 2002a: 307–338).

12.4.3 *Spongy Hyperostosis and Cribra Orbitalia*

In pre-Columbian America, these lesions have been associated with megaloblastic anaemia due to folic acid, vitamin B and iron deficiencies of a diet based mainly on maize and the lack of meat consumption. Maize is low in iron and contains phytates that inhibit the absorption of minerals, contributing to aggravate the problem. These lesions of similar aetiology seem to have a higher incidence during the delactation stage. Based on the evidence of infections in infant and child skeletons (Mensforth et al. 1978, Stuart-Macadam and Kent 1992, Stuart-Macadam 1995), the occurrence of spongy hyperostosis has also been associated with infectious and parasitic problems. The synergistic effect between poor nutrition and infection has been demonstrated in cases of diarrhoea in the weaning period as a cause of infant mortality in several contemporary populations (Scrimshaw et al. 1968; Scrimshaw 2000; Walker 1986: 345–354).). Nutritional deficiency weakens the immune system, which in turn affects the body's ability to cope with infectious agents. On the other hand, infections sometimes inhibit as well the ability of the digestive system to absorb the nutrients required during diarrhoea, generating a vicious circle. The combination of these factors has been identified as the leading cause of child mortality in contemporary developing nations.

Mensforth et al. (1978) suggest recording the lesions according to their stage, taking into account the cases with bone remodelling, focusing specially in the co-occurrence pattern and the affected ages. Thus, infections occur earlier and hasten the nutritional deficiency of anaemia. According to Storey (1992: 198), cases with bone remodelling in subadults and adults represent survivors of anaemia and infection problems.

Results. Spongy hyperostosis was identified in the four series (Table 12.4), with frequencies ranging from 27.4% in Jaina (17/62) to 89.1% in Palenque (49/55), and Cribra (lesion in the orbit roof) showed the highest frequency of 80.8% in Palenque (21/26) and lowest of 25% in Colonial Xcaret (12/48). In the latter group results are 35.7% in men, 25% in women and 21.4% in subadults. Considering the age groups, children between 0 and 4 years old and women of childbearing age (25 to 29 years old) reached the highest frequencies (Márquez Morfín et al. 2002a). It is important to mention that the relation of this lesion with that produced by periosteal reactions is generally caused by gastrointestinal infections and parasitosis, which prevent the use of nutrients such as folic acid, vitamin B, or minerals such as iron. In Palenque, 98.9% of individuals living in the area had Cribra while half of the skeletal remains in Copán and Jaina showed the injury. In spite of being rich in marine resources and complemented by other terrestrial foods, the type of diet may not have been used biologically due to recurrent infections, which explains these results. In other sites in the Maya area, different values of spongiosum hyperostosis have been recorded (Table 12.5) showing the wide variability in which the ancient Maya of Yucatan were affected by feeding problems and especially by anaemia. The percentage of this indicator is very high in the series of the sacred cenote of Chichen Itza, which is explained by the social identity of the people sacrificed in that place, perhaps slaves or prisoners of war (Márquez Morfín 2010). Values in coastal sites such as Chac Mool, Playa del Carmen and Xcambo range from 31% to 44%.

Table 12.4 Number of cases and percentages of health indicators for the four series

	Copán		Jaina		Palenque		Xcaret Colonial	
Indicators	Cases/N	%	Cases/N	%	Cases/N	%	Cases/N	%
Spongy hyperostosis	41/69	59.4	17/62	27.4	49/55	89.1	37/74	50.0
Cribra orbitalia	40/100	40.0	21/56	37.5	21/26	80.8	12/48	25.0
Incisor hypoplasia	138/151	91.4	27/27	100.0	32/45	71.1	15/37	40.5
Periosteal reactions tibia	78/159	49.1	39/78	50.0	91/92	98.9	64/85	75.3
Systemic periosteal reactions	33/134	24.6	19/46	41.3	109/114	95.6	51/88	58.0

Source: Calculations from Márquez Morfín and Hernández Espinoza database (2002 and 2003)

Table 12.5 Distribution of spongy hyperostosis in other Maya population of Yucatan

Site	Chronology	Cases/N	Percentage	Reference
Chac Mool	Late Classic period	23/45	44	Márquez Morfín and Hernández Espinoza (2006a: 127)
Chac Mool	Postclassic period	15/22	68	Márquez Morfín and Hernández Espinoza (2006a: 127)
Playa del Carmen	Postclassic period	10/23	43.4	Márquez Morfín et al. (1982: 114)
Chichen chultun	Postclassic period	51/75	68	Márquez Morfín (2010: 275)
Sacred Cenote 1 Sacred Cenote 2	Postclassic period	24/36 52/54	66 96	Márquez Morfín (2010: 275) Márquez Morfín (2010: 275)
Xcambó	Classic period		31.3	Cetina Bestida and Sierra Sosa (2005: 269)

12.4.4 Enamel Hypoplasia in Incisors

Dental enamel hypoplasia is a deficiency in the thickness of the enamel matrix due to the cessation of enamel formation by visible ameloblasts, such as bands, grooves or transverse lines in dental crowns. Once formed, these marks are unalterable over time, so they have been considered as chronological memories of nutritional stress or disease episodes during the development of individuals. Enamel hypoplasia is related to nutritional deficiency problems, especially in critical periods of ontogenetic development of individuals. The incidence of these indicators is altered in direct relation with the socioeconomic status and, therefore, nutritional condition since the incidence of hypoplasia in people of poor social classes is higher (Goodman et al. 1987; Goodman and Rose 1990). On the other hand, the susceptibility between teeth is differential, being higher in maxillary central incisors than in maxillary canines. All nutritional indicators are closely related to the disease, i.e. causes for the appearance of these indicators in the bones and teeth are due to the synergic effect between nutrition problems (hypoalimentation) and infectious diseases, mainly gastrointestinal.

This indicator was evaluated in 27 individuals in Jaina. All had the injury, which would explain, in turn, the high infant mortality in this group due to malnutrition and health problems. Copan individuals showed a similar situation, since they had feeding problems, or infectious and parasitic diseases, as shown by the results of the percentage analysis of this indicator. Individuals with one (47%), two (47%) or more lines were registered among young men. Forty-eight per cent of women had one line and 35% registered two or more lines. For subadults, 60% had one line and 26% had two or more lines (Márquez Morfín et al. 2002a: 293). Hypoplasia was identified in 71% of cases in Palenque.

12.4.5 Periosteal Reaction

The periosteal reaction is constantly related to infections, although this is not the only cause. These infections can be acute and localized, chronic, or systemic. Periosteal processes are strong indicators of disease endemic status among the population, whether produced by bacterial or mycotic process, by the excessively adverse environment in which individuals live and develop, or by opportunistic diseases that appear when the immune system is depressed by malnutrition (Martin et al. 1991).

The infectious diseases of the pre-Hispanic world include those caused by staphylococci and streptococci, viruses, tuberculosis, treponematosis and fungi (Mensforth et al. 1978, Ubelaker 1982), in addition to the effects of parasites that were probably very common in tropical zones. Some of these diseases were more common among infants and young children than in adults. For example, gastroenteritis is one of the most common diseases in these early ages of pre-industrial societies. How can we know which of these diseases may have affected the ancient settlers and which groups were most at risk? According to the incidence by age groups, we conclude that children and adolescents are the most susceptible. According to the pattern of incidence of spongiosum hyperostosis, the infections found are not related to nutrition problems, as we explained above and correspond to an infectious process, which are impossible to identify specifically. Some researchers have associated the increase in infections with a high demographic density and the type of urban life that causes high unhealthy conditions, as it would be the case of all the groups under analysis (Cohen and Armelagos 1984; Steckel and Rose 2002; Cohen 2007).

12.4.6 Periosteal Reaction in Tibia

In Palenque, 98.9% of evaluable cases have mild periosteal reactions in tibia, with very few severe cases. Taking into account the differences by gender, 32.2% corresponds to women and 39.7% to men. Among those under 15 years, 10.7% have this injury in Palenque, 75.3% in Xcaret, 50% in Jaina and 49.1% in Copán.

12.4.7 Systemic Periosteal Reaction

Systemic infections evaluated in several parts of skeletal remains reached 95.6% in Palenque, 58% in Xcaret, 41.3% in Jaina and 24.6% in Copán. In the first three sites this indicator was registered in its initial or mild stage while in Copán the severe type predominates. The possible interpretation of these results is related to a life in a large city, where living conditions are not adequate, with severe health problems and perhaps a low-protein diet.

12.4.8 Comparison of Results Through Prevalence

One of the most frequent and important problems researchers face about the health of old populations is to make their data comparable, which is a complex task when the series to be compared differ in their age-structure, a very common method in osteological work. The standardization method proposed by Waldron (2016, see chapter 4) allows us to overcome these methodological obstacles. First, it is based on the calculation of specific prevalence rates (as per the acronym in Spanish, TEP) by age and gross prevalence rates (as per the acronym in Spanish, PBR). The latter summarize statistically the impact of a condition on a population. TEPs are obtained by dividing the affected individuals in an age group among the total of individuals that make up the age group under analysis and multiplied by 100 or 1000, depending on the size of the sample, or the scale of the inferences required. When the total of individuals affected by a condition is divided by the total population, a gross rate is obtained, which can also be multiplied by 100 or thousand inhabitants (Waldron 2016).

Table 12.6 shows an example of the calculation of specific prevalence rates for orbitalia cribra per 100 inhabitants, where the highest values are observed in Copán and Palenque in all their age groups. Note the discrepancies between the age-structures of each sample. For the age groups with fewer individuals, as in the case of ≥50 in Jaina and Xcaret, TEPs increase, producing a false image of the impact level of this indicator among older adults. The purpose of the standardised preva-

Table 12.6 Specific and gross prevalence rates for the indicator

Cribra orbitalia						
Copán				Jaina		
Age	*n*	*N*	[a]Rate/10^2	*n*	*N*	[a]Rate/10^2
15–29	17	21	80.95	2	5	40.00
30–49	27	32	84.38	6	17	35.29
50 and +	35	42	83.33	2	2	100.00
	79	95	**83.16**		24	**41.67**
	Proportional rate		**2.00**	Proportional rate		**0.50**
Palenque				Xcaret		
	n	N	[a]Rate/10^2	n	N	[a]Rate/10^2
15–29	10	13	76.92	5	15	33.33
30–49	21	32	65.63	7	18	38.89
50 and +	1	3	33.33	1	2	50.00
	32	48	**66.67**	13	35	**37.14**
	Proportional rate		**1.79**	Proportional rate		**0.56**

Source: Calculations of the authors. *N* = totals of individuals by age group; n: number of individuals with cribra orbitalia for each age group. The specific rates are shown in the third and sixth columns. The gross prevalence rate for both populations was obtained from total n/total N of each population. The relation between both rates indicates the comparative level of prevalence (*proportional rate*)

[a]Specific rate

lences is to obtain several readings (Table 12.6). The standardised proportions rate between the Copán and Jaina gross rates is 2.00, which means that Copán has twice as many individuals affected with orbitalia as Jaina. When reversing the proportion, Jaina has only 0.50 individuals affected by this injury compared to Copán.

To eliminate the effect produced by the differences in the number of individuals by age group in each sample, Waldron (2016: 83-86) suggests applying some method that will make allowances for the different structures of the two groups and produce overall prevalences that are directly comparable. This is achieved by some form of standardization. In direct standardization, the age-specific rates of the populations to be compared are applied in turn to a standard population to produce a standardised rate, sometimes called a comparative mortality figure. There are several potential choices for the standard population: it can be (1) entirely artificial; (2) a real population that may or may not be related to those under study; (3) one of the two groups being studied; or (4) both groups combined. For palaeoepidemiological purposes, the most convenient standard would be a combination of the two assemblages being compared. It should be clear from what has gone before that standardization can be used only to compare two populations; if one wishes to compare the prevalence in more than two, then this has to be done in a pair-wise fashion comparing A with B, A with C, and B with C, for example. This can be rather cumbersome if several prevalences are to be compared, and the results are then usually best presented in the form of a matrix.[4]

The age structures of the series compared for interpretation should be taken into account (Table 12.6). Copán has the best representation of individuals in the three age groups, while the other three series have an evident absence of people ≥50 years, which is a bias for simple percentages. Upon obtaining Copán standardization, all indicators show lower values except for enamel hypoplasia. Xcaret and Palenque have 2.0 and 1.87 times more hypoplasia than Copán (see percentages in Table 12.4). When comparing the behaviour of indicators with Jaina's age-structure, values remain low, unlike Palenque where it is evident that Copán and Jaina have 2.3 times more periosteal reaction in tibia than that registered in Palenque. Systemic injuries are 2.6 times more frequent in Copán when compared to Palenque's age-structure, so that readings of the indicator impact is different from the one discussed when describing the results of percentages (Table 12.4): Xcaret Colonial has a greater presence of spongy hyperostosis and enamel hypoplasias while Jaina and Copán of cribra orbitalia and periosteal reactions (Table 12.7).

12.5 Discussion

The importance of reconstructing the living conditions of population sectors that inhabited different environments and had different social, economic and political organization is to highlight the effect that could have several factors, such as the

[4]The chapter has a page limit set, so we are unable to describe the entire procedure. See Waldron, 2016, Chapters 4 and 5.

Table 12.7 Directly standardised prevalences matrix for biological disruption indicators

Cribra Orbitalia				
	Copán	Jaina	Palenque	Xcaret
Copan	–	1.41	0.64	1.23
Jaina	1.01	–	0.58	1.42
Palenque	1.98	1.46	–	3.16
Xcaret	0.67	0.57	0.36	–
Espongy hyperostosis				
Copán	–	1.16	0.67	0.41
Jaina	1.05	–	0.58	0.46
Palenque	1.97	2.05	–	0.30
Xcaret	2.28	2.17	1.13	–
Enamel hypoplasia in canine				
Copán	–	1.45	1.47	2.00
Jaina	0.55	–	0.73	1.18
Palenque	0.83	1.84	–	1.87
Xcaret	0.46	0.91	0.57	–
Periostitis in tibia				
Copán	–	0.79	0.37	0.47
Jaina	1.00	–	0.37	1.18
Palenque	2.31	2.34	–	1.27
Xcaret	1.87	1.80	0.79	–
Periostitis in other bones				
Copán	–	0.73	0.33	0.48
Jaina	1.44	–	0.51	0.73
Palenque	2.64	1.59	–	1.37
Xcaret	1.95	1.24	0.72	–

Source: Calculations from the authors

demographic density, and its distribution in health, epidemiological and mortality profiles, since public health systems were inadequate and most likely similar regardless of the types of natural and cultural environment (Walker 1986).

The evaluation on the presence of spongy hyperostosis and cribra orbitalia reveals that its frequency is much higher in the two Maya cities studied here. Regardless of the differences in population size and density of Palenque and Copan, the relationship of the presence of these indicators and the natural environment lies in the intensity of transmission of parasites and diseases, which increases in hot and humid climates. On the other hand, the infant diet has great influence on the variety of parasites that can be transmitted, which increases in those associated with marine resources whose diversity is greater and are not part of the most common type of parasites in cities in jungle areas inland. An important finding in assessing the frequency of lesions using the Waldron method was to verify the high prevalence rates of spongiosum hyperostosis in the Xcaret series hidden under an age structure with underreporting in the three age groups considered.

Regarding the hypoplasia lines, Maya populations of Xcaret and Palenque have more affected individuals with this indicator, and almost duplicated in the Copán individuals. Storey and collaborators (2002: 293) performed an analysis for the Xcaret pre-Hispanic series showing that hypoplasia remained low, and visibly increased at the time of contact with Europeans, as seen in the Xcaret Colonial data. Therefore, it can be stated that there was a deterioration in the quality of life of its inhabitants. The feeding habits of the Xcaret (Rodríguez Suárez 2004) and Chac Mool inhabitants (Berriel 2002), another site on the eastern coast, was based on the consumption of marine resources in accordance with the analysis results of trace elements, among which high strontium levels were found associated with plants and marine species. In addition to fish, molluscs and other marine foods, remains of dogs were found on the site, which were used in the diet. Finally, the low frequency of magnesium in bone samples could be associated with low maize consumption (Márquez Morfín and Hernández Espinoza (2006a).

Infections are more frequent in the Palenque series, but individuals from the other Maya centres were not exempt from periosteal reactions, since their percentages were close to 50%. Furthermore, we consider that the explanation of these results consists, in addition to the social, economic and political factors, in the natural environment of each site and in the sanitary and hygienic conditions that foster the proliferation of these diseases. The percentages of infections are lower in Copán despite the fact that population size and demographic density were high. We should bear in mind that Copán was settled in a 24 km^2 valley inhabited by 27,000 individuals in its peak period (Webster et al. 2000), while Palenque, in its golden age, had 5000 inhabitants in a 16 km^2 area (Liendo 2004). Regardless of whether the cities have had urban equipment that included sanitary structures such as drainage, storage wells and garbage dumps, these were not the only determinants. According to our results, the site climate and natural environmental conditions defined the epidemiological profiles of their populations, together with social, economic and political circumstances, which had a differential impact on each society. An important factor when considering the quality of life of those individuals, as mentioned, refers to the socioeconomic and political problems that characterized the populations of Copan and Palenque at the end of the Classic period. Both populations, despite belonging to groups considered as elite, were in a situation of social and political conflict with a breakdown of the control system, which should have impacted the economic aspects, food production and distribution, and the possibility of having sufficient resources (Webster and Freter 1990). In Palenque, individuals buried in the central area and in some of the housing units should have belonged to the upper social stratum. Indicators such as periosteal reactions related to infections are found in high percentages in almost all skeletal remains, although with signs were mild. There were few cases of severe conditions that, according to Weston (2012), should be revaluated in view of new research perspectives, since this injury, in its non-severe forms, can be associated with the recovery of a different morbid process and not necessarily due to an infection. Their presence is related to sanitary conditions and parasitosis all Palenque inhabitants were exposed to almost without exception.

Hernández Espinoza (2006) states in an evaluation of Palenque's demographic profile that one of the serious problems of public hygiene was bodily waste and garbage, which undoubtedly contributed to the virus and bacteria proliferation contaminating water and food, hence the high frequency of periosteal reactions in the skeletal remains evaluated. Cetina Bastida and Sierra Sosa (2005: 667–668) observed in the Xcambó series analysis that infectious diseases show a high frequency. The presence of inactive periostitis in 60% of the cases reveals a strong physiological resistance and survival to the pathological state in an environment characterized by severe conditions of extreme stress during the Classic period. The presence of light periosteal reactions was 47.90% in Xcambó while severe cases were scarce. The frequency of this injury was similar to that found in the Jaina (48%) and Copán (40.80%) settlements of the Classic period (Whittington 1989; Storey 1999).

Palenque presents the highest percentages in almost all the indicators revealing malnutrition and infectious problems. The relation between the spongiosum hyperostosis indicators and periosteal reactions reinforces this approach, since results point to parasitosis and other types of gastrointestinal problems causing that, even though food consumption was adequate, its biological use was inappropriate. Outdoor fecalism, the lack of hygiene in the food preparation, and the consumption of contaminated or in poor condition products, especially in times of drought, are elements that allow us to understand the infectious and nutritional problems. In each of these cases, results are interpreted in an integral manner, taking into account the aforementioned factors. We consider that there were no unique causes, rather it was due to the interaction of the physical environment with the socioeconomic and political problems, responsible for the differences in the quality of life and in the epidemiological and demographic profiles of individuals that inhabited these sites. The wide value variability among pre-Hispanic Maya forces to make specific interpretations for each population, considering comprehensively the different determinants of the living conditions and health. Results show the situations that Maya faced to adapt and survive, which depended on factor multiplicity and multicausality.

References

Benavides Castillo A (2007) Jaina en el contexto de las poblaciones del Clásico en el occidente peninsular. In: Hernández Espinoza PO, Márquez Morfín L (eds) La población prehispánica de Jaina. Osteobiografía de 106 esqueletos. Escuela Nacional de Antropología e Historia-Promep, México, D.F., pp 13–32

Berriel RE (2002) Paleodieta de los mayas de Chac Mool, Quintana Roo. División de Posgrado. Escuela Nacional de Antropología e Historia, México, D. F.

Camacho Martínez MA (2018) Las tendencias de la mortalidad en menores de cinco años en la parroquia del Sagrario, Zacatecas, entre 1835–1865: un estudio de antropología demográfica. División de Posgrado. Escuela Nacional de Antropología e Historia, Ciudad de México

Cetina Bastida A, Sierra Sosa T (2005) Condiciones de vida y nutrición de los antiguos habitantes de Xcambó, Yucatán. Estud Antropol Biol XII:661–678

Cohen MN (2007) Ancient health: skeletal indicators of agricultural and economic intensification. University of Florida Press, Gainsville

Cohen MN, Armelagos GJ (1984) Paleopathology at the origins of agriculture. Academic Press, New York

Con MJ (1991) Informe del Proyecto Xcaret. Cuarta y quinta temporada. Instituto Nacional de Antropología e Historia, México, D. F., pp 1990–1991

Cucina A (2011) Maya subadult mortality and individual physiological frailty: an analysis of infant stress by means of linear enamel hypoplasia. Childhood Past 4:105–116

Cucina A, Tiesler Blos V (2007) Nutrition, lifestyle, and social status of skeletal remains from non-funerary and "problematical" contexts. In: Tiesler Blos V, Cucina A (eds) New perspectives on human sacrifice and ritual body treatments in ancient Maya society. Springer, New York, pp 251–262

Cucina A, Cantillo CP, Sierra Sosa T et al (2011a) Carious lesions and maize consumption among the Prehispanic Maya: an analysis of a coastal community in northern Yucatan. Am J Phys Anthropol 145:560–567

Cucina A, Tiesler Blos V, Sierra Sosa T et al (2011b) Trace-element evidence for foreigners at a Maya port in Northern Yucatan. J Archaeol Sci 38:1878–1885

Danforth ME (1994) Stature change in prehistoric Maya of the Southern Lowlands. Lat Am Antiq 5:206–211

Danforth ME (1997) Late classic Maya health patterns: evidence from enamel microdefects. In: Reed DM (ed) Bones of the Maya studies of ancient skeletons. Smithsonian Institute, Washington, pp 116–125

Danforth ME (1999) Coming up short: stature and nutrition among the ancient Maya of the southern lowlands. Reconstructing Ancient Maya Diet. University of Utah Press, Salt Lake City

Dávalos E, Romano A (1973) Estudio preliminar de los restos osteológicos encontrados en la tumba del Templo de la Inscripciones, Palenque. In: Ruz A (ed) El Templo de las Inscripciones. Instituto Nacional de Antropología e Historia, México, pp 253–254

Faulhaber J (1976) Investigación longitudinal del crecimiento. Instituto Nacional de Antropología e Historia, México D. F.

Frenk J, Bobadilla JL, Stern C et al (1991) Elementos para una teoría de la transición en salud. Salud Publica Mex 33:448–462

Glassman DM, Garber JF (1999) Land use, diet, under effects on the biology of the prehistoric Maya of northern ambergris Caye, Belize. In: White C (ed) Reconstructing ancient Maya diet. University of Utah Press, Salt Lake City, pp 119–132

Goodman AH, Martin DL (2002) Reconstructing health profiles from skeletal remains. In: Steckel RH, Rose JC (eds) The backbone of history. Health and nutrition of the western hemisphere. Cambridge University Press, New York, pp 11–60

Goodman AA, Rose JC (1990) Assessment of systemic physiological perturbations from dental enamel hypoplasias and associated histological structures. Am J Phys Anthropol 33:50–110

Goodman AA, Allen LH, Hernandez GP et al (1987) Prevalence and age at development of enamel hypoplasias in Mexican children. Am J Phys Anthropol 72:7–19

Haviland WA (1967) Stature at Tikal, Guatemala: implications for Ancient Maya Demography and Social Organization. Am Antiq 32:316

Hernández Espinoza PO (2006) La regulación del crecimiento de la población en el México Prehispánico. México: Instituto Nacional de Antropología e Historia

Hernández Espinoza PO, Márquez Morfín L (2007) La población prehispánica de Jaina. Estudio osteobiográfico de 106 esqueletos. Publicaciones de los Cuerpos Académicos. Primera ed. ENAH - INAH – Promep, México, D. F.

Hernández Espinoza PO, Márquez Morfín L (2015) Maya Paleodemographics: what do we know? Am J Hum Biol 27:747–757

Herrera Flores DA, Götz CM (2013) La alimentación de los antiguos mayas de la Península de Yucatán: Consideraciones sobre la identidad y la cuisine en la época prehispánica. Estud Cultura Maya 43:71–98

Liendo Stuardo R (2000) Reyes y campesinos. La población rural de Palenque. Arqueol Mexicana 45:34–37

Liendo R, Vega F (2000) Técnicas agrícolas en el área de Palenque: inferencias para un estudio sobre la organización política de un señorío maya del Clásico. Arqueología. Segunda Época 23:3–25

Liendo Stuardo R (2004) *El paisaje urbano de Palenque: una perspectiva regional*. México: Instituto Nacional de Antropología e Historia / Universidad Nacional Autónoma de México

Livi-Bacci M (2002) Historia mínima de la población mundial. Editorial Ariel, S.A., Barcelona

López Alonso S, Serrano Sánchez C (1997) Los entierros humanos de Jaina, Campeche. In: Malvido E, Pereiya G, Tiesler V (eds) El cuerpo humano y su tratamiento mortuorio, 1st edn. Instituto Nacional de Antropología e Historia, México, D. F., pp 145–160

López Alonso S, Serrano Sánchez C (2007) Estatus social y contexto funerario durante el Clásico, en Jaina, Campeche. In: Hernández Espinoza PO, Márquez Morfín L (eds) La población prehispánica de Jaina. Estudio osteobiográfico de 106 esqueletos, 1st edn. Escuela Nacional de Antropología e Historia, México, D. F., pp 77–110

López Bravo R (1995) El grupo B de Palenque, Chiapas. Una unidad habitacional maya del Clásico Tardío. Dissertation, Escuela Nacional de Antropología e Historia

López Bravo R (2000) La veneración de los ancestros en Palenque. Arqueol Mexicana 45:38–43

Magennis LA (1999) Dietary change of the lowland Maya site of Kichpanha, Belize. In: White CD (ed) Reconstructing ancient Maya diet. University of Utah Press, Salt Lake City, pp 133–150

Márquez Morfín L (1981) Spongy hyperostosis and criba orbitalia in a Maya subadult sample. Paleopathol Newsl 35:13–15

Márquez Morfín L (1982) Distribución de la estatura en colecciones óseas mayas prehispánicas. Estud Antropol Biol I:253–269

Márquez Morfín L (1984) Osario infantil en un Chultun en Chichen Itza. In: SMd A (ed) XVIII Mesa Redonda de Antropología. Sociedad Mexicana de Antropología, San Cristobal de Las Casas, pp 89–103

Márquez Morfín L (1987) Qué sabemos de los mayas peninsulares a partir de us restos óseos. In: México UNA (ed) Primer Coloquio Internacional de Mayistas. Universidad Nacional Autónoma de México, México, D. F., pp 42–59

Márquez Morfín L (1989) Evidencias óseas sobre la dieta maya durante el Postclásico. Segundo Congreso Internacional de Mayistas. Centro de Estudios Mayas, Uiversidad Nacional Autónoma de México, San Cristobal de las Casas

Márquez Morfín L (1991) La dieta maya prehispánica en la costa yucateca. Estud Cultura Maya XVIII:354–394

Márquez Morfín L (1996) Paleoepidemiología en las poblaciones prehispánias mesoamericanas. Las enfermedades calan hasta los huesos. Arqueol Mexicana IV:4–11

Márquez Morfín L (2010) Morir por los dioses …y uno que otro humano. ¿Sacrificio de niños en Chichén Itzá o práctica funeraria? In: Márquez Morfín L (ed) Los niños actores sociales ignorados. Levantando el velo, una mirada al pasado. Escuela Nacional de Antropología e Historia-Promep, México, D. F., pp 253–282

Márquez Morfín L, Del Angel A (1997) Height among prehispanic Maya of the Yucatán peninsula: a reconsideration. In: Whittington SL, Reed DM (eds) Bones of the Maya. Studies of ancient skeletons. Smithsonian Institution Press, Washington, pp 51–61

Márquez Morfín L, Hernández Espinoza PO (2002) Base de datos en SPSS de la series osteológicas mayas. Laboratorio del Posgrado de Antropología Física, División de Posgrado, Escuela Nacional de Antropología e Historia

Márquez Morfín L, Hernández Espinoza PO (2003) Indicadores de salud y nutrición en poblaciones prehispánicas mesoamericanas: la colección prehispánica de Chac Mool, Quintana Roo. Escuela Nacional de Antropología e Historia, México, D. F.

Márquez Morfín L, Hernández Espinoza PO (2004) Health and society among prehispanic Maya from Chac Mool, Quintana Roo, México. Paper presented at 69th annual meeting of the Society of American Archaeology, Montréal, April 1–4

Márquez Morfín L, Hernández Espinoza PO (2006a) La transición al Posclásico y su efecto en la salud, nutrición y condiciones de vida de algunos pobladores de Chac Mool, Quintana Roo. In: Márquez Morfín L, Hernández Espinoza PO, González Licón E (eds) La población maya costera de Chac Mool. Análisis biocultural y dinámica demográfica en el Clásico Terminal y el Posclásico. CONACULTA-INAH-Promep, México, D. F., pp 113–160

Márquez Morfín L, Hernández Espinoza PO (2006b) Los mayas prehispánicos. Balance de salud y nutrición en grupos del Clásico y cl Posclásico. In: Márquez Morfín L, Hernández Espinoza PO (eds) Salud y Sociedad en el México Prehispánico y Colonial. CONACULTA-INAH-Promep, México, D. F., pp 73–102

Márquez Morfín L, Hernández Espinoza PO (2006c) ¿Privilegios en la salud? Testimonio osteológico de un sector de la elite de Palenque. In: Márquez Morfín L, Hernández Espinoza PO (eds) Sociedad y salud en el México Prehispánico y Colonial. CONACULTA-INAH–Promep, México, D. F., pp 265–290

Márquez Morfín L, Hernández Espinoza PO (2013) Los mayas del Clásico Tardío y Terminal. Una propuesta acerca de la dinámica demográfica de algunos grupos prehispánicos: Jaina, Palenque y Copán. Estud Cultura Maya 42:55–86

Márquez Morfín L, Hernández Espinoza PO (2016) La esperanza de vida en la Ciudad de México (siglos XVI a XIX). Rev Secuencia 96:6–44

Márquez Morfín L, Jaén MT (1997) Una propuesta metodológica para el estudio de la salud y nutrición de poblaciones antiguas. Estud Antropol Biol VIII:47–63

Márquez Morfín L, Peraza ME, Miranda T et al (1982) Playa del Carmen: una población de la costa oriental en el Postclásico (un estudio osteológico). Instituto Nacional de Antropología e Historia, México, D. F.

Márquez Morfín L, Hernández Espinoza PO, González Licón E (2001) La salud en las grandes urbes prehispánicas. Estud Antropol Biol X:291–313

Márquez Morfín L, Hernández Espinoza PO, Gómez Ortíz A (2002a) La población urbana de Palenque durante el Clásico Tardío. In: Tiesler Blos V, Cobos R, Robertson MG (eds) La organización social entre los mayas, 1st edn. Instituto Nacional de Antropología e Historia-Universidad Autónoma de Yucatán, México, D. F., pp 13–34

Márquez Morfín L, Jaén Esquivel MT, Jiménez López JC (2002b) Impacto biológico de la colonización en Yucatán. La población de Xcaret, Quintana Roo, México. Antropol Física Latinoamericana 3:25–42

Martin DL, Goodman AH, Armelagos GJ et al (1991) Black Mesa Anasazi health: reconstructing life from patterns of death and disease. Southern Illinois Univeristy, Carbondale

Masur LJ (2011) Stature trends in ancient Maya populations: re-examining studies from Tikal and altar de Sacrificios. Totem: the University of Western Ontario. J Anthropol 17:13–22

Meindl RS (2003) Current methodlogical issues in the study of prehistoric demography. Estud Antropol Biol XI:679–692

Meindl RS, Russell KF (1998) Recent advances in method and theory in paleodemography. Annu Rev Anthropol 27:375–399

Meindl RS, Mensforth RP, Lovejoy OC (2008) Comentarios sobre los principales errores del trabajo paleodemográfico: el cálculo de la mortalidad promedio, la estructura por edad y la tasa de crecimiento anual. Un ejemplo del Ohio prehistórico, en Estados Unidos. In: Hernández Espinoza PO, Márquez Morfín L, González Licón E (eds) Tendencias Actuales de la Bioarqueología en México, 1st edn. Escuela Nacional de Antropología e Historia–Promep, México, D. F., pp 15–36

Mensforth RP, Lovejoy OC, Lallo JW et al (1978) The role of the constitutional factors, diet and infectious disease on the etiology of Porotic hyperostosis and periosteal reactions in prehistoric infants and children. Med Anthropol 2:1–59

Ortega Muñoz A (2007) Los mayas prehispánicos de El Meco. La vida, la muerte y la salud en la costa oriental de la península de Yucatán. Instituto Nacional de Antropología e Historia, México, D. F.

Peña Reyes ME, Hernández Espinoza PO, Márquez Morfín L (2007) Estatus de crecimiento y condiciones de salud en los niños Jaina. In: Hernández Espinoza PO, Márquez Morfín L (eds) La población prehispánica de Jaina. Estudio osteobiográfico de 106 esqueletos. ENAH - INAH – Promep, México, D. F., pp 139–164

Piña Chan R (1968) Jaina. La casa en el agua. Instituto Nacional de Antropología e Historia, México, D. F.

Rodríguez Suárez R (2004) Paleonutrición de poblaciones extinguidas en Mesoamérica y Las antillas: Xcaret y el Occidente de Cuba. División de Posgrado. Escuela Nacional de Antropología e Historia, México, D. F.

Ruz Luihllier A (1958) Exploraciones arqueológicas en Palenque: 1953–1956. Anales del Instituto Nacional de Antropología e. Historia 10:69–299

Ruz Luihllier A (1973) El Templo de las Inscripciones, Palenque. Instituto Nacional de Antropología e Historia, México, D. F.

Saul FP (1972) The skeletal remains from altar de Sacrificios. An osteobiographic analysis. Harvard University, Cambridge

Schrimshaw NS, C.E. Taylor y JE. Gordon (1968) *Interactions of Nutrition and Infection*. Ginebra: World Health Organization

Scrimshaw NS (2000) Infection and nutrition: synergistic interactions. En: The Cambrisge World History of Food. Kenneth F. Kiple and K. C. Ornelas, eds. pp. 1397–1411. Cambridge, UK: Cambridge University Press

Steckel RH, Rose JC (2002) The backbone of history. Health and nutrition in the western hemisphere. Cambridge University Press, New York

Storey R (1992) The children of Copan: issues in paleopathology and paleodemography. Anc Mesoam 3:161–167

Storey R (1999) Late classic nutrition and skeletal indicators at Copan, Honduras. In: White C (ed) Reconstructing ancient Maya diet. University of Utah Press, Salt Lake City, pp 169–182

Storey R, Márquez Morfín L, Smith V (2002) Social disruption and the Maya civilization. In: Rose J, Steckel R (eds) The backbone of history. Health and nutrition in the western hemisphere. Cambridge University Press, Cambridge, pp 286–306

Stuart-Macadam, Patricia y Susan Kent, eds. (1992) *Diet, demography and disease. Changing perspectives in anemia*. New York: Aldine de Gruyter.

Stuart-Macadam, Patricia (1995) Breastfeeding in Prehistory. In: *Breastfeeding. Biocultural Perspectives*. Patricia Stuart-Macadam and Katherine A. Dettwyler, eds. pp. 75–100. Evolutionary Foundations of Human Behavior. New York: Aldine de Gruyter.

Tiesler Blos V (2001) La estatura entre los mayas prehispánicos. Consideraciones bioculturales. Estud Antropol Biol X:257–273

Tiesler Blos V, Cucina A (2004) Janaab' Pakal de Palenque. Vuda y muerte de un gobernante Maya. Universidad Nacional Autónoma de México - Universidad Autónoma de Yucatán, México, D. F.

Tiesler Bloss V, Cucina A (2007) New perspectives on human sacrifice and ritual body treatments in ancient Maya society. Springer, New York

Tiesler Bloss V, Cucina A, Sierra Sosa T et al (2005) Comercio, dinámicas biosociales y estructura poblacional de asentamiento costero de Xcambo, Yucatan. Los investigadores de la Cultura Maya 13(II):365–372

Ubelaker D (1982) The development of American paleopathology. In: Spencer F (ed) A history of American physical anthropology 1930–1980. Academic Press, New York, pp 337–351

Waldron T (2016) Paleoepidemiology. The measure of disease in the human past. Routledge, New York

Walker PL (1986) Porotic hyperostosis in a marine-dependent California Indians. Am J Phys Anthropol 69:345–354

Webster D, Freter A (1990) Demography of late Clasic Copán. In: Culber PT, Rice D (eds) Precolumbian population history in the Maya lowlands. University of Nuevo Mexico Press, Albuquerque, pp 37–61

Webster D, Freter AC, Golin N (2000) Copan. The rise and fall of an ancient Maya kingdom. Wadswoth Group—Thomson Learning, Belmont
Weiss KW (1973) Demographic models for anthropology. Society for American Archaeology Memoir No. 27, Washington
Weston DA (2012) Nonspecific infection in paleopathology: interpreting periosteal reactions. In: Grauer AL (ed) A companion to paleopathology. Wiley, West Sussex, pp 492–512
Whittington SL (1989) Characteristics of demography and disease in low status Maya from classic period Copan, Honduras. Dissertation. The Pennsylvania State University
Whittington SL, Reed DM (1997) Commoner diet at Copan: insights from stable isotopes and porotic hyperostosis. In: Whittington SL, Reed DM (eds) Bones of the Maya. Smithsonian Institution Press, Washington, pp 157–170
Wright LE (1997) Ecology or society? Paleodiet and the collapse of the pasión Maya lowlands. In: Whittington SL, Reed DM (eds) Bones of the Maya. Studies of ancient skeletons. Smithsonian Institution Press, Washington, pp 181–195
Wright LE, White CD (1996) Human biology in the classic Maya collapse: evidence from paleopathology and paleodiet. J World Prehistory 10:147–198

Chapter 13
Crossing the Threshold of Modern Life: Comparing Disease Patterns Between Two Documented Urban Cemetery Series from the City of Mérida, Yucatán, Mexico

Vera Tiesler, Julio Roberto Chi-Keb, and Allan Ortega Muñoz

Abstract This study compares two cemetery series together with their civil records from the city of Mérida, Yucatán, Mexico, which spotlight changes in lifestyle, life expectancy, and health during the twentieth century. To this end, we scored health indications in a skeletal series from the Central Cemetery of Mérida ($N = 104$; collected during the beginning of last century), and a recent population from the Xoclán Cemetery of Mérida, collected last decade ($N = 174$). The latter materializes living conditions towards and during the turn of the twenty-first century. The records under study include basic life and socioeconomic information, obtained from the civil records, along with skeletal data of age at death, sex, benign tumors, nonspecific stress markers, arthritis, and osteopenia. Our results, once age-corrected, indicate a rise in almost all analyzed indications towards the turn of the present century, which we will discuss in terms of pharmaceutical advances, public sanitation and longevity, changes in lifestyle and nutrition.

13.1 Introduction

There is a growing need in our globalized academic world of biological anthropology to systematize, standardize, and make available information about identified modern skeletal collections from different areas of the world, which are not only important for developing and testing osteological methods but also provide references for international forensic identification and medical work. In these efforts, detailed and validated background information on the individuals that make up each

V. Tiesler (✉) · J. R. Chi-Keb
Laboratorio de Bioarqueología e Histomorfología, Facultad de Ciencias Antropológicas, Universidad Autónoma de Yucatán, Mérida, Yucatán, México

A. Ortega Muñoz
Department of Physical Anthropology, Centro INAH Quintana Roo, Instituto Nacional de Antropología e Historia, Chetumal, Quintana Roo, México

H. Azcorra, F. Dickinson (eds.), *Culture, Environment and Health in the Yucatan Peninsula*, https://doi.org/10.1007/978-3-030-27001-8_13

collection is crucial, as only then, the skeletal data can be used as reliable references. Unfortunately, there are still very few identified modern reference collections curated in Mexico (see Gómez-Valdés et al. 2011, 2012; Menéndez et al. 2011; Talavera et al. 2006) and none of them stems from the vast stretches of southeastern Mexico. It is therefore the purpose of this communication to present to the scientific community a new and growing modern human skeletal series from this part of Mexico, namely the municipal cemetery of Xoclán, in Mérida, capital of the state of Yucatán, Mexico. This is the urban capital of some two million mostly Maya and mestizo inhabitants of the state of Yucatán, Mexico.

This study compares the civil records and skeletal stress indicators in still another documented cemetery population from the city of Mérida, Yucatán. The earlier series one comes from the Central Cemetery of Mérida (CeM; $N = 104$) and was collected by Harvard University during 1927 and materializes the living conditions of locals prior to 1925 (Tiesler et al. 2015; Fig. 13.1). The more recent cemetery population from the Xoclán Cemetery (XoM) of Mérida, collected between 2003 and to date ($N = 174$). The latter materializes the conditions towards and during the turn of the twenty-first century. By putting the observed trends into context, we spotlight the shifts of lifestyle, life expectancy, and health problems in Mérida, Yucatán, showcasing the benefits but also the historically persistent negative heath burdens of industrialization, coupled with modernization and globalization in Latin America.

Fig. 13.1 Museum facilities of the Peabody Museum, Harvard University, Cambridge (photo, Laboratorio de Bioarqueología e Histología, UADY)

Fig. 13.2 Recovery of the burial series from the Xoclán cemetery (photo, Laboratorio de Bioarqueología e Histología, UADY)

The skeletal collection which anchors this contribution is being collected, inventoried, and housed at the School of Anthropological Sciences of the Autonomous University of Yucatán, at Mérida, Mexico (Fig. 13.2). The specific information on each individual was obtained from the municipal civil records of Mérida. Subsequent population profiles were elaborated using the 12th national and municipal census from the year 2000 and the second national population counts from 2005 (INEGI 2013a, b [http://www.beta.inegi.org.mx/temas/mortalidad/]). For this chapter, we profiled the cause of death, provenience, sex and age at death, and the social background of the individuals that make up the series. We discuss the representativeness of the skeletal cohort and validate its specific uses as a resource for ongoing and prospective skeletal research.

13.2 Validating the Two Burial Series

The modern skeletal series comes from the urban municipal Cemetery of Xoclán, Mérida, Yucatán, Mexico. It was recollected during six recovery periods between 2003 and 2018 after signing an inter-institutional agreement between the UADY and the Municipal Government. It is made up of 192 mostly well-preserved skeletons with individualized information on names, last residences, genders, causes of and ages at death. The age and sex information of the civil records was validated with osteological discriminant function analyses in those individuals whose preservation and adult age allowed application. The socioeconomic status, as inferred from the neighborhoods of last residence (INAEGI n/d), concurs with the poorer segments of urban population although the average income inferred for those who make up the series is similar to the average income level of the state of Yucatán. The individuals that constitute this documented skeletal series were born between 1900 and 1990 and died between 1995 and 2010, with a peak in the year 2001. The great majority comes from the town itself and its surrounding rural areas in Mexico's Yucatán peninsula.

The individuals that make up the second, historic collection, died some 80 years before those of the modern series and represent the living conditions during and past the years of the Caste War in the Yucatán (Patch 1991; Reed 2001). There is no systematic individual record of these individuals, whose skeletons were collected by Harvard University during 1927. To overcome this lack of information, we recorded systematically Mérida's historical records of 800 individuals dying during the early years of the twenties of last century. Both sets of records, i.e., the entries collected at the civil agency from the years of death of the historic and the modern skeletal collection, include basic life and socioeconomic information.

In a subsequent step, we compared these entries to the skeletal data. To validate the reliability and accuracy of the modern municipal civil registry, the information on sex and age at death was tested osteologically in each skeleton, using conventional osteological techniques and measurements of the pelvic bone as described in Buikstra and Ubelaker (1994) and Bruzek (2002). This validation process had the sole purpose of endorsing the information of the civil registry and was confirmed largely by validating osteological discriminant function analyses in those individuals whose preservation and adult age allowed application (Bruzek 2002), except for two discrepancies in the 80+ years of age-at-death cohort, which might be attributable to senile bone deformation. Skeletal age estimates were obtained using the auricular surfaces (Lovejoy et al. 1985) and the pubic symphyses of adults (Brooks and Suchey 1990). The age at death of the only subadult with preserved teeth was determined by stage of dental maturation, using the taxonomy described in Ubelaker (1989). The remaining 13 skeletons were either too deteriorated to allow scoring or presented a fused pelvic girdle. A total of 73 adult individuals allowed metric sex discrimination in the pelvic bone, including three skeletons that were too deteriorated to allow age determination. The skeletal data were entered on a separate sheet in the database that holds the information on each individual, as it was retrieved from the municipal registry of Mérida.

In the case of the adult skeletons that make up the second, historic collection, we proceeded analogously, although in this case, no direct confrontation with the civil registry was possible, because no personal information accompanied the skeletons curated at the Peabody Museum. Apart from sex and age-at-death estimations, we recorded and compared among the two collections, three nonspecific stress markers: porotic hyperostosis in the skull vault and in the orbital roofs, periostoses in lower extremities, and the frequencies of depression fractures in the frontal bone and appendicular diaphyses. For the purposes of this study, we used conventional scoring techniques, as described in Buikstra and Ubelaker (1994), Ortner (2003: 206–207), and Schultz (1988).

13.3 Age-at-Death Profiles

Despite the 80 years that lie between the recording of the two sets of civil records (the twenties of last century versus the two first decades of the twenty first century), both death entries include equivalent basic life and socioeconomic information, which we were able to compare to the skeletal data of age at death, sex, nonspecific stress markers, arthritis, and osteopenia. In the following, we present the ages at death and causes of death among the civil records that correlated the cemetery registry. A noticeable increase in ages at death and shifts in causes of death is evident when the historic registry is compared to the modern one from 1990–1999 (Figs. 13.3 and 13.4; Chi-Keb et al. 2013). The overall preservation of the remains is good, although there is variation in the degree of preservation of the bone surfaces. Some appear eroded, due either to diagenesis or specific curation procedures in our protocol, namely disinfecting the remains with diluted chloride and peroxide. Some of the front teeth fell out of the sockets of some individuals during decomposition and could not be retrieved during recollection.

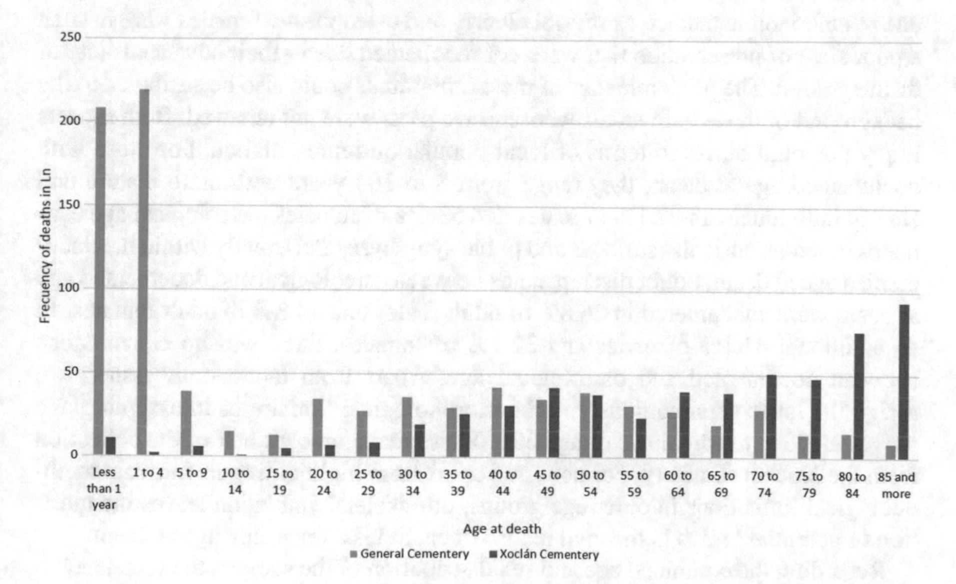

Fig. 13.3 Distribution of recorded ages at death by total of deaths; (**a**) municipal registry of Mérida, Yucatán 1924–1925, in light shade; (**b**) survey of recent distribution of ages at death (1990–1999, in dark shade)

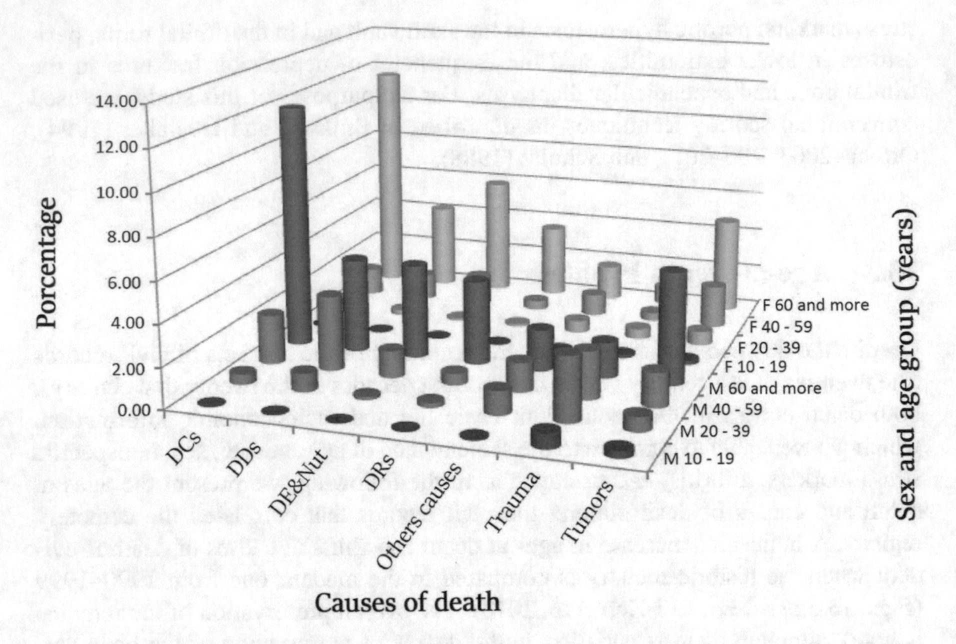

Fig. 13.4 Distribution of recorded causes of death among the males and females of the Xoclán skeletal series according to age group

The skeletal collection is composed of 110 males and 49 females. The under-representation of the female segment (the sex ratio is 2:1 males vs. females) may be due to taphonomic factors, as several elderly and osteoporotic females were in such a poor state of preservation that were not recollected due to their advanced state of disintegration. The predominance of male individuals could also be attributed to the background of those individuals whose grave plots were not renewed. Both aspects imply potential biases in terms of local population representation. For those with documented age at death, they range from 8 to 104 years with more mature and elderly individuals. In 32.1% of males and 53.6% of females, osteological age estimates based on auricular surfaces and pubic symphyses fall exactly within the documented age at death. Slight discrepancies between osteological and documented age at death were encountered in 26.8% of adult males and 14.3% of adult females. In an additional 41.1% of males and 32.1% of females, there was no concordance between documented and osteological age. Apart from taphonomic issues, we assign the latter parsimoniously to the morphological limitations in assigning age ranges to elderly individuals of age 60+. These make up over half of the collection from the Xoclán cemetery. For now, and apart from highlighting its intrinsic methodological limitations in older age groups, our skeletal validation leaves the question of potential biases in the civil record open, at least regarding age at death.

Regarding the combined age and sex distribution of the sample, the female average age at death exceeds the male average. Also, the median values are dissimilar when both sexes are compared (Tables 13.1 and 13.2). Age at death is more uni-

Table 13.1 Chi-square values between sexes by causes of death

	1990–1999	2000–2008
Causes of death	Chi-square	Chi-square
DCs	94.05*	154.10*
DDs	352.31*	644.69*
DE&Nut	266.73*	95.08*
DRs	29.12*	25.56*
Others causes	9.66*	106.54*
Trauma	1455.27*	1663.12*
Tumors	79.19*	56.12*

DCs Diseases of the circulatory system, *DDs* Diseases of the digestive system, *DE&Nut* Diseases of the endocrine glands, nutrition, metabolism and disorders of immunity, *DRs* Diseases of the respiratory system
*$P > 7.815$ $a = 0.005$ (gl. = 3)

Table 13.2 Chi-square values between sexes and decades (1990–1998 and 2000–2008) by causes of death and age group

Age group (years)	Causes death	Chi-square	Age group (years)	Causes of death	Chi-square
10–19	DCs	1.76	20–39	DCs	31.08*
	DDs	1.05		DDs	268.09*
	DE&Nut	3.96*		DE&Nut	124.14*
	DRs	2.39		DRs	20.62*
	Others causes	4.48*		Others causes	75.95*
	Trauma	291.31*		Trauma	1608.15*
	Tumors	15.03*		Tumors	9.36*
40–59	DCs	202.28*	≥60	DCs	13.03*
	DDs	724.00*		DDs	3.87*
	DE&Nut	9.01*		DE&Nut	224.71*
	DRs	12.37*		DRs	19.30*
	Others causes	29.16*		Others causes	6.62*
	Trauma	1018.92*		Trauma	200.01*
	Tumors	96.66*		Tumors	14.25*

DCs Diseases of the circulatory system, *DDs* Diseases of the digestive system, *DE&Nut* Diseases of the endocrine glands, nutrition, metabolism and disorders of immunity, *DRs* Diseases of the respiratory system
*$P > 3.841$ $a = 0.005$ (gl. = 1)

formly represented in the male cohort, although there is an absence of individuals below 14 years and between 40–44 and 50–59 years. Conversely, the female cohort includes subadults in the 5–9-year-old category, but a relative absence of individuals in the young, middle, and mature adult groups (15–19, 25–34, and 50–64 years) is noteworthy.

All individuals were born between 1900 and 1990. The cohort's mode corresponds to the decade surrounding 1922 and its combined median to the year 1929.

Sex differences are apparent despite the small sample sizes. Overall, women were born years before their male counterparts, with men dying at a younger age than women. The difference is apparent when comparing median years of birth, which indicate that half of the men were born by 1931, while half of the women were born by the year 1926, 5 years before the male median. Regarding the year of death, the individuals that make up the skeletal series died between 1995 and 2004, with a peak in the year 2001 and an average and median of 2001 and 2002, respectively. When the years of death are compared between the sexes, there are minor or no discrepancies.

When the mortality profiles of the skeletal collection are compared to contemporary mortality statistics from 1995–2007 from urban Mérida, 5–14, 35–39, and 45–49-year-old individuals are overrepresented in the skeletal sample. The sample interred in the cemetery of Xoclán represents the 0.06% to the 0.49% of all municipal deaths per year from 1995 to 2007. Eight age groups of the collected cohort fit closely to the corresponding municipal mortality profile. Its overall pattern indicates a low mortality rate in Mérida, and a relatively high life expectancy, which in 2007 averages 72.6 and 77.4 years, respectively, for Yucatecan men and women (Consejo Nacional de Población 2013). The skeletal sample shows a similar tendency, especially in the older age group, although it is noteworthy that the younger age groups are underrepresented in the cemetery cohorts. In general, considering all of the above and considering the limitations imposed by the fact that unclaimed individuals compose the collection, the skeletal series is still likely much representative of the age-at-death distribution of present-day Meridians, at least related to specific sectors of the society.

13.4 Stress Markers

Despite the biases introduced by longer life spans in recently deceased Yucatecans from our series, our comparison of three nonspecific stress markers between the two collections is quite revealing. The frequencies of porotic hyperostosis (in the skull vault and in the orbital roofs) are generally related to childhood anemia—whether due to malnutrition, gastrointestinal ailments, or a combination of nonspecific factors (Macadam 1989). The frequencies of porotic hyperostosis were slightly lower on average in the historical CeM collection, with 41.17%, when compared to the 44.68% presence in the XoM series (Fig. 13.5a). Because the development of this condition is limited to subadult age, the almost exclusively adult population represented in both series suggests that any biases related to the differences between average ages at death should be negligible.

The second set of nonspecific skeletal stress markers relates to the periostosis/osteomyelitis complex, which identifies a range of infectious disorders and/or healed hemorrhagic episodes, which lead to pathological remodeling and accumulations of bone on top of the diaphyseal shafts. The frequencies of the lesioned segments that are related to this condition are similarly predominant among the

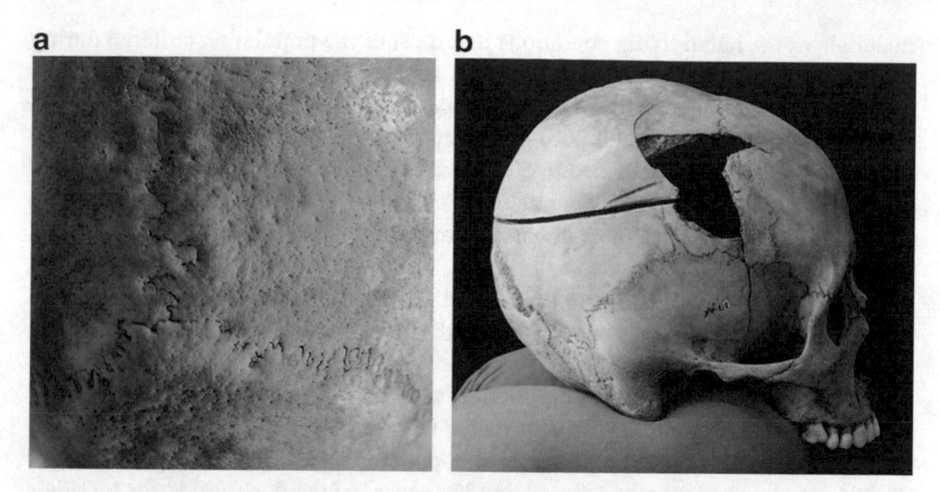

Fig. 13.5 (**a**) Healed porotic hyperostosis in an adult individual from the Xoclán cemetery, (**b**) unhealed fracture on left side of an autopsied adult neurocranium (photo, Laboratorio de Bioarqueología e Histología, UADY)

skeletonized modern deceased and their historic counterparts, given their percentual ratios of 30.20% [XoM] versus 30% [CeM] of affected individuals (Fig. 13.4b).

As regards trauma in bone, these may express interpersonal violence, accidents (and specifically vehicle accidents), and surgical (mal)practice. Healed or unhealed trauma frequencies per preserved anatomical segment were compared between the Xoclán series and the set of skeletons recollected during the first part of the twentieth century (N = 104). In the group of complete frontal segments from XoM and CeM, used in this study, show much lower frequencies of healed trauma on the frontal bone among the historic cemetery population [5.88%], when compared to the higher rates in the XoM series [60%] and especially the male segment (Fig. 13.4b). In the cemetery series from the early twentieth century, lesions were scored in only very few of the long bones of arms (4 percent of which show evidence of trauma; $N = 99$) and legs (with 2 percent of visible trauma; $N = 100$); these frequencies constitute only a tenth to a quarter of those displayed among the modern population from Xoclán (Fig. 13.5b).

13.5 Discussion and Conclusions

At less than a century, the comparison between both samples demonstrates a sharp rise in life expectancy, as shown by the shift in age at death. The individuals commissioned during 1927 by Harvard's, had died some 80 years before those of the modern series and come to materialize a pre-antibiotic, pre-automobilist lifestyle (Tiesler et al. 2015; see also Cova 2010; Milner 2013). In the local context of Mérida and the Yucatán, the low average of age at death among these remains most probably materializes the harsh living conditions that most of the population suffered during the aftermath of the so-called Caste War in Yucatán (1847-1901), which had initiated with a revolt of native Maya people against the privileged, European-descended sectors of local society (Reed 2001). This conflict was officially terminated at the beginning of the twentieth century, but its continued effects were suffered by exploited and poor indigenous populations subsisting across the Yucatán, which in this study made us expect signs of harsh living conditions among the skeletal series from this time (Fig. 13.6).

Although we cannot compare the ages at death of the historical collection with those of the present series we do know that the increase in lived years is the consequence of improved sanitary conditions and advances in the pharmaceutical care, namely the introduction of antibiotics, which have been available locally since many decades. This trend is partly attributed to larger life spans and a higher degree of surgical intervention among the recent population from Xoclán. Important shifts in time are also evident when comparing the causes of death recorded for Mérida's population still during the twenties of last century (high on infectious diseases), and today's death causes, among which degenerative and especially metabolic diseases predominate. Nine percent are attributed directly to diabetes in the modern death

Fig. 13.6 Late sisal plantation from Yucatán, the production of which included forced labor of the indigenous Maya population in Yucatán (photo, Fototeca Guerra, UADY)

entries. Unfortunately, at present there are still few systematic studies on the specific effects that this chronic metabolic disorder has on the skeleton (Brickley and Ives 2008; Waldron 2009). However, the frequent occurrence of diffuse inflammatory ossifications of connective tissues among the elderly from the Xoclán cemetery and the high ratio of amputated lower legs do invite a broader systematic coverage of this series and others in view of a more factual and diachronic evaluation of this metabolic collective killer (Fig. 13.7).

Accidents and violence in the statistics are expressed in the skeletal series in the form of high-impact (automobilistic) and weapon-related trauma. Therefore, the series from Xoclán appears to materialize some of the activity trends and shifts in life style that have been experienced by local Yucatecans during the second half of past century. From the high rate and the distinctive frequencies and lesion types between the sexes, we infer that trauma distribution must mirror broader social and logistical aspects of emerging modern life styles, especially the way of living sustained among the poorer urban sectors. In fact, the frequencies recorded in the skeletal collection from Xoclán roughly echoes the age and sex distributions in trauma-related mortality, which during the decade of 1990 to 1999 appears as the main cause of death among males between 10 and 39 years (but not among analogously aged females).

Fig. 13.7 Bottled, artificially sweetened soft drinks, offered to the urban population of Mérida (photo, Laboratorio de Bioarqueología e Histología, UADY)

In older males, degenerative disease and tumor-related deaths gained in importance similar to that of women. It is of note that past the age of 60, females were more prone to trauma-related deaths (0.31% to 0.86%, probably related to osteopenia), while elderly males retain a stable specific mortality rate of 1.92% (below 40–60) and 1.65% (over 60), respectively. Similar trauma-related mortality rates have been recorded for the first decade of the twenty-first century indeed.

So, what do we learn from confronting biographic data and skeletal remains at a distance of some 80 years? If the conditions documented in the earlier series still echo those of colonial and pre-Hispanic Yucatecan locals (Tiesler et al. 2012, 2015), the present populations have changed dramatically both in terms of age at death and in a host of degenerative, metabolic, and traumatic damage, which we were unable to document in the earlier cohort. Erosive arthroses become prominent, including younger age groups. These conditions stand most probably in direct relationship with a more sedentary lifestyle and automobile transport, a longevity induced pharmaceutically and surgically, together with an unfavorable nutritional regimen highlighting the dark side of globalization. The latter marks the effects of a last nutritional transition, a "food globalization" mediated culturally, socially, and economically, without translating automatically into improved dietary conditions for all. Cheap, highly processed products have flooded the food market, replacing the former local produce (Fig. 13.8). Direct or indirect consequences of this new form of malnutrition are elevated rates of obesity, diabetes, and metabolic syndrome among the population of Yucatán, including minors. Sadly, only recently has awareness grown regarding the damage caused by what is commercialized by the food industry as "prestigious goods," such as soda beverages, processed bread and cookies.

In concluding, we may say that the trends shown in this study bring out into the open that humankind is striding new territories in terms of collective diet, demography, and health. These do not necessarily imply improvements for all, especially when life styles and food regiments are called upon. This at least comes to mind after reviewing the century-wide gap in life span and well-being among the disadvantaged of Mérida and the Yucatán. Although global technical and scientific progress percolates economically, socially, and culturally, it does not automatically

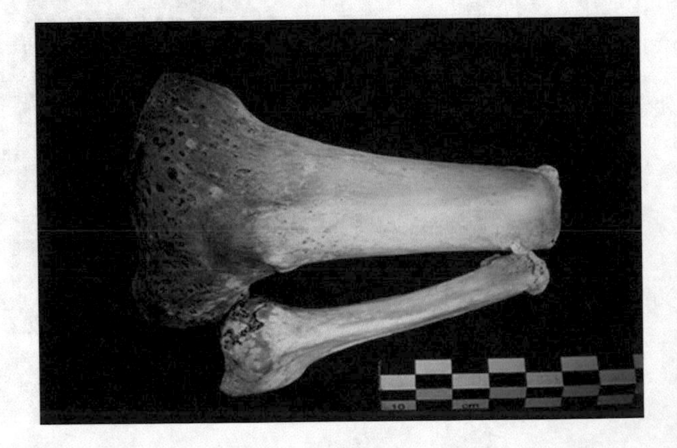

Fig. 13.8 Healed stump of an amputated lower leg of a terminally diabetic individual, Xoclán collection (photo, Laboratorio de Bioarqueología e Histología, UADY)

improve the living conditions for all. This dark side of globalization explains recent dietetic shifts towards highly processed, industrial foods. Like never before, the present dietary "revolution" has alienated humanity from its evolutionarily roots, dietary needs, and habits, mostly at the expense of the poor (Lindeberg 2010). Much inverted from the past is today's relationship between socioeconomic sector and body weight. Most visible are the effects on the urban poor and uneducated, like many a *Meridano*, whose daily dietary intake is made up largely of prepackaged snacks. Junk food is sold and bought over the counter, mistaken by the consumer as a cheap but safe and healthy choice that replaces traditional nourishment, just as artificially sweetened soft drinks are accepted as an alternative to water. As for the Yucatán Peninsula with its vast fertile stretches of semi-tropical lands and gardens, its fish-laden coastlines, we may ask ourselves if the industrially imposed "dietary revolution" is really necessary? In view of a more historically reinforced awareness, we hope that factual surveys like this one may prepare the grounds of a safer crossing, without the grave dietary and health tokens we pay as part of our modern lives.

Acknowledgements Funding was received through the projects "Demografía y enfermedad de los antiguos habitantes de Yucatán a través del análisis histomorfológico de sus restos óseos" (CONACYT, no. 37743-H), and "Nuevas referencias histomorfométricas sobre edad a la muerte, morbilidad y condiciones de vida entre los antiguos mayas (CONACYT, no. 49982, no. 152105), along with UC MEXUS Colaborative Grant VT/KT 2015-2017). We are indebted to our kind colleagues of Harvard University's Peabody Museum for providing access and help throughout the study of Mérida's historic cemetery collection from the *Cementerio Central*, to the Facultad de Ciencias Antropológicas de la Universidad Autónoma de Yucatán, and the *Dirección de Panteones y Cementerios* de la Ciudad de Mérida, Yucatán (municipal government), for lending continued logistical and financial support for our recovery and recording efforts at the Xoclán cemetery. Collection and curation of the skeletal series was made possible by an agreement between the Autonomous University of Yucatán and the Municipal Government of Mérida (represented by Manuel Jesús Fuentes Alcocer and Leandro Martínez García). The support we received from Dr. Christopher Goetz, Araceli Hurtado, Margarita Valencia, Raúl López Pérez and Saúl Chay Vela whose initial study of the Xoclán collection has set the pace of our present endeavors, to the cemetery keepers was instrumental in collecting and identifying the skeletal individuals.

References

Brickley M, Ives R (2008) The bioarchaeology of metabolic bone disease. Academic Press, Oxford

Brooks ST, Suchey JM (1990) Skeletal age determination based on the os pubis: a comparison of the Ascari-Nemeskéri and Suchey Brooks methods. Hum Evol 5:227–238

Bruzek J (2002) A method for visual determination of sex, using the human hip bone. Am J Phys Anthropol 117:158–167

Buikstra JE, Ubelaker D (1994) Standards for data collection form human skeletal remains. Arkansas Archaeological Survey Research Series no. 44, Fayetteville

Chi-Keb J et al (2013) A new reference collection of documented human skeletons from Mérida, Yucatan, Mexico. Homo 64(5):366–376

Consejo Nacional de Población (2013) Indicadores demográficos básicos 1990–2010, México http://www.conapo.gob.mx/es/CONAPO/De_las_Entidades_Federativas_1990-2010. Accessed 9 May 2010

Cova C (2010) Cultural patterns of trauma among 19th-century-born males in cadaver collections. Am Anthropol 112:589–606

Gómez-Valdés J et al (2011) Discriminant function analysis for sex assessment in pelvic girdle bones: sample of the contemporary Mexican population. J Forensic Sci 56:297–301

Gómez-Valdés J et al (2012) Comparison of methods to determine sex by evaluating the greater sciatic notch: visual, angular and geometric morphometrics. Forensic Sci Int 221:156.e1–156.e7

INEGI 2013a XII Censo de Población y Vivienda de México (México), Instituto Nacional de Estadística y Geografía http://www.inegi.org.mx/est/contenidos/Proyectos/ccpv/cpv2000/. Accessed 9 May 2010

INEGI 2013b. II Conteo de Población y Vivienda de México, Instituto Nacional de Estadística y Geografía http://www.inegi.org.mx/est/contenidos/proyectos/ccpv/cpv2005/. Accessed 9 May 2010

Lindeberg S (2010) Food and western disease. Health and nutrition from an evolutionary perspective. Wiley-Blackwell, New York

Lovejoy C et al (1985) Chronological metamorphosis of the auricular surface of the ilium: a new method for the determination of adult skeletal age-at-death. Am J Phys Anthropol 68:15–28

Menéndez A, Gómez J, Sánchez G (2011) Comparación de ecuaciones de regresión lineal para estimar estatura en restos óseos humanos en población mexicana. Antropo 25:11–21

Milner G (2013) Trauma in the medieval to early modern Sortebrode from Odense, Denmark. In: Lozada M, O'Donnabh B (eds) The dead tell tales. Essays in honor of Jane E. Buikstra. Cotsen Institute of Archaeology Press, University of California, Los Angeles, pp 172–185

Ortner DJ (2003) Identification of pathological conditions in human skeletal remains. Smithsonian Institution Press, Washington, D.C.

Patch RW (1991) Decolonization, the agrarian problem and the origins of the caste war, 1812–1847. In: Brannon JT, Gilbert MJ (eds) Land, labor, and capital in modern Yucatan. Essays in regional history and political economic. The University of Alabama Press, Tuscaloosa, Alabama, pp 51–82

Reed N (2001) The caste war of Yucatán. Stanford University, Stanford. (Revised edition)

Schultz M (1988) Paläopathologische diagnostik. In: Knußmann R (ed) Anthropologie, wesen und methoden der anthropologie, vol 1. Gustav Fischer Verlag, Stuttgart, pp 480–496

Talavera A et al (2006) Catálogo San Nicolás Tolentino: una colección osteológica contemporánea mexicana. Instituto Nacional de Antropología e Historia, México

Tiesler V et al. (2012) What we learn from the past: health, life style and urbanism in prehispanic, colonial and modern populations in Yucatán. In: Meeting supplement of the 54° international congress of Americanists 54, Viena, Austria

Tiesler V et al (2015) Qué aprendemos del pasado: salud, estilo de vida y urbanismos en poblaciones prehispánicas, coloniales y contemporáneas en la península de Yucatán. Estudios de Antropol Biol 17(1):11–30

Ubelaker DH (1989) Human skeletal remains, 12th edn. Taraxacum, Washington, D.C.

Waldron T (2009) Trauma. In: Barker G (ed) Palaeopathology. Cambridge, manuals in archaeology. Cambridge University Press, New York, pp 138–167

Part III
Environment and Health

Chapter 14
Health and Well-being in the Yucatan Peninsula Revisited with a Human Ecology Perspective

Hugo Laviada-Molina, Oswaldo Huchim-Lara, and Nina Méndez-Domínguez

Abstract Human health and well-being are highly dependent of the individual and collective human activities and behaviors, but they are also tied to dynamic ecological determinants in a constantly changing society. These interactions can seem simple or obvious, as the relation between high sugar consumption with diabetes, diving and decompression illness or mosquito bites with febrile infections. But when it comes to a human ecology approach to health problems, the study of the aspects underlying those simple associations become less simplistic. Complex health-related questions can be thus analyzed within the interdisciplinary approach that characterizes the human ecology. In this chapter, we invite you to familiarize yourself with our inclusive approach and to have an insight of the health conditions in the Yucatan peninsula, with emphasis in those that are both endemic and frequent in the region.

14.1 Introduction

Health is defined as an optimal human status in which individuals can develop their capacities and express them in an appropriate, desirable way. When this welfare is present, people can actively engage in productive, sociocultural activities and interrelate with their environment with freedom.

However, just as health depends on the degree of involvement of human beings with their environment, also the physical, sociocultural practices, cultural traits and

H. Laviada-Molina
Escuela de Ciencias de la Salud, Universidad Marista de Mérida, Mérida, Yucatán, México
e-mail: hlaviada@marista.edu.mx

O. Huchim-Lara
Unidad Experimental Marista, Universidad Marista de Mérida, Mérida, Yucatán, México
e-mail: rhuchim@marista.edu.mx

N. Méndez-Domínguez (✉)
Escuela de Medicina, Universidad Marista de Mérida, Mérida, Yucatán, México
e-mail: nmendez@marista.edu.mx

H. Azcorra, F. Dickinson (eds.), *Culture, Environment and Health in the Yucatan Peninsula*, https://doi.org/10.1007/978-3-030-27001-8_14

behaviors that prevail in a certain society can directly influence on human health. Contemporary people living in the Yucatan peninsula, therefore, face challenges to their health that are very particular. As Yucatecan society has its roots in the culture of the peninsular Maya and people have preserved greatly these ancestral traditions, at the same time, they have also incorporated practices from other cultures on their lifestyle.

In the productive and occupational environment, the Yucatecans have shown their creativity by incorporating innovations that have allowed them to exploit the natural resources of the region. Finally, the climatic conditions, the coexistence of the population of Yucatan with the native or imported animal or plant species, generate health profiles in a certain sense, like those of other tropical regions, but at the same time different by the way in which the population, stakeholders, and the authorities address them.

In this chapter, we will address the health and disease profiles of the Yucatecan population from the human ecology perspective, so that we will unravel the impact of human habits and behavior; of economic activities and of the coexistence with pathogenic species applied in the understanding of three health problems in force in the Yucatan: (a) type 2 diabetes mellitus, (b) decompression sickness, and, finally, (c) diseases transmitted by *Aedes* mosquitoes.

Our interest in addressing three pathologies of different nature lies mainly in shedding into understanding, from a broad perspective, how the communicable (infectious), noncommunicable diseases (chronic, degenerative) and the occupational diseases of epidemiologic relevance in Yucatan can be understood with precision when approached from the human ecology perspective.

14.2 Human Ecology of the Highly Prevalent Chronic Non-transmissible Health Conditions

14.2.1 Approach of Metabolic Diseases Related to Nutrition from a Medical, Cultural, and Environmental Perspective

The interactions between social, economic, and environmental determinants of lifestyle play a vital role in chronic diseases development (Egger and Dixon 2014). Yucatan is an example of how variations in lifestyle may have contributed in modifying the prevalence of obesity and diabetes.

Obesity and diabetes represent two related conditions. In turn, obesity and type 2 diabetes are the source of other comorbid entities, among which are high blood pressure, lipid disturbances, cardiovascular and cerebrovascular disease, as well as chronic complications such as renal failure, blindness, and limb amputations.

The prevalence of obesity and type 2 diabetes mellitus in Mexico has significantly increased in recent years. Obesity is a risk factor for numerous non-communicable chronic diseases including type 2 diabetes (Pi-Sunyer 2002). There is limited information about the underlying causes that brought about this epidemiologic profile. The National Survey of Health and Nutrition of 2016 explored the state of various chronic diseases in Mexico, including diabetes in population older than 20 years of age. It was found that the prevalence of diabetes in the country increased from 9.2% in 2012 to 9.4% in 2016, this based on a previous diagnosis of the disease (Rojas-Martínez et al. 2018). Women report higher values of diabetes (10.3%) than men (8.4%). Nationwide, this trend is observed both in urban locations (10.5% in women and 8.2% in men) and in rural areas (9.5% in women, 8.9% in males). In this section, we will focus on the rural area of Yucatán, with a predominance of Mayan population, in which the change in the prevalence of these chronic conditions has been experienced in a more dramatic way.

In a recent study, the rural Maya of Yucatan were found to have a high body mass index (BMI) average and a 10.6% diabetes prevalence. The latter places them among the highest in Mexican aboriginal groups (Loria et al. 2018).

There are no data before 1997 for any indigenous group except for the report of Chávez et al. (1963) performed in rural Yucatan in 1962. At that time, the diagnosis of diabetes was made by a blood glucose assay using the Folin-Wu method, measured 1.5 h after a 100 g load of glucose in subjects with trace positivity in at least one of the following two screening tests for glycosuria: glucose oxidase tape method and Benedict test. Despite the low sensitivity of the 1962 screening tests, it did not play a major role in the low rate of diabetes rating 2.3% in women and 0% in men. This is supported by data on weight and height from the 1962 survey (Chávez A, unpublished data), i.e., body weight was measured at that time using a mechanical scale with a precision of ±0.5 kg, and body height with a metallic metric tape fixed to a wall, with subjects without shoes, wearing light clothing, and standing on a flat surface. Females ($N = 139$) had higher BMI than the 141 males (25.3 vs. 22.0 kg/m^2). Also, in agreement with low diabetes prevalence and low BMI values in 1962, Chavez and Pimentel (1963) reported a widespread adult pellagra and malnutrition, associated to the consumption of a low energy and low protein diet, in this same Yucatan population.

In 2000, after 38 years, in San Rafael and Uci, two communities of the same rural Maya zone, our group reported a prevalence of nearly 11% of type 2 diabetes. Despite improved access to health services in the early twenty-first century, it is to be noted that no one had been previously diagnosed (Loria et al. 2018).

The 1962 average BMI was clearly lower than the 2000 BMI of more than 28 kg/m^2 in males and females. The differences between 1962 and 2000 show a drastic change in the prevalence of obesity and diabetes in the rural Maya of Yucatan during almost four decades. The rapid change in obesity and diabetes may have been influenced by the low prevalence in 1962 due to the chronic undernutrition suffered during five centuries of external and internal colonization of Yucatan and other regions of Mexico (Bracamonte 2007).

Henequen (*Agave fourcroydes*) planting had been the main crop of the Yucatan Peninsula since the middle of the nineteenth century. The leaves of *A. fourcroydes* yield a fiber also called *henequen*, which is suitable for rope and twine, used in agricultural and naval activities. For many decades this activity was decisive in the lifestyle and occupation of the Mayan population settled in the rural areas of Yucatan.

In the publication of Loria et al. (2018), we shared an analysis of the potential causes of these abrupt changes experimented regionally in lifestyle since the early 1960s, influenced by relevant socioeconomic modifications (see Chap. 2):

(a) Tourist industry detonation in the Eastern Coast of Yucatan Peninsula. Since the mid-1970s, large investments transformed Cancun from a small village of a few hundred people to a city of more than 800 thousand. The urban and touristic development attracted labor force for the construction and services industries, with temporal or permanent patterns of migration into the zone which have influenced the economy of the Mayan communities into a predominantly cash economy. This internal migration has resulted in a cash flow to the families that have remained in their communities (Iglesias 2011).

(b) Decline of henequen production. Due to the introduction of synthetic fiber in the hard-fiber market in 1960, there started a steady decline in henequen production. In the early 1970s the Federal Government took direct control of the whole industrialization process leading to a subsidized economy of the rural sector through the capital inflow provided by the Federal Government to the henequen industry (Escalante 1988). The Government subsidized the rural henequen growers leading to increased food availability and to less strenuous physical activity. But these official actions did not prevent the final collapse of the henequen agroindustry at the beginning of the 1980s.

(c) Changes in physical activity and nutrition patterns appear to be the main transformations linked to the increasing prevalence of obesity and diabetes in rural Yucatan. By 2000, the population of San Rafael and Uci consumed calorie-dense diets. Loria findings agree with the dietary studies performed by Daltabuit Godas et al. (1997), Leatherman and Goodman (2005), and Leatherman et al. (2010) in the community of Yalcoba, located in the corn-producing area of the Yucatan Peninsula and in relative proximity to Cancun. These authors reported a transformation from self-production to a market economy in the area. These changes took place earlier in the henequen-producing areas in the Central and Western parts of the Peninsula, since subsidies from the state-owned henequen industries preceded the Cancun tourist development. The socioeconomic transition initiated in the 1960s appears to have modified the traditional pattern of food consumption in rural Yucatan, to a more varied and energy-dense diet with more fried than cooked ingredients when compared with other Mexican groups, including the corn-producing area of rural Yucatan described by Leatherman and Goodman (2005). In addition, physical activity was reduced drastically, associated with the change in the type of productive activity recently adopted.

(d) Migration to urban areas and expansion of the assembly, "maquila," industry. New activities have also contributed to the transition of Yucatan from a predominantly rural society, dependent on agriculture, into an increasingly urban population concentrated in Mérida and other few cities, laboring in "maquila industry" and services (Iglesias 2011). Similar changes were experienced by the two counties where San Rafael and Uci are located, which experienced a decrease in primary activities from above 60% to below 20% in the lapse 1970–2000.

(e) Free trade agreements implementation. A more recent macroeconomic factor may be the establishment of NAFTA (North American Free Trade Agreement) since 1994 (Arroyo et al. 2000). However, it is accepted that Yucatan would be less affected by NAFTA, since it is in the Southeastern tip of the country farthest than any other Mexican State from the US border.

The near absence of diabetes and the high prevalence of nutritional deficiencies of the rural Mayan population in the early 1960s in comparison with the high prevalence of obesity and diabetes in 2000 and thereafter are determined by social, economic, and cultural factors generating lifestyle changes that occurred in the last 5 or 6 decades in rural Yucatan.

14.2.2 Self-Efficacy and empowerment of Metabolic Control: Experience in Using Peer-Support Interventions in Diabetes

In Yucatán, 62.7% of the residents are identified as indigenous of Maya origin (Conapo, 2018). This population faces challenges in healthcare. Rates of poverty among them are approximately 50% higher than in the mainstream Mexican population, a factor that hinders health access (Inegi, 2018). Furthermore, western medicine often emphasizes individual behaviors and ignores the influence of sociocultural determinants of health. In Yucatán, the legacy of institutional discrimination against Mayan communities currently affects communication and trust between the community and healthcare providers (Frank and Durden 2017). Thus far, conventional care has done little to address diabetes or inequities in health in this vulnerable population. The lack of formally trained diabetes educators is another important factor that makes patient education a huge challenge.

Diabetes education: Advancing towards a model that integrates and respects the Mayan culture and their own social networks.

Diabetes self-management education (DE) is the process of facilitating the knowledge, skill, and ability for diabetes self-care. DE should incorporate the needs and life experiences guided by evidence-based standards. The objectives of diabetes education are: support informed decision-making, self-care behaviors, problem-solving and active collaboration with the healthcare team and to improve health status, and quality of life. Diabetes has evolved from primarily didactic expositions to more theorctically based empowerment model. Programs incorporating behavioral, cultural, and psychosocial strategies demonstrate improved outcomes (Funnell et al. 2009). Additional studies show that culturally appropriate programs improve outcomes and that group education is effective. The failure of traditional educational approaches for ethnic groups may be due to a lack of cultural competency on the part of providers and failure to address issues of relevance for the population (Nam et al. 2012). Most studies agree on the importance of social networks in providing support during the relevant events that occur to people with chronic disease. In the case of patients with type 2 diabetes, other authors report that having company and a well-established family and social network significantly increases patients' perception on their individual capacity to manage treatment. It provides a greater sense of emotional well-being and increases patients' health empowerment (Juárez-Ramírez et al. 2015).

Traditional models of diabetes care may need to include the cultural and social determinants of health as aspects that influence adherence to treatment and consider beliefs about illness and the emotional aspect of chronic disease. Studies about chronic diseases have shown that strengthening mutual support at these levels of social relationships positively influences patients' therapeutic behavior. Peer support is seen as a promising viable and culturally appropriate approach for diabetes care. It provides the ability for participants to support each other throughout their regular self-management treatment strategies. Peer support has been defined as the provision of support from an individual, member of the community, with experiential knowledge based on sharing similar life experiences that can be provided through various modalities: group meetings, home visits, face-to-face contact, text messaging, and other technologic approaches (Funnell 2009). The impact of peer support on diabetes control has differed based on the type of interaction, preparation time and the actual level and time of involvement of the peer leaders (Qi et al. 2015).

A recent longitudinal study in two communities of Yucatan, Komchen and Conkal (Castillo-Hernández 2017) shows that peer support added to diabetes self-management education improves diabetes self-care behaviors, diabetes control, program attendance, and some diabetes-related quality of life measures; the results in this underserved population with so many socioeconomic, cultural, and linguistic challenges are encouraging, representing an improvement in metabolic biochemical results but also in quality of life measurements. This approach may be of interest to places with limited human resources such as diabetes educators and/or in countries where conditions of economic and cultural disadvantage predominate. The cost-benefit analysis of this model would provide additional information to consider its adoption as a public health strategy in vulnerable areas where the challenge of diabetes management is even greater.

14.3 Diving-Related Health Issues Among Fishermen Divers

Coastal communities, fishing activity, and health-related issues. Fishing is the main economic activity and source of protein for millions of people around the world, and approximately 56 million of the people work in fisheries, and 85% of those are in the small-scale or artisanal fisheries type (FAO 2016). Definition of small-scale fisheries or artisanal is complex, because what is considered small in one context or region could be large-scale in another; however, generally involves small vessels (25 ft.), one day trips in with low technology and usually near to the coast (Huchim-Lara et al. 2016). In Mexico, as well, fishing is an important economic activity and fisheries capture contributes with 85% of the fish. In terms of employment, from the 271,000 Mexican people working in fisheries, 79% work in capture fisheries in 76,096 vessels, which 97% of these vessels are small-scale. (FAO 2017) According to the fatal and non-fatal injuries reported, fishing is one of the most dangerous occupations, and even more if we consider the number of accidents that are not registered. Among Mexican fishers, the rate of accidents is 4.6 × 100 workers and mortality are estimated to be 3.6 × 10,000 workers. The risk of accidents could be influenced by the fishing method, which depends of the target species. Among the variety of methods used in small-scale fisheries, those used to harvest benthonic species, as the lobster (*Panulirus argus*) and sea cucumber (*Isostichopus badionotus*), are particularly a source of risk for fishers' health (Huchim-Lara et al. 2016). In these particular fisheries, autonomous (SCUBA) and semiautonomous (hookah) diving are the elected method to access to the marine resources (Huchim-Lara et al. 2015; Huchim-Lara and Seijo 2018).

However, considering the advantages of the hookah diving as the amount of time the diver can spent in the water, the depth reached and the relative low cost of the equipment, is the first option for many fishers' communities worldwide. Basically, hookah diving utilizes a gasoline compressor which take air from the environment and compress it into a volume tank; the air is breathed by the diver through a plastic hose, usually of 100 m length. This equipment sits on wooden planks in the middle of the deck. It is usually built locally and the knowledge regarding the assembly is transmitted by peers or from a family member.

Breathing compressed gas under increased pressure has been related to decompression illness (DCI) that has been reported among recreational divers and military with a lower incidence, but among fishers, the incidence, severity, amount of deaths, and socioeconomic burden are significantly higher. The risk factors to develop DCI related to the diving technique are the time spent underwater, the depth while diving, and the speed during the ascent to the surface. But, there are other factors inherent to the fishers that can increase the risk for DCI; among them are obesity, asthma, dehydration, wounds, alcohol, and tobacco consumption, as well as patent foramen oval.

Signs and symptoms of DCI are fatigue, skin itch, pain, dizziness, vertigo, skin rash, paralysis, difficulty urinating, and unconsciousness. The treatment for DCI is the recompression at a hyperbaric chamber (Huchim-Lara et al. 2017); however, first aid with early oxygen while transportation to the chamber is important to reduce symptoms and have a better prognosis. In the Yucatan state, there are no chambers along the coastline.

Among the fishermen divers of the Yucatan, DCI is the main health problem. Every lobster fishing season (July–August) more than a hundred cases are treated, but when the sea cucumber season starts (opens), the number of cases increases dramatically. This increase is because for the latter species, anyone who has a permit can participate into the fishery, however, for lobster, only those who belong to a fishing cooperative participate (Huchim-Lara et al. 2017).

Carbon monoxide (CO) is an odorless, tasteless, colorless, and non-irritating gas that results of the oxidation or combustion of fossil fuel or organic matter. In the environment, the concentration of CO is less than 10 parts per million (ppm). In the human blood, CO binds to hemoglobin and displaces oxygen; hence, higher levels of the gas can cause injuries as overwhelm an adaptive response and lead to neurocognitive disorders (Chin et al. 2015). Symptoms of CO poisoning are joint pain, nausea, vomiting, headache, syncope, weakness, and tachycardia; but this nonspecific set of symptoms suggests a wide range of possible health problems and difficult a correct diagnosis (Vann et al. 2011). The treatment for CO poisoning is breathing pure oxygen through a mask or hyperbaric oxygen therapy at a hyperbaric chamber (Huchim et al. 2017).

In the Yucatan state, it is estimated that 1484 fishers work in the lobster fishery, concessional to fishers' cooperatives. In the northeastern coast, where hookah diving is the only fishing method to harvest the Caribbean spiny lobster (*Panulirus argus*), approximately close to 500 divers work. When sea cucumber fishery open, and participate the private sector, the number increases to an estimate of 1701 divers, although the real number could be two or three times higher, as this phenomenon can be easily underdiagnosed. The predominant age group among the fishers is 40–44, followed by 45–49. However, the main age group of fishers treated at a single hyperbaric chamber was 30–39 during a lobster fishing season, and during a sea cucumber season the mean age was 38.8 (±9.46) (Huchim et al. 2017).

Among the risk factors prevailing among the fishers' population we found the following prevalence: overweight (47%), obesity (42%), high blood pressure (20.9%), and high glucose level (19%) (Mendez et al. 2018). Between 2012 and 2016, four hyperbaric facilities in the Yucatan provide a total of 1298 attentions because of DCI (Huchim et al. 2017). Death among fishers has been related to the inexperience of the occasional fishers for diving (Mendez et al. 2017) and also to the diving equipment, as it is related to the air quality; an average diving equipment (of artisanal fishermen divers) usually contains up to 15 times the ppm allowed by the safety at work agencies. To understand the practice of diving in fisheries besides the health risks involves the comprehension of the social, cultural, and economic issues of the fishers, but also the link between the fishers with the sea.

Mitigating the risk. Reducing the risk of CO poisoning because the diving equipment in the lobster fishery by cooperative fishers is a good example of successful educational intervention when the users are involved in planning, implementation, and evaluation. Fishers from Rio Lagartos and Celestun were invited to participate in focus groups during 2017, in a community-based participatory research. In researcher-led demonstrations and training sessions, different options to improve the hookah equipment have proven to significantly reduce the CO levels. Improvement options cover important issues as feasibility, low cost, and practicality. The elected option was installed in the hookah equipment with previous and later measures of air quality. Fishers witnessed the process where CO levels decreased up to 80%. Months after the intervention, many fishers make the improvement on their own equipment (Chin et al. 2016).

To reduce the risk of DCI, agencies have been training fishers but when the diving type developed by fishers is not recreational or military, the success of diving training turns complicated. Hence, understanding the diving behavior of fishers can help to reduce the risk to DCI. Fishers diving profile was followed by 4 years and the results help to feedback fishers. They usually try to follow the recommendations for diving safety; however, the availability of the target marine resources makes fishers dive deeper and far from the coast. Comparing maps of the fishing ground visited in the 1990s and the current fishing ground, now they fish farther from the coast. Also, the skills to fish influence the time spend underwater and also the selection of the fishing ground (Huchim-Lara et al. 2016).

Considering the number of DCI events in the region, that there are no hyperbaric chambers in the coastline and that fishing is an important activity for coastal communities, building capacities of diving fishers become relevant. Since, as described in literature (Chin et al. 2016), oxygen supply can help with symptoms and have better prognosis, fishers where trained to supply oxygen in case of DCI to their peers during transportation. For training in choke, recovering position, and Cardiopulmonary Resuscitation, fishers were moved to a medical simulation center in the capital city of Yucatan. A course was tailored for fishers using high fidelity medical simulation technology. Knowledge regarding first aid was evaluated before and after the course. Knowledge and competences increased, and training was well accepted, arguing that hands on help to understand the process (Huchim-Lara et al. 2018).

In the case of fishers' inherent risk factors as chronic disease, congenital disease related to diving and knowledge regarding symptomatology of DCI, implementation of medical checkups was developed periodically to diagnose and early treatment. Infographic campaigns of DCI risk factors and symptoms were delivered between the cooperatives to increase fishers' awareness, because results of studies in the area showed a low-risk perception regarding DCI and risk factors associated.

It is important to mention that the different activities performed with fishers was a result of collaboration of different researchers from different knowledge areas, including human ecologists, anthropologist, medical doctors, hyperbaric medicine specialists, fisheries biologists, and physical therapy and medicine students who were involved to conduct the different studies. Understanding the environment and the sociocultural context of the economic activities on the coastal communities is important to develop effective strategies to reduce the health risk factors among the fishers.

14.4 *Aedes*-Borne Viral Diseases in Yucatan

Vector-borne diseases are communicable ailments that derive from human contact with invertebrates or nonhuman vertebrates which transmit infective organisms from one host to another. Mosquitoes are members of the family *culicidae*, *Aedes* being a genus of this family that includes the mosquitoes that are capable of transmitting viruses that cause infection and disease to human hosts.

Aedes aegypti is a common vector in the American continent that transmits infections of clinical importance. The capacity of *A. aegypti* as a transmitter of diseases derives from its ability to feed repeatedly on one or several hosts until their dietary needs are met. *A. aegypti* flight range depends on how close or far they need to fly to have access to feed from human blood. Under certain situations in which *A. aegypti* has access to nearby hosts, its flight range is short and therefore it is recognized as a peri domiciliary vector (a vector that remains near to homes that are inhabited by humans).

Since 2015, the Yucatan peninsula has been facing the co-circulation of three viruses that are transmitted by vectors of the genus *Aedes*, all of them of clinical and epidemiological importance. Given the abundance of the vector and the re-emergence of the chikungunya virus and the emergence of zika, the capacities of the health services and the communities for the prevention, treatment of acute symptoms, and the mitigation of harm was challenged.

Dengue was first reported in the state of Yucatan in 1979, and since then, the four dengue serotypes have circulated periodically, in such a way that some specialists have considered dengue as a hyperendemic disease in the region (Loroño et al. 1993).

For over 30 years, dengue virus infection has been the main threat of disease causing several epidemic periods in association to the four serotypes of virus and, despite all efforts, a sustained reduction in the *Aedes* abundance has not yet been achieved (Gómez-Carro et al. 2017).

The appearance of the chikungunya virus in 2015 and the zika virus in 2016 in Yucatan posed new challenges for physicians and regional public health due to the superposition of clinical characteristics and coexistence of three co-circulating viruses. In Fig. 14.1, we present the number of confirmed dengue, chikungunya, and zika cases in Yucatan between 2014 and 2016. Please note that only 5% of all chikungunya, dengue, and zika cases were confirmed by the health system during outbreaks exclusively, meaning that out of outbreak season diagnosis is made basically on clinical manifestations or passive surveillance and therefore not reported). (DGCES 2018).

Dengue. Since its first appearance in Yucatan in 1979, dengue virus caused major epidemics in the State for the following 5 years with circulation of three of the four existing serotypes. With the arrival of serotype 4 infections to Yucatan, the incidence of severe hemorrhagic manifestations of dengue increased significantly. According to the antibodies enhancement theory, the first contact of an individual with any serotype of dengue tends to unleash a mild self-limited infection, but

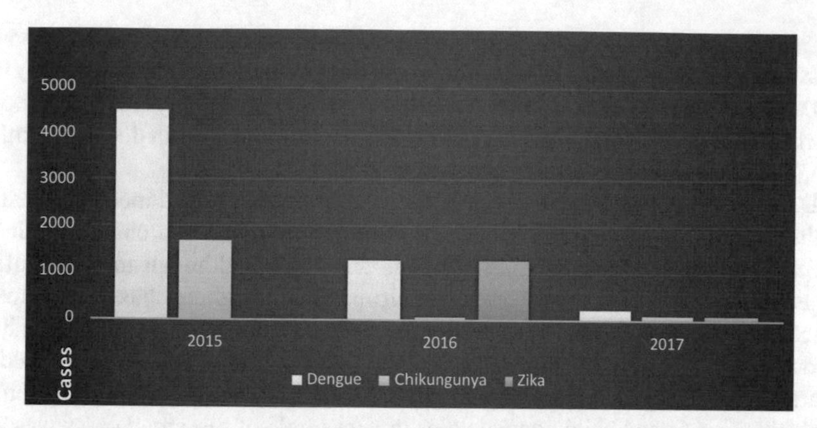

Fig. 14.1 Dengue, chikungunya, and zika laboratory confirmed cases in Yucatán

subsequent infections (with other serotypes) tend to typically cause increasing severity infections; for that reason the co-circulation of all serotypes in Yucatan has historically caused not only dengue fever but also an increase of cases with hemorrhagic manifestations or even a life-threatening shock. Fatalities have been reported due to severe dengue infections, particularly among young children or seropositive infants that are exposed to infected vectors and thus develop subsequent and enhanced immune response (Mendez-Dominguez et al. 2016).

The dengue serologic surveys conducted in Yucatan during the 1980s showed an overall seropositivity close to 70% of the population residing in urban environments. Later surveys showed an overall seropositivity to dengue infection among adult population of 59.94% in 1996 and 81.52% in 2006 (Gómez-Carro et al. 2017). The latest survey was conducted in 2015 and showed that the highest seroprevalence was observed in adults older than 50 years, 83.4% (Pavía-Ruz et al. 2018) of whom had been infected with dengue at least once.

Chikungunya. There is historical evidence by Halstead that chikungunya is not an emerging disease in America, but a reemergence of a previously circulating virus whose origins go back to the late 1820s and whose re-entry to the American continent in 2013 marked the beginning of outbreaks in those regions where there is presence of the transmitting vector (Weaver and Forrester 2015).

Prior to reappearance in America, the chikungunya virus was the cause of the outbreak in Kenya during 2004, later reaching numerous islands of the Pacific, among which was the Meeting, where a total of 46,000 cases and 284 deaths were reported. For 2006, imported cases of this infection were reported in Europe (Amraoui and Failloux 2016).

The first case of recent local transmission in America was reported in the islands of Saint Martin and Martinique in 2013. For 2014, cases were reported in other islands of the Caribbean, Central and South America. In that same year, the first case imported into Mexico was confirmed. For the epidemiological week number 45 of the year 2014, 11 imported and 14 autochthonous cases had been reported (in the state of Chiapas, Mexico) and later also in other states of the country. In the state of Yucatán, the first two cases were confirmed in the month of May 2015, and the peak incidence was observed in the month of September 2015 (DGCES 2018).

During 2015, there were 12,588 confirmed cases of chikungunya fever in Mexico, of which 1669 occurred in the state of Yucatán, confirming 1 death due to this infection (DGCES 2018; Mendez-Dominguez et al. 2016).

During the outbreak, more than a half of the patients received medical attention in the capital city (Méndez et al. 2017). Pain and inflammation characteristically affected the patients' small joints in hands and ankles, although they also affect larger joints. In 40–50% of the cases maculo-papular exanthema has been found in the chest and facial edema. The persistence of the arthralgias for 4 months has been reported in 33% of the infected patients. The average time for patients affected by arthralgia to seek specialized treatment for rheumatology was, on average, 8 months. Beside the confirmed chikungunya cases, a subsample of discarded cases was also analyzed for diagnosis that let to identify that patients with chikungunya were significantly more prone to develop pruritus and polyarthralgia when compared to dengue patients (Mendez-Dominguez et al. 2017).

Chikungunya is a virus that like the one of the dengue and the yellow fever initiated its cycle of urban transmission by amplificating in the presence of many infected mosquitoes that reached humans. The residents of Yucatan had no immunological defenses against chikungunya, because it was a reemergent disease; therefore, the population was naïve, susceptible to infection.

The rapid dispersion of chikungunya showed that it is necessary to continuously maintain not only the passive but also the active surveillance measures, as well as the importance of awareness about the strategies of individual, household, and school that centered protection against the mosquito bites, and also for the health department to be able to anticipate future outbreaks in a timely manner and act effectively against new viruses transmitted by the same vector.

The fact that 83% of the patients during the chikungunya outbreak in Yucatan had had outpatient management also points that if attention had been optimized in the first contact health centers, patients could have managed from their nearest medical unit without having to be referred to a general hospital, as has been done in certain places where there have been recent outbreaks.

Zika. Zika is the name of the viral diseases caused by a flavivirus that was originally isolated in 1947 from the blood of a sentinel rhesus monkey (*Macaca mulatta*) in Uganda (Dick et al. 1952). Zika is a disease of recent appearance, and since the initial reports of human infection in 2007, it has become of high importance in clinical and epidemiological terms. Zika has been the causal agent of recent outbreaks and associated with the unprecedented increase in two (among others) rare and lethal health conditions: congenital microcephaly and Guillain Barre syndrome incidence (Rodrigues 2016).

Microcephaly is a severe condition of pathologic in utero development in infants born to women that were infected with zika during the first trimester of pregnancy, while Guillain Barre syndrome referrers to an acute autoimmune inflammatory neuritis, where demyelination occurs in peripheral nerves and nerve roots, causing progressive weakness, loss of sensation, and loss of deep tendon reflexes affecting respiratory musculature and paralysis.

Even when severe, microcephaly and Guillain-Barre syndrome are uncommon compared to the cases in which zika occurs as a self-limited febrile disease. This infection is characterized by symptoms that can last for one week, with a clinical presentation like that of other arbovirus infections such as chikungunya and dengue, which include mild fever, rash, arthralgia, arthritis, myalgia, headache, conjunctivitis, and edema (Kuno et al. 1998; Pech-Cervantes et al. 2017).

14.4.1 The Context of Aedes-Borne Transmission in Yucatan

The amplification and rapid transmission of *Aedes*-borne diseases could be explained by the fact that, in Yucatan, almost half of the population of the state is concentrated in Mérida, the capital, where the demographic density favors the transmission of the *Aedes*-borne viruses, while intra and interstate migration precedes the arrival of the disease to inner state communities.

Aedes-borne viral diseases, after infecting a human host, cause a lifelong immunity against the virus. For the case of dengue, as there are four serotypes, immunity is serotype-specific, which means that any person, after being infected with a specific serotype, will not be susceptible to future infections caused by the same serotype, but can still develop dengue if in contact with any of the other three serotypes. Unlike dengue, chikungunya and zika, as emergent or reemergent diseases, met a totally naïve, susceptible, population when first identified in the Yucatan; for that reason, the outbreaks were the events that evidenced that the measures that were taken by the population and the health authorities-held community-based strategies to avoid the abundance of the vector or the transmission of the disease were suboptimal.

The rapid dispersion of both recent viral diseases showed a need to continuously maintain the active vector surveillance measures, as well as the importance of spreading the strategies of individual self-care to be able to anticipate future outbreaks in a timely manner and act effectively against new viruses transmitted by the same vector.

Conventional vector-control actions carried out by governmental public health entities in Mexico include measures against the adult vector and its hatcheries, because females of *A. aegypti* deposit their eggs on the walls of containers containing crystalline fresh water, but also the cleaning of public lands, proper management of urban drainage systems and disposal of garbage in areas free of puddles or standing water that could mean the formation of *A. aegypti* hatcheries.

Fumigation for the control of adult mosquitoes is usually practiced during outbreaks. In relation to this control measure, it should be considered not only the cost and the difficulty to cover the different rural and urban areas that each new case of chikungunya, zika, or dengue need to be controlled, but also the possibility of resistant varieties of mosquitos that could translate in ineffective fumigations.

14.4.2 *Future, Promising or Controversial Strategies*

The release of the vaccine against the four dengue serotypes in 2015 was promising for the population of Yucatan; however, shortly after its commercial release, the formulation showed not to be safe enough for its application to open population, besides being a specific measure for a single virus. The chikungunya and zika outbreaks showed that needed measures cannot be limited to a virus-specific immunological protection, because as long as the vector remains abundant and evolution of pathogens and migration continues as expected, preventive strategies need to be diverse.

Upcoming measures to reduce the vector abundance rely on the open liberation of genetically modified mosquitoes that are infertile, with the purpose of limiting the reproduction of *A. aegypti* vectors in Yucatan. The World Health Organization has developed general recommendations before the liberation of infertile laboratory mosquitoes in any open region, including the need of the development of specific guidelines to regulate the laboratory use of modified mosquitoes and the eventual liberation in an open ecosystem that include environmental contingency plans, as the impact on any ecosystem is still unknown.

Previous studies have shown that the social and cultural role of women in the context of Yucatan implies caring for the sick in the family context, but when the population is not aware of the need for isolation, taking care of an infected relative also implies becoming infected. During outbreaks, in Yucatan, it has not been uncommon that when a person becomes infected, the virus is transmitted to the relatives that cohabit with the infected patient, extending the transmission to the surroundings of the domicile and neighboring sites, generating the so-called clusters, that are characterized by several cases that coincide geographically and temporally; intradomiciliary chikungunya transmission studies have suggested the presence of a gender-related transmission (Bargas-Ochoa et al. 2017).

To date, all epidemiologic data, research studies, and clinical series point that women are more prone to become infected, which, as mentioned before, could be culturally driven. There is also a possibility that men as main providers and culturally designed as the economic pillar of the households avoid medical consultations, in an intent to keep working even when sick, and opting for self-medication. There is, to our opinion, a need in the region to develop campaigns to prevent Aedes-borne diseases based on gender-specific recommendations (Alonzo-Ancona et al. 2017).

The measures that can modify the transmission of the viruses are mainly the immunological defense against the viruses and their serotypes, i.e., the vaccination, but even more, the prevention, not only to avoid the contact with the transmitting vectors of the virus, but also with the isolation and distancing of infected patients.

The preventive measures in the Yucatan peninsula have historically been centered on the reduction of vectoral load, by means of the elimination of mosquito breeding places by the health authorities; however, these measures of fumigation and elimination of breeding places by means of chemicals have been ineffective in reducing the abundance of vectors, resulting, in the long run, in the greater dissemination and dispersal of mosquitoes in rural and urban environments. Even when community participation has been considered in government-centered preventive strategies, it has been focusing only on the elimination and disposal of water containers in the household open spaces, but to our knowledge, broader campaigns to promote the social responsibility for isolation of infected patients have not yet been applied.

Infected patients may be carriers of the disease, which in turn would continue to be transmitted through the contact of the vectors with the infected patients and the possible contact of the vectors with healthy susceptible patients. There is evidence (Graham 2017) that indicates that individual protection measures can reduce the transmission of the virus to individuals, individual protection measures should ideally be used by patients infected during the acute period and by patients prone to develop the infection.

The protection of homes and school classrooms through the use of mosquito nets and fumigation along with individual use of repellent has shown to reduce intra-school transmission. In previous studies related to the transmission of dengue, it has been shown that dengue seropositivity becomes more frequent in children with each grade of their school life and that the history of having been exposed to dengue is higher among children that attend schools with a higher density of students. Interventions directed to school age children and teachers have shown to be efficiently translated in an improvement in the knowledge, attitudes, and practices to prevent mosquito-borne diseases in the school community (Bargas-Ochoa et al. 2017). We consider that those measures are important and should be continued as part of a holistic prevention program.

14.5 Concluding Remarks

By presenting these three characteristic scenarios of the Yucatan, where human biology, ecological and socioeconomic conditions collide to contribute to an increased human health risk, we also wanted to invite our readers to consider the possibility that as in the cases of diabetes, decompression illness or *Aedes*-borne viral infections of the Yucatan, mitigation strategies might be achievable, not only for health problems but also for any human ecology phenomenon. Even when the mitigation strategies might not absolutely vanish all endemic health problems for good, and even when community-based interventions might have discrete or gradual results over time, we believe that as soon as the problems are sufficiently identified, interdisciplinary efforts should be directed to inform communities and favor their active participation in problem solving thru awareness, empowerment, training, or health education.

References

Alonzo-Ancona A, Coba Y, Gomez-Carro S et al (2017) Probable intradomiciliary transmission of chikungunya virus during the 2015 outbreak in Yucatan, Mexico. ACPM Annual Scientific Meeting., Portland, Oregon. Available via https://www.eventscribe.com/2017/posters/acpmannual/SplitViewer.asp?PID=MTEyMzQzMDQwNjQ

Amraoui F, Failloux A-B (2016) Chikungunya: an unexpected emergence in Europe. Curr Opin Virol 21:146–150. https://doi.org/10.1016/j.coviro.2016.09.014

Arroyo P, Loria A, Fernández V et al (2000) Prevalence of pre-obesity and obesity in urban adult Mexicans in comparison with other large surveys. Obes Res 8:179–185. https://doi.org/10.1038/oby.2000.19

Bargas-Ochoa M, Espadas-Torres M, Quintana A et al (2017) School-based education for individual protection against Aedes transmitted diseases during outbreak season in an endemic community. ACPM Annual Scientific Meeting, Portland, Oregon. https://www.eventscribe.com/2017/posters/acpmannual/SplitViewer.asp?PID=MTEyMzQ3MjgwMzY

Bracamonte P (2007) Una deuda histórica. Ensayo sobre las condiciones de pobreza secular entre los mayas de Yucatán. CIESAS, Mérida. ISBN: 9707019875, 9789707019874

Castillo-Hernández K (2017) Efectos de un programa de intervención educativa con apoyo entre pares para pacientes con diabetes mellitus tipo 2 implementado en una comunidad semiurbana del Estado de Yucatán. Universidad Autónoma de Yucatán. Dissertation

Chavez A, Pimentel R (1963) The epidemiology of pellagra in a rural area. Bol Ofic Sanit Panamer 55:398–404

Chávez A, Balam G, Zubirán S (1963) Estudios epidemiológicos de la diabetes en tres comunidades de la zona henequenera del estado de Yucatán. Rev Investig Clin 15:333–344

Chin W, Huchim O, Wegrzyn GH et al (2015) CO and CO_2 analysis in the diving gas of the fishermen of the Yucatan Peninsula. Undersea Hyperbar Med 42:297–305. Available via https://www.ncbi.nlm.nih.gov/pubmed/26403015

Chin W, Huchim-Lara O, Salas S (2016) Decreasing carbon monoxide in the diving air of artisanal fishermen in the Yucatán peninsula by separation of engine exhaust from compressor intake. Undersea Hyperbar Med 43:411–419. Available via https://europepmc.org/abstract/med/28763170

Conapo (2018) Consejo Nacional de Población. Datos abiertos Consulta en línea Ultima actualización: abril de 2018. https://datos.gob.mx/busca/dataset/

Daltabuit Godás M, Bérrio M, Garzón L (1997) Conducta reproductiva e ideales de fecundidad en una comunidad maya de Yucatán. Estudios de Antropol Biol 6. Available via http://revistas.unam.mx/index.php/eab/article/view/42727/38818

DGCES (2018) Anuario de Morbilidad. Dirección General de Calidad y Educación en Salud. Available via http://www.epidemiologia.salud.gob.mx/anuario/html/anuarios.html

Dick G, Kitchen S, Haddow A (1952) Zika virus (I). Isolations and serological specificity. T Roy Soc Trop Med H 46:509–520. https://doi.org/10.1016/0035-9203(52)90042-4

Egger G, Dixon J (2014) Beyond obesity and lifestyle: a review of 21st century chronic disease determinants. Biomed Res Int 2014. https://doi.org/10.1155/2014/731685

Escalante R (1988) The state and henequen production in Yucatán, 1955–1980 ISA Occasional Papers:1–44. Available via https://sas-space.sas.ac.uk/4543/1/B56_-_The_State_and_Henequen_Production_in_Yucatan_1955-1980.pdf

FAO (2016) Food and Agriculture Organization of the United Nations. The State of World Fisheries and Aquaculture. Contributing to food security and nutrition for all. Rome. Available via http://www.fao.org/3/a-i5555e.pdf

FAO (2017) Food and agriculture Organization of the United Nations. El buceo en la Pesca y la acuicultura en América Latina y el Caribe Orientaciones operativas, legislativas, institucionales y de política Para garantizar condiciones de empleo decente. Organización de las Naciones Unidas para la Alimentación y la Agricultura, Santiago, Chile. Available via http://www.fao.org/3/a-i7331s.pdf

Frank SM, Durden TE (2017) Two approaches, one problem: cultural constructions of type II diabetes in an indigenous community in Yucatán, Mexico. Soc Sci Med 172:64–71. https://doi.org/10.1016/j.socscimed.2016.11.024

Funnell MM (2009) Peer-based behavioural strategies to improve chronic disease self-management and clinical outcomes: evidence, logistics, evaluation considerations and needs for future research. Fam Pract 27:i17–i22. https://doi.org/10.1093/fampra/cmp027

Funnell MM, Brown TL, Childs BP et al (2009) National standards for diabetes self-management education. Diabetes Care 32(S1):S87–S94. https://doi.org/10.2337/dc10-S089

Gómez-Carro S, Méndez-Domínguez N, Mendez-Galván J (2017) Dengue seropositivity in a randomly selected sample from Yucatan analyzed in the context of dengue cases reported between 1996 and 2006. J Epidemiol Res 3(23). https://doi.org/10.5430/jer.v3n1p23

Graham S (2017) Overview of mosquito-borne diseases and prevention strategies. Pract Nurs 28(5):214–220. https://doi.org/10.12968/pnur.2017.28.5.214

Huchim O, Rivas-Sosa F, Rivera-Canul N et al (2017) 350 years of hyperbaric medicine: historic, physiopathologic and therapeutic aspects. Gac Med Mex 153:854–860. https://doi.org/10.24875/GMM.M18000102

Huchim-Lara O, Seijo JC (2018) Risk perception in small-scale fishers and hyperbaric personnel: a risk assessment of hookah diving. Undersea Hyperb Med 45:313–322. Available via https://europepmc.org/abstract/med/30028918

Huchim-Lara O, Salas S, Chin W et al (2015) Diving behavior and fishing performance: the case of lobster artisanal fishermen of the Yucatan coast, Mexico. Undersea Hyperb Med 42(4):285–296. Available via https://europepmc.org/abstract/med/26403014

Huchim-Lara RO, Salas S, Fraga J et al (2016) Fishermen's perceptions and attitudes toward risk diving and management issues in small-scale fisheries. Am J Human Ecol 5:1–10. https://doi.org/10.11634/216796221504760

Huchim-Lara O, Chin W, Salas S et al (2017) Decompression sickness among diving fishermen in Mexico: observational retrospective analysis of DCS in three sea cucumber fishing seasons. Undersea Hyperb Med 44:149–156. Available via https://europepmc.org/abstract/med/28777905

Huchim-Lara O, Cetina-Sauri G, Méndez-Domínguez N et al (2018) First-aid training through medical simulation technology: the case of small-scale fishers. Int J Comm Dev 6:21–27. https://doi.org/10.11634/233028791503932

Iglesias E (2011) Las nuevas migraciones yucatanenses: Territorios y remesas. Migración y desarrollo 9(17):69–90. Available via http://www.scielo.org.mx/scielo.php?script=sci_arttext&pid=S1870-75992011000200003

Inegi (2018) Banco de Indicadores. Instituto Nacional de Información Geográfica y Estadística Ultima actualización: abril de 2018. http://www.beta.inegi.org.mx/app/indicadores/

Juárez-Ramírez C, Théodore FL, Villalobos A et al (2015) Social support of patients with type 2 diabetes in marginalized contexts in Mexico and its relation to compliance with treatment: a sociocultural approach. PLoS One 10(11):e0141766. https://doi.org/10.1371/journal.pone.0141766

Kuno G, Chang G-JJ, Tsuchiya KR et al (1998) Phylogeny of the genus Flavivirus. J Virol 72(1):73–83. Available via https://www.ncbi.nlm.nih.gov/pubmed/9420202

Leatherman TL, Goodman A (2005) Coca-colonization of diets in the Yucatan. Soc Sci Med 61:833–846. https://doi.org/10.1016/j.socscimed.2004.08.047

Leatherman TL, Goodman AH, Stillman T (2010) Changes in stature, weight, and nutritional status with tourism-based economic development in the Yucatan. Econ Hum Biol 8(2):153–158. https://doi.org/10.1016/j.ehb.2010.05.008

Loria A, Arroyo P, Fernandez V et al (2018) Prevalence of obesity and diabetes in the socioeconomic transition of rural Mayas of Yucatan from 1962 to 2000. Ethnic Health:1–7. https://doi.org/10.1080/13557858.2018.1442560

Loroño MA, Farfán AA, Rosado EP et al (1993) Epidemic dengue 4 in the Yucatán, México, 1984. Rev Inst Med Trop Sao Paulo 35:449–455. https://doi.org/10.1590/S0036-46651993000500011

Méndez N, Baeza-Herrera L, Ojeda-Baranda R et al (2017) Perfil clinicoepidemiológico de la infección por Chikungunya en casos hospitalarios atendidos en 2015 en Mérida, México. Rev Panam Salud Publica 41:e91. Available via https://www.scielosp.org/article/rpsp/2017.v2041/e2091/

Mendez N, Huchim-Lara O, Rivera-Canul N et al (2017) Fatal cardiopulmonary decompression sickness in an untrained fisherman diver in Yucatan, Mexico: a clinical case report. Undersea Hyperb Med 44:279–281. Available via https://europepmc.org/abstract/med/28779584

Mendez N, Huchim O, Chin W et al (2018) Body mass index in association to decompression sickness events: cross-sectional study among small-scale fishermen-divers in Southeast Mexico. Undersea Hyperb Med 45:439–445. Available via https://europepmc.org/abstract/med/30241124

Mendez-Dominguez N, Achach-Asaf J, Basso-García L et al (2016) Septic shock secondary to non-congenital chikungunya fever in a young infant: a clinical case. Rev Chil Pediatr 87:143–147. Available via https://europepmc.org/abstract/med/27032486

Mendez-Dominguez N, Janssen-Aguilar R, Pacheco-Tucuch F et al (2017) Chikungunya fever in clinically diagnosed patients: a brief report of comparison between laboratory confirmed and discarded cases. Arch Clin Infect Dis 12(4):e12980. https://doi.org/10.5812/archcid.12980

Nam S, Janson SL, Stotts NA et al (2012) Effect of culturally tailored diabetes education in ethnic minorities with type 2 diabetes: a meta-analysis. J Cardiovasc Nurs 27(6):505–518. https://doi.org/10.1097/JCN.0b013e31822375a5

Pavía-Ruz N, Rojas DP, Villanueva S et al (2018) Seroprevalence of dengue antibodies in three urban settings in Yucatan, Mexico. Am J Trop Med Hygiene 98:1202–1208. https://doi.org/10.4269/ajtmh.17-0382

Pech-Cervantes CH, Lara-Romero EG, Haas-Solis EG et al (2017) Syndrome in patients with zika: clinical series with laboratory confirmation in the state of Yucatan. Am J Trop Med Hygiene 97(Suppl 5):555. Available via https://www.astmh.org/ASTMH/media/2017-Annual-Meeting/ASTMH-2017-Abstract-Book.pdf

Pi-Sunyer FX (2002) The obesity epidemic: pathophysiology and consequences of obesity. Obes Res 10:97S–104S. https://doi.org/10.1038/oby.2002.202

Qi L, Liu Q, Qi X et al (2015) Effectiveness of peer support for improving glycaemic control in patients with type 2 diabetes: a meta-analysis of randomized controlled trials. BMC Public Health 15:471. https://doi.org/10.1186/s12889-015-1798-y

Rodrigues LC (2016) Microcephaly and Zika virus infection. Lancet 387:2070–2072. Available via https://www.ncbi.nlm.nih.gov/pubmed/26993880

Rojas-Martínez R, Basto-Abreu A, Aguilar-Salinas CA et al (2018) Prevalence of previously diagnosed diabetes mellitus in México. Salud Publ Mex 60:224–232. https://doi.org/10.21149/8566

Vann RD, Butler FK, Mitchell SJ et al (2011) Decompression illness. Lancet 377:153–164. https://doi.org/10.1016/S0140-6736(10)61085-9

Weaver SC, Forrester NL (2015) Chikungunya: evolutionary history and recent epidemic spread. Antivir Res 120:32–39. https://doi.org/10.1016/j.antiviral.2015.04.016

Chapter 15
Hair Mercury Content in an Adult Population of Merida, Yucatan, Mexico, as a Function of Anthropometric Measures and Seafood Consumption

Sally López-Osorno, Flor Árcega-Cabrera, José Luís Febles-Patrón, and Almira L. Hoogesteijn

Abstract Acute or chronic mercury exposure may cause adverse effects to health and development. Total mercury concentration in hair of a general population ($n = 90$) between 18 and 45 years of age was studied in the city of Merida, Yucatan, Mexico. Subjects provided data on age, sex, body weight, height, body mass index (BMI), and seafood consumption frequency. Mean mercury concentration ranged from 0.136 to 4.383 μg/g, with a mean of 1.119 ± 0.854 μg/g. Forty-two percent of the sampled individuals had total mercury hair concentration over 1 μg/g which corresponds to the U.S. EPA reference dose. No statistically significant differences were found regarding age, weight, body fat percentage, and BMI. Women had lower total mercury than men, although the difference was not statistically significant ($p > 0.05$). There was a positive correlation between height and hair mercury concentration ($p < 0.05$). A multiple linear regression analysis showed a significant correlation of total mercury content in hair with respect to the frequency of seafood consumption per month ($p < 0.001$). Mercury exposure can be dangerous; there is a critical need to understand mercury distribution in the environment and its human health impacts under prevailing cultural and social conditions, to mitigate

S. López-Osorno
Facultad de Medicina, Universidad Autónoma de Yucatán, Mérida, Yucatán, México
e-mail: sally.lopez@correo.uady.mx

F. Árcega-Cabrera
Unidad de Química Sisal, Facultad de Química, Universidad Nacional Autónoma de México, Sisal, Yucatán, México
e-mail: farcega@unam.mx

J. L. Febles-Patrón · A. L. Hoogesteijn (✉)
Departamento de Ecología Humana, Cinvestav-Mérida, Mérida, Yucatán, México
e-mail: jose.febles@cinvestav.mx; almirahoo@cinvestav.mx

H. Azcorra, F. Dickinson (eds.), *Culture, Environment and Health in the Yucatan Peninsula*, https://doi.org/10.1007/978-3-030-27001-8_15

the potential effect of pollutants on health of all living beings. Seafood consumption advisory should be developed for the region, especially for women in childbearing age.

15.1 Introduction

During our lives we intuitively develop a sense of toxicology that guide our day-to-day decisions, e.g., we have learned to moderate consumption of alcohol or caffeine, using the most basic principle of toxicology: dose-response. Humans are not aware about the hazardous chemicals that are part of daily life. Global production of chemicals rose from a million tons in 1930 to over 400 million tons in the year 2000 (European Commission 2001), the European Union chemical exports (pharmaceuticals excluded) reached €146.2 billion in 2016. In 2015 Latin America imported €19.2 billion from the United States and €9.4 billion from EU worth of chemicals (CEFIC 2018). To put things in perspective, we illustrate the immense ignorance related to chemicals in the environment and our health, using the American-Chemistry-Council data: the United States Federal Chemical Regulation Law known as the Toxic Substance Control Act (TSCA) required the U.S. Environmental Protection Agency (U.S. EPA) to keep a list of all chemicals manufactured or processed in the United States. The TSCA list included roughly 84,000 chemicals—but The American-Chemistry-Council states that this number does not reflect chemicals that were actually produced and used (American Chemistry Council 2018). The website asserts that currently the total numbers of chemicals produced and used in commerce are closer to 7700. In June 2016, the Lautenberg Chemical Safety for the 21st Century Act was passed. Prioritization is the new mandate of the Act; by December 2019 EPA should have prioritized 20 high-risk chemicals and 20 low-risk chemicals to undergo risk evaluation (EPA 2017). That an Act was passed, demanding prioritization as an alternative to solve the lack of knowledge related to chemicals used, is very troubling; that a rough calculation shows that only 0.51% ($n = 7700$) of the chemicals currently in commerce are going to be assessed is even more troubling. This only shows that the substances that are in the products we use every day in our food, cosmetics, cleaning products, clothes, or furniture are *terra incognita*. The current approach to chemicals is a huge experiment; women and children are especially at risk, because of different and more vulnerable systems.

At the Rio Summit in 1992, world governments agreed that the best way to protect the environment and humans from pollution was to act in a preventive measure by applying the precautionary principle (in face of scientific uncertainty of innocuousness, discretionary use should be enforced). The precautionary principle has not been respected, the chemical industry is strong, and overrules precaution. Most countries lack trained human resources and adequate facilities to monitor exposure, pollution and its impact, such systems simply do not exist. In the current legal situation of Mexico, industry is not legally obliged to prove that the chemicals it produces, and the consumer articles they end up in, are in fact safe. The current state of the affair forces the consumer to prove that the substance is detrimental to human health and the environment before any action is taken. A good example of this situation is the current concerns regarding the massive disposal of one-time-use-plastics and how the citizens of the world, not the government and certainly not the industry, are trying to clean up the mess we have created.

The negative effects of chemicals are only investigated after chemical catastrophes, high incidence of disease, or very strong evidence of an immediate link. However, certain health impacts emerge only after decades or even generations of exposure. Such an accident, which also alerted us to the relationship between the environment and human health, is illustrated in the tragedy, which occurred to residents that lived in the region of the Minamata Bay in Japan. In the 1930s, a plastic-producing chemical plant began discharging mercury into the Minamata Bay. In the 1950s, cats and pigs in the area began showing abnormal behavior; the condition was described as the disease of the dancing cats. In the 1960s, villagers began to show symptoms such as sensory impairment, hearing loss, ataxia, and speech impairment. Approximately 6% of the infants born near Minamata had cerebral palsy. By 1980 over 50 people died and 700 were left permanently paralyzed, the cause: methylmercury. It is estimated that over 600 tons of mercury were discharged from the plant between 1932 and 1968. Fish was a staple food of the people living in the Minamata Bay (Rodier and Zeeman 1993).

Mercury is an element with chemical symbol Hg, atomic number 80 and atomic weight 200.59. Once released, mercury circulates in the environment between air, water, sediments, soil, and biota in various forms, three of which are important to this chapter: (1) Elemental or metallic mercury (Hg^0) is liquid and volatilizes at room temperature, it is the major form in the air and is scarcely soluble in water, it is the familiar metal we find in thermometers; (2) inorganic mercury in the form of salts when it combines, for example, with chlorine, oxygen, or sulfur, as either monovalent (Hg^+) or divalent (Hg^2), it is sparingly soluble in water, in the atmosphere it associates with particles, and (3) organic mercury in the form of methylmercury (CH_3Hg^+) which is soluble in water.

Although most of the mercury released is inorganic, a fraction of it is converted into organic mercury (mostly methylmercury) by bacteria, this is important because methylmercury accumulates in the food web so effectively that most fish consumption is the primary exposure pathway for methylmercury in humans and wildlife (Harris et al. 2003). Because methylmercury is stored throughout the life of aquatic organisms it is transferred up the food web and results in the highest concentration in larger long-lived, predatory/carnivorous species such as swordfish, pike, and ocean tuna. The bulk of methylmercury is stored in the muscle. Fish, shellfish, and marine mammal consumption are the primary source of human exposure (Gilbert 2004). Elemental mercury is used in electric lamps and switches, gauges and controls (e.g., thermometers, barometers, thermostats), battery production, and nuclear weapons. Because mercury has a high affinity for gold and silver, it is used in precious metal extraction from the ore (Gilbert 2004). Mercury is used in folk remedies and Santeria with unknown health implications (Newby et al. 2006).

How does mercury reach the trophic web? The Earth's crust is the ultimate source of all metallic elements found in the environment. Some of the mercury found in the environment is product of rock leaching and erosion processes, volcanic activity and forest fires. Metals including mercury are not created or destroyed by humans but are redistributed through mining and industrial activity. Generally environmental expected levels of inorganic mercury tend to be very low, between 10 and 20 ng/m^3 have been measured in urban outdoor air or less than 5 ng/L in surface water, these levels are below safe levels to breath or drink (Alimonti and Mattei 2008).

Mercury is freed to the atmosphere as a particulate in the vapors, which originate from coal combustion, waste incineration, chlor-alkali facilities (to produce chlorine and caustic soda), vinyl chloride monomer production, and mining. Total anthropogenic emissions of mercury to the atmosphere in 2010 are estimated at 1960 tons. However, the emissions estimate still has large associated uncertainties, giving a range of 1010–4070 tons per year (UNEP 2013).

Annual anthropogenic mercury releases into the water total 1000 tons at a minimum. Annual releases have been calculated for different human activities: power plants and factories release approximately 185 tons; old mines, landfills, waste disposal locations release 8–33 tons; artisanal and small-scale gold mining total around 800 tons. Deforestation and forest fires mobilize 260 tons into rivers and lakes (UNEP 2013). Current anthropogenic sources are responsible for 30% of annual emissions of mercury to air. Another 10% comes from natural geological sources, and the rest 60% is from re-emissions of previously released mercury (UNEP 2013).

Mexico is a mercury-mining country; the Secretariat of Economy reported 314 active mines. The potential reserve was calculated in 42,000 metric tons (Castro Díaz 2013). Between 2007 and 2009 the average annual production was estimated in 13 tons, informal miners using low-technology operations accomplished extraction (Camacho et al. 2016). There are gaps of knowledge related to mercury cycling and mercury concentration in water, sediments, and the lower food web in the Gulf of Mexico (Harris et al. 2012). This premise applies to the coastline of the State of Yucatan. The presence of most of the pollutants used in a range of industrial activity manifests within the coastal regions in the proximity of oil refineries and the plants used for mining, metallurgy and the production of agrochemicals and untreated domestic discharge (Botello et al. 2018). Arcega-Cabrera and Fargher (2016) report that even sociocultural and economic patterns have an influence on the release and exposure to mercury, and other metals on children in Yucatan State. Dissolved non-anthropic concentrations in the world range between 0.4 and 2 ng/L. In the Mexican east coastline mercury was found up to a concentration of 75 μg/L in heavily industrialized areas during the 70s (Núñez Nogueira 2005). The Global Mercury Observation System established a monitoring station in the town of Celestún, Yucatan State. The annual average air concentration was calculated to be 1.047 ± 0.271 ng/m^3. The study concluded that local sources did not influence significantly the total gas mercury variations (Velasco et al. 2016; Sprovieri et al. 2017). In simulations made by Harris et al. (2012), total mercury depositions for the coast of Yucatan are between 12 and 21 μg/m^2/yr. There is a need to identify which mercury sources are most important to the coast, how mercury is redistributed, and how sources vary amongst locations (Harris et al. 2012).

When small amounts of metallic mercury are swallowed (for example a broken thermometer) less than 0.01% is absorbed through a healthy intestine. However, when metallic mercury vapors are inhaled, most of it (up to 80%) enters the blood stream directly from the lung and rapidly goes to every part of the body including brain and kidneys (ATSDR 1999). After entering the brain mercury oxidizes and will not transfer back across the blood–brain barrier. Mercury in all its chemical forms is a potent neurotoxicant. The pathology focuses in the cerebellum (granule cells), calcarine cortex, and dorsal root ganglia. The lesions correlate with the neurological signs and symptoms, namely ataxia, constriction of the visual field, and sensory disturbances (Chang and Cockerham 1994). Most of the metallic mercury concentrates in the kidney, but eventually leaves the body through urine and feces (ATSDR 1999). Mercury salts are steady at room temperature and do not easily evaporate, therefore it is less likely to inhale mercury in this form. When ingested up to 10% can be absorbed. Very small amounts can pass the skin. When inorganic mercury enters the bloodstream, it reaches organs and tissues, it accumulates mainly in the kidneys, but does not pass the blood–brain barrier as easily as metallic or methylmercury (ATSDR 1999). Methylmercury is readily absorbed by the digestive tract, up to 95%; it also evaporates easily entering the body through inhalation. Dimethylmercury passes the skin barrier. Methylmercury leaves the body slowly over periods of several months, mostly through the feces. Methylmercury crosses the placental barrier; levels in the fetal brain are about 5–7 times that of maternal blood (Cernichiari et al. 1995).

The general toxic mechanism for mercury may be outlined as follows. Mercury, especially methylmercury, has a high affinity for sulfhydryl (–SH) ligands. It forms complexes with rich –SH amino acid cysteine, which acts as a carrier for mercury in the blood and into the nervous system. In the nervous cell mercury binds with other –SH rich components, specifically membranous structures within the cell such as endoplasmic reticulum, mitochondria, and Golgi system. The cysteine–methylmercury complex also resembles another amino acid, methionine that is important in the initiation of the polypeptide chain in protein synthesis. Thus, mercury not only disrupts the energy metabolism but also the protein synthesis. Demethylation may occur to some extent within the cell. The cleavage of the carbon–mercury bond generates methyl free radicals, which promote lipid peroxidation. Methylmercury also disrupts microtubules, damaging the cytoskeleton contributing to cellular dysfunction. This is particularly important in developing neurons, especially when they are active differentiating and migrating, as is the case in fetuses and children (Chang and Cockerham 1994). According to current knowledge toxic metals also possess endocrine-disrupting properties (Dyer 2007). High levels of mercury in blood have been positively associated with metabolic syndrome and obesity, and the association was dose dependent across metabolic and weight phenotypes (Lee 2013; Bae et al. 2016), with visceral adipose tissue in healthy adults (Park et al. 2017), and with high cholesterol in children (Fan et al. 2017) damaging pancreatic beta cells, causing hyperglycemia (González Villalva et al. 2016).

Several studies have demonstrated an association between dietary intakes of fish and the accumulation of mercury in the body (Iwasaki et al. 2003; Knobeloch et al. 2007; Miranda et al. 2011). In Mexico the average adult consumes approximately 12 kg of seafood per year (CONAPESCA 2015). We were unable to find data on average seafood consumption in the State of Yucatan; however, a surprisingly low proxy measurement indicated that the yearly expenditure in seafood was the third lowest in the country with $ 6.27 US per year (Zacatecas with the lowest annual expenditure $ 5.20 US and Nayarit the highest with $ 28.18 US, exchange rate 18.71 pesos/U.S. dollar, 1 October 2018) (INEGI 2013). Regardless of the level of seafood consumption, mercury accumulates in growing scalp hair (Table 15.1). Concentrations in hair are proportional to simultaneous concentrations in blood but are about 250 times higher. They are also proportional to concentrations in the target tissue, the brain (Cernichiari et al. 1995).

Legal norms in Mexico (Norma Oficial Mexicana-NOM) regulate admissible concentrations of mercury for fresh, tinned, frozen, and refrigerated fish. Said norms have not yet established the permissible levels of mercury in human hair; however,

Table 15.1 Total mercury hair concentration reported in selected studies of residents from Mexico and Florida, United States

Location	Sampled population	Total mercury concentration, mean (±sd)	*n*	Reference
Indian River Lagoon, Florida, USA	Adults[a]	1.53 ± 1.89 μg/g	135	Schaefer et al. (2014)
Indian River Lagoon, Florida, USA	Adult men[a]	2.02 ± 2.38 μg/g	73	Schaefer et al. (2014)
Indian River Lagoon, Florida, USA	Adult women[a]	0.96 ± 0.74 μg/g	62	Schaefer et al. (2014)
Yucatan, Mexico	Adults (dentists)	3.32 ± 3.75 ppm (0.10–15.77)[b]	20	Mendiburu-Zavala et al. (2011)
Yucatan, Mexico	Adults (≥4 amalgams)	1.66 ± 1.65 ppm (0.0004–5.50)[b]	18	Mendiburu-Zavala et al. (2011)
Yucatan, Mexico	Adults (0 amalgams)	1.68 ± 2.37 ppm (0.05–8.98)[b]	13	Mendiburu-Zavala et al. (2011)
Jalisco, Mexico	Adult women	2218.0 ± 14707.5 ppb	92	Trasande et al. (2010)
Jalisco, Mexico	Adult women	686.4 ± 720.2 ppb	91	Trasande et al. (2010)
Veracruz, Mexico	Adults	1.48 ± 0.86 μg/g (0.36–3.36)	27	Guentzel et al. (2007)
Veracruz, Mexico	Children (≤18)	1.12 ± 0.62 μg/g (0.18–2.57)	20	Guentzel et al. (2007)

[a]Total hair mercury concentration was statistically significantly associated with total seafood consumption ($p < 0.01$)
[b]Range

maximum permissible mercury levels in hair have been established by international regulatory instances. One of the most common cutoff points used is 1 μg/g, as presented by the U.S. EPA, the National Academy of Sciences (NAS) and the United Nations Environment Program (UNEP) (Nutall 2006).

The aim of this study was to determine concentrations and the influence of seafood consumption on total mercury concentrations in hair for a population at the city of Merida. Potential factors such as age, sex, body weight, height, and BMI were also assessed.

15.2 Materials and Methods

15.2.1 Samples

The study was conducted in Merida, Yucatan, the major city in southeast Mexico from August 2011 to January 2012 as part of an ongoing monitoring project of heavy metal pollution in the city of Merida and its zone of influence. Sample size was calculated using Minitab® (version 16.1, State College, Pennsylvania). Mean was used as a parameter for normal distribution; the standard deviation was obtained from reviewed literature (0.85), with a margin of error of 0.25. By these calculations, sample size should be ≥73 people.

The Ethics Committee (Cobish, Spanish acronym) of Cinvestav approved the investigation. The purpose of the study was explained to all individuals interested in participating, who also gave a signed consent. Volunteers provided head hair in exchange for the analysis results. Pregnant women were excluded from the study because of the risk of bioimpedance analysis to the concept. People with artificial hair color or perm were excluded from the study since some hair dyes and perm lotions contain heavy metals that do not wash out. Two grams of hair were cut as close as possible to the scalp of the occipital area. Collected hair samples were coded and stored in clean polyethylene bags until mercury analysis. Data on age, sex, weight, height, fat percentage, and BMI were recorded at the same time the hair sample was taken (Table 15.2). Smoking habits, alcoholic beverage consumption, exposure to broken light bulbs or thermometers, blood transfusions, and general health were recorded through a questionnaire (Table 15.3).

15.2.2 Dietary Study

Participants were asked about their seafood consumption habits, including number of seafood meals consumed per month, and type of seafood (fish, shellfish, octopus, cuttlefish, etc.) reflecting their average intake. Frequency of fish consumption was classified into five categories: 1: never/month; 2: 1–7 times/month; 3: 8–14 times/month; 4: 15–21 times/month; and 5: >21 times/month (Table 15.2).

Table 15.2 Total hair mercury concentration (μg/g) by demographic, anthropometric, and seafood consumption (days/month) variables in Merida, Yucatan

Participants	*n*	Percentages	Mean ± SD	Range	*p*-value[a]
Sex					
Men	45	50			
Women	45	50			
Age					
Men	45	50	29.24 ± 7.66	19–44	
Women	45	50	27.38 ± 5.85	18–39	
BMI					
Men	44	50.5	27.7 ± 5.5	18.9–42.8	<0.01
Women	43	49.4	24.8 ± 4.8	18.8–38.5	
Body fat %					
Men	44	50.5	24.2 ± 9.5	7.3–51.3	<0.05
Women	43	49.4	29 ± 8.6	12.1–49.1	
Total hair mercury (μg/g)					
Men	45	50	1.24 ± 0.966	0.153–4.383	>0.05
Women	45	50	0.983 ± 0.713	0.136–3.172	
Total	90	100	1.119 ± 0.854	0.136–4.383	
Total seafood consumption days/month					
Never	7	7.8			
1–7	46	51.1			
8–14	29	32.2			
15–21	3	3.3			
>21	5	5.6			

[a]*p*-value from independent sample *t*-test using square root of mercury

Table 15.3 Exposure status and percentages of pathological and non-pathological records of adult population that participated in this study ($n = 90$)

Record	Status	*n*	%
Accidental exposure (broken thermometers, light bulbs, batteries)	Exposed	23	25.6
	Not exposed	67	74.4
Cigarette	Smokers	17	18.9
	Non-Smokers	73	81.1
Alcohol consumption	Exposed	60	66.7
	Not exposed	30	33.3
Cardiac disease (diagnosed)	With	6	6.7
	Without	84	93.3

15.2.3 Mercury Determination

Using ultrasound-cleaning equipment (BRANSON, Bransonic®, CT 06813-1961 USA), hair was washed with 100 mL acetone for 5 min and rinsed three times with deionized water for 5 min (Diez et al. 2008; Llorente et al. 2011; Marrugo 2008). Washed samples were frozen at −67 °C for 1 h, and lyophilized (LABCONCO®, Kansas City, Missouri 64,132, USA) at −0.22 mBar, −53 °C for 24 h to eliminate water. One gram of hair was digested with concentrated HCl (0.5 mL), HNO_3 (5 mL), H_2O_2 (2 mL) and H_2SO_4 (5 mL) by a pre-established method, using a microwave system (Anton Par®, Graz, Austria) and close Teflon recipients (EPA 1996). Digested samples were stored at 4 °C until analysis. Samples were analyzed at the Biogeochemistry and Environmental Quality Unit, Sisal, Yucatan of the School of Chemistry of the National Autonomous University of Mexico. Total mercury (THg) determination was done using a cold vapor atomic absorption spectrometry method (FIAS-400, Perkin Elmer®, CT 06484-04794, USA), coupled to an atomic absorption spectroscopy equipment (AAnalyst-800, Perkin Elmer®, MA 02451, USA). Total mercury was measured at 253.7 nm. The accuracy of the method was calculated by a certified reference material from the Analytical Quality Control Services, IAEA, Vienna, Austria (IAEA-085®). Accuracy was estimated by a reference material analysis with values of 99.44 ± 11.24% (range 81.29–115.26) ($n = 8$) from the certified value. The limit of detection (three times the standard deviation of the blank) was 0.22 μg g^{-1} and the limit of quantification was 18.86 μg g^{-1} evaluated as the lowest analyzed concentration on the calibration plot. Each sample was analyzed in triplicate.

15.2.4 Statistical Analysis

Data distribution of mercury concentration, age, weight, height, BMI, and fat percentage in men and women were evaluated by the Kolmogorov–Smirnov test. Only three variables presented normal distribution: (1) age in women, (2) height, and (3) fat percentage. Even though most variables did not present a normal distribution, the central limit theorem indicates that in such cases in which the sample size n is equal or higher than 30, it is valid to assume that the mean of said sample reasonably reaches a normal distribution (Daniel 2002). Under the mentioned premise a comparative analysis of the means was made with the Student's t test. We used Pearson correlation test to evaluate associations between variables (Triola 2009). We developed a linear regression model were the response was total hair mercury concentration (total mercury was square root transformed to achieve normal distribution and homogeneous variance) with seven explicative variables (frequency of seafood consumption/month, BMI (kg/m^2), body fat percentage, fat-free mass, age in years, number of alcoholic beverages consumed/month, number of cigarettes smoked/month). The level of significance was set as $p < 0.05$ for all analysis. All statistics were performed with Statistica® version 6.0 (StatSoft, Inc. Tulsa Oklahoma).

15.3 Results

Seafood consumption, body measurements, and hair samples were provided by a total of 90 people, 45 men and 45 women. Anthropometric data, hair mercury concentrations, and seafood consumption frequency are shown in Table 15.2. Most of the participants were volunteers recruited from two higher education study centers (65.6% Cinvestav, Merida Unit $n = 59$, 33.3% Autonomous University of Yucatan $n = 30$) but not exclusively students. Consequently, the sample was not random, but drawn primarily from people who volunteered. It was not a sample that concentrated on mercury labor-exposure, but represented the average population. Weight and height data were missing for three people (3.3%). The mean age for men was 29.24 years (Min.–Max. 19–44); the mean age for women was 27.38 (Min.–Max. 18–39). BMI and fat percentages were different between sexes; men had a higher BMI and lower fat percentage than women ($p < 0.01$, $p = 0.01$ respectively). In this sample 47.1% presented some degree of overweight or obesity.

Women presented lower mercury concentration in hair than men, but it was not statistically significant ($p > 0.05$). Forty percent had mercury levels above 1 µg/g hair. The study participants appeared to be healthy and some of them confirmed specific exposure to mercury or had diagnosed cardiac problems (Table 15.3).

We did not find evidence of correlation between age, weight, body fat percentage, BMI, and total mercury among the whole group tested ($p > 0.05$). However, there was a significant correlation between height and total mercury ($p = 0.05$). There was no significant correlation between age and seafood consumption ($p > 0.05$).

A multivariate linear regression model was developed for seven variables (seafood consumption frequency, BMI, body fat %, fat free mass, age in years, number of alcoholic beverages consumed in 1 month, number of cigarettes smoked in 1 month) on the square root of total mercury hair concentration (to achieve normality). After eliminating nonsignificant variables ($p > 0.05$) the final theoretical linear regression model was $\hat{Y} = (\beta_0 + \beta_1 X_1)^2 + \varepsilon$, where the only explicative variable for total mercury concentration in hair was seafood consumption frequency/month. The regression model adjusted to the predictive variable was: total mercury concentration in hair = (0.809803 + 0.0244192 ∗ (monthly frequency of seafood consumption)^2 (Fig. 15.1). For this model we obtained an F-coefficient of 20.66 ($p < 0.001$). Because in an ANOVA table the value is lower than 0.001, we can conclude there is a statistically significant relationship with a 99% confidence between variables. With an r^2 value = 0.1901 and an r^2 adjusted = 0.1809, the model explains 18.09% of the mercury concentration variability after transformation. The correlation coefficient of r was 0.436 ($p < 0.001$), showing a weak but statistically significant correlation between variables. Standard error estimation was 0.334 and residual vs. fit graphs verified assumptions in the model (normal distribution). We also applied the Durbin–Watson test (1.848, $p > 0.05$) to ascertain significant correlation based on the order of the data. Because the p value was larger than 0.05 there is no indication of serial correlation between residuals.

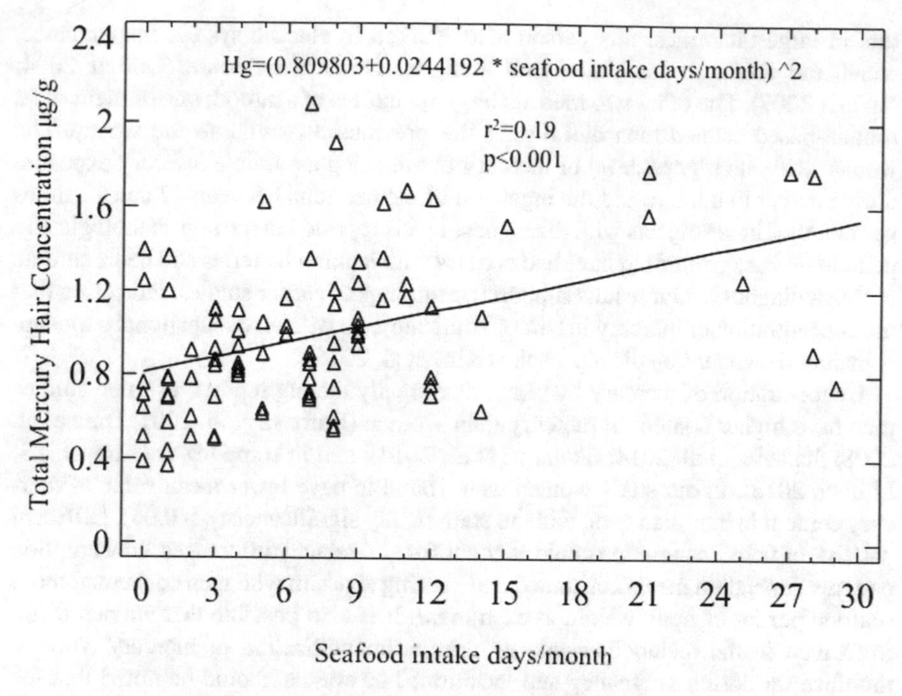

Fig. 15.1 Multiple linear regression analysis graph of seafood consumption frequency and the square root of total hair mercury concentration

15.4 Discussion

Study participants had a mean hair mercury concentration of 1.119 µg/g; this concentration is in accordance with findings in other parts of Mexico and Florida, USA (Table 15.1). Our data showed values slightly higher than those suggested by international organizations, 42.2% of the samples obtained from participants exceeded levels deemed as safe (1 µg/g). From the 45 tested women, 27% ($n = 12$) had concentrations above the U.S. EPA benchmark dose for maternal exposure level related to child developmental impairment (1.2 µg/g). Because these levels have been associated with neurological deficits like child cognition and visual recognition, verbal performance, language, attention and memory deficiencies (Mahaffey et al. 1997), prospective mothers in this group should seek medical advice in case of pregnancy.

Seafood consumption was the strongest and best predictor of total mercury in the studied population ($p < 0.001$). We examined possible exposure sources that could contribute to hair mercury concentration (Table 15.3); however, this study was not designed to make an exhaustive account of possible exposures. We observed that when eliminating extreme values, the highest mercury levels were present in the people with the highest seafood intake frequency. Three men and one woman had levels above 3 µg/g; one man had more than 4 µg/g. The person with the highest hair mercury concentration consumed fish at least 7 times/month but always apex predators in large quantities, this person also worked in electric system maintenance, which the literature associates with a higher mercury exposure (Gilbert 2004; NIOSH 2007). The other two men ate large quantities of seafood; one of them did a protein-based canned tuna diet during the previous 3 months to the survey. The woman with the highest level of mercury commented a possible mercury exposure during a visit to a mine, and the ingestion of canned tuna between 17 and 21 times per month. Three subjects with the highest levels reported in the non-pathologic and pathologic backgrounds to have had contact with leaking batteries and had a cardiac problem diagnosis. Our results support the findings of earlier studies, indicating that the concentration of mercury in hair of adult subjects with no occupational exposure is mainly dependent on dietary intake (Kim et al. 2016).

Concentration of mercury by sex varies greatly among reports; in some studies men have higher content of mercury than women (Barbosa et al. 2001; Diez et al. 2008; Schaefer et al. 2014; Skalnaya et al. 2014), and in some lower (Airey 1983; Li et al. 2013). In our study women were found to have lower mean value of mercury content in hair than men, with no statistically significance ($p > 0.05$). Different patterns of fish consumption could account for sex-related differences, consumption patterns depend on meals consumed and serving sizes, maybe men consumed more seafood per kg of body weight at each meal. It is also possible that women differ from men in the metabolism and fat compartmentalization of mercury. Women mobilize fat during pregnancy and lactation; however, it should be noted that the translactional barrier is more effective to mercury acquisition than the transplacental barrier (Dórea 2004).

Because of the lack of correlation ($p > 0.05$) between mercury hair concentration and age, weight, body fat percentage, and BMI we conclude that in this study we did not find a modulatory effect of these variables on mercury metabolism. Additional research is required to estimate the dynamics of toxic trace elements in overweight and obesity (Skalnaya et al. 2014). We only found a positive statistically significant correlation between heights and mercury hair concentration ($p < 0.05$). The reviewed literature does not explain about the reason for this situation; however, our findings are not exclusive, other authors found similar results (Diez et al. 2008; Palacios-Torres et al. 2018).

The results of the multiple linear regression analysis are consistent with other published research in which seafood exposure is the main factor that contributes to body mercury burdens (Mahaffey et al. 2009; Smith et al. 2009; Park et al. 2017; Lincoln et al. 2011). Because 42.2% of the sampled population presented levels higher than those estimated safe (1 μg/g), a local advisory should be designed to reduce exposure through the seafood intake pathway, especially in women of childbearing age. Such an advice should include canned tuna and species already identified with high mercury burdens (top predators), but it would also be important to emphasize the beneficial nutritive qualities of seafood. The challenge should be to find a balance between positive and negative effects of seafood consumption.

In summary, there is a need for more exhaustive regional estimates of mercury exposure among individuals who consume seafood at a higher frequency than the general population. These data provide insights into regional mercury exposure where access and consumption of seafood has not been deemed high by authorities (INEGI 2013). Given that other states in Mexico apparently spend more money in seafood purchases and therefore consumption, further studies are warranted to confirm the body burdens of Mexicans especially women in childbearing age. Further studies are needed to understand the relation between mercury and increased body mass index, visceral abdominal fat and height, especially considering the obesity epidemics in the region.

Acknowledgments The authors acknowledge the support of the Consejo Nacional de Ciencias y Tecnología of Mexico (CONACyT), Sally López, Grant Number 245024, and MSc Felipe García López for his statistical advice.

References

Airey D (1983) Total mercury concentration in human hair from 13 countries in relation to fish consumption and location. Sci Total Environ 31:157–180

Alimonti A, Mattei D (2008) Biomarkers for human biomonitoring. In: Conti ME (ed) Biological monitoring: theory and applications. WIT Press, Boston, pp 163–211

American Chemistry Council (2018) Debunking the myths: Are there really 84,000 chemicals? https://www.chemicalsafetyfacts.org/chemistry-context/debunking-myth-chemicals-testing-safety/. Accessed 12 Sept 2018

Arcega-Cabrera F, Fargher LF (2016) Education, fish consumption, well water, chicken coops, and cooking fires: using biogeochemistry and ethnography to study exposure of children from Yucatan, Mexico to metals and arsenic. Sci Total Environ 568:75–82

ATSDR (1999) Toxicological profile for mercury. Agency for Toxic Substances and Disease Registry, Atlanta

Bae S, Park SJ, Yeum KJ et al (2016) Cut-off values of blood mercury concentration in relation to increased body mass index and waist circumference in Koreans. J Investig Med 64:867–871

Barbosa AC, Jardim W, Dórea JG et al (2001) Hair mercury speciation as a function of gender, age, and body mass index in inhabitants of the Negro River Basin, Amazon, Brazil. Arch Environ Contam Toxicol 40:439–444

Botello AV, Villanueva FS, Rivera RF et al (2018) Analysis and tendencies of metals and POPs in a sediment core from the Alvarado Lagoon System, Veracruz, Mexico. Arch Environ Contam Toxicol 75:157–173

Camacho A, Van Brussel E, Carrizales L et al (2016) Mercury mining in Mexico: I. Community engagement to improve health outcomes from artisanal mining. Ann Glob Health 82:149–155

Castro Díaz J (2013) An assessment of primary and secondary mercury supplies in Mexico, 1st edn. Commission for Environmental Cooperation, Montreal

CEFIC (2018) Extra—EU chemicals trade balance. In: CEFIC-Chemdata International. http://fr.zone-secure.net/13451/451623/?startPage=14-page=21. Accessed 10 Sept 2018

Cernichiari E, Brewer R, Myers GJ et al (1995) Monitoring methylmercury during pregnancy: maternal hair predicts fetal brain exposure. Neurotoxicology 16:705–710

Chang LW, Cockerham LG (1994) Toxic metals in the environment. In: Cockerham LG, Shane BS (eds) Basic environmental toxicology, 1st edn. CRC Press, Boca Raton, pp 109–132

CONAPESCA (2015) Anuario estadístico de acuacultura y pesca. Comisión Nacional de Acuacultura y Pesca, SAGARPA, Sinaloa

Daniel W (2002) Bioestadística. Base para el análisis de las ciencias de la salud. Limusa-Wiley, México. D. F.

Diez S, Montuori P, Pagano A et al (2008) Hair mercury levels in an urban population from southern Italy: fish consumption as a determinant of exposure. Environ Int 34:162–167
Dórea JG (2004) Mercury and lead during breast feeding. Br J Nutr 92:21–40
Dyer CA (2007) Heavy metals as endocrine-disrupting chemicals. In: Gore AC (ed) Endocrine-disrupting chemicals: from basic research to clinical practice, 1st edn. Humana Press, New Jersey, pp 111–133
EPA (1996) Microwavc assisted acid digestion of siliceous and organically based matrices, Method 3052. https://www.epa.gov/sites/production/files/2015-12/documents/3052.pdf. Accesed 12 Sept 2018
EPA (2017) Prioritizing existing chemicals for risk evaluation. In: Assessing and managing chemicals under TSCA. Environmental Protection Agency. https://www.epa.gov/assessing-and-managing-chemicals-under-tsca/prioritizing-existing-chemicals-risk-evaluation. Accessed 12 Sept 2018
European Commission (2001) White-paper strategy for a future chemicals policy. Brussels
Fan Y, Zhang CL, Bu J (2017) Relationship between selected serum metallic elements and obesity in children and adolescent in the US. Nutrients 9(2):104. https://doi.org/10.3390/nu9020104
Gilbert SS (2004) A small dose of toxicology. CRC Press, Boca Raton
González-Villalva A, Colín-Barenque L, Bizarro-Nevares P et al (2016) Pollution by metals: is there a relationship in glycemic control? Environ Toxicol Pharmacol 46:337–343
Guentzel JL, Portilla E, Keith KM et al (2007) Mercury transport and bioaccumulation in riverbank communities of the Alvarado Lagoon System, Veracruz state, Mexico. Sci Total Environ 388(1–3):316–324. https://doi.org/10.1016/j.scitotenv.2007.07.060
Harris R, Krabbenhoft D, Mason R et al (2003) Introduction. In: Harris R, Krabbenhoft D, Mason R et al (eds) Ecosystem responses to mercury contamination, indicators and change, 1st edn. CRC Press, Boca Raton, pp 1–11
Harris R, Pollman C, Hutchinson D et al (2012) A screening model analysis of mercury sources, fate and bioaccumulation in the Gulf of Mexico. Environ Res 119:53–63
INEGI (2013) Encuesta nacional de gastos del hogar. Instituto Nacional de Estadística y Geografía, México D. F.
Iwasaki Y, Sakamoto M, Nakai K et al (2003) Estimation of daily mercury intake from seafood in Japanese women: Akita cross-sectional study. Tohoku J Exp Med 200(2):67–73
Kim KH, Kabir E, Jahan SA (2016) A review on the distribution of Hg in the environment and its human health impacts. J Hazard Mater 306:376–385
Knobeloch L, Gliori G, Anderson H (2007) Assessment of methylmercury exposure in Wisconsin. Environ Res 103(2):205–210
Lee K (2013) Blood mercury concentration in relation to metabolic and weight phenotypes using the KNHANES 2011–2013 data. Int J Occup Environ Health 91:185–193
Li J, Cen D, Huang D et al (2013) Detection and analysis of 12 heavy metals in blood and hair sample from a general population of Pearl River Delta Area. Cell Biochem Biophys 70:1663–1669
Lincoln RA, Shine JP, Chesney EJ et al (2011) Fish consumption and mercury exposure among Louisiana recreational anglers. Environ Health Perspec 119(2):245–251
Llorente B, Vírseda C, Peral J et al (2011) Metilmercurio en el cabello de población infantil. Sanidad Militar 67:299–303
Mahaffey KR, Swartout J, Rice GE (1997) Mercury study report to Congress Volume IV: An assessment of exposure to mercury in the United States. https://www3.epa.gov/airtoxics/112nmerc/volume4.pdf. Accessed 12 Sept 2018
Mahaffcy KR, Clickner RP, Jeffries RA (2009) Adult women's blood mercury concentrations vary regionally in the United States: association with patterns of fish consumption. Environ Health Perspec 117:47–53
Marrugo N (2008) Distribution of mercury in several environmental compartments in an acuatic ecosystem impacted by gold mining in northern Colombia. Arch Environ Contam Toxicol 55:305–318

Mendiburu-Zavala CE, Acereto-Escoffie PO, López-Villanueva ME et al (2011) Mercurio total en cabello de cirujanos dentistas de práctica general del estado de Yucatán, México. Rev Odontol Latinoamericana 3:11–16

Miranda ML, Edwards S, Maxson PJ (2011) Mercury levels in an urban pregnant population in Durham County, North Carolina. Int J Environ Res Public Health 8(3):698–712

Newby A, Riley DM, Leal-Almeraz TO (2006) Mercury use and exposure among Santeria practitioners: religious versus folk practice in Northern New Jersey, USA. Ethnic Health 11:287–306

NIOSH (2007) NIOSH pocket guide to chemical hazards National Institute for Occupational Safety and Health. Centers for Disease Control and Prevention, National Institute for Occupational Safety and Health, Cincinnati

Núñez Nogueira G (2005) Concentration of essential and non-essential metals in two shark species commonly caught in the Mexican (Gulf of Mexico) coastline. In: Botello AV, Rendón von Osten J, Gold-Bouchot G et al (eds) Golfo de Mexico, contaminación e impacto ambiental: Diagnóstico y tendencias, 2nd edn. Universidad Autónoma de Campeche, Universidad Nacional Autónoma de México, Instituto Nacional de Ecología, Mexico, pp 452–473

Nutall KL (2006) Interpreting hair mercury levels in individual patients. Ann Clin Sci 36:248–261

Palacios-Torres Y, Caballero-Gallardo K, Olivero-Verbel J (2018) Mercury pollution by gold mining in a global biodiversity hotspot, the Choco biogeographic region, Colombia. Chemosphere 193:421–430

Park JS, Ha KH, Ka H et al (2017) Association between blood bercury level and visceral adiposity in adults. Diabetes Metab J 41:113–120

Rodier DJ, Zeeman MG (1993) Ecological risk assessment. In: Cockerham LG, Shane BS (eds) Basic environmental toxicology, 1st edn. CRC Press, Boca Raton, pp 581–604

Schaefer AM, Jensen EL, Bossart GD et al (2014) Hair mercury concentrations and fish consumption patterns in Florida residents. Int J Environ Res Public Health 11:6709–6726

Skalnaya MG, Tinkov AA, Demidov VA et al (2014) Hair toxic element content in adult men and women in relation to body mass index. Biol Trace Elem Res 161:13–19

Smith KM, Barraj LM, Kantor M et al (2009) Relationship between fish intake, n-3 fatty acids, mercury and risk markers of CHD [National Health and Nutrition Examination Survey 1999–2002]. Public Health Nutr 12:1261–1269

Sprovieri F, Pirrone N, Bencardino M et al (2017) Five-year records of mercury wet deposition flux at GMOS sites in the northern and southern hemispheres. Atmos Chem Phys 17(4):2689–2708. https://doi.org/10.5194/acp-17-2689-2017

Trasande L, Cortes JE, Landrigan PL et al (2010) Methylmercury exposure in a subsitence fishing community in Lake Chapala, Mexico: an ecological approach. Environ Health 9(1). https://doi.org/10.1186/1476-069X-9-1

Triola M (2009) Estadística. Pearson Educación, Ciudad de México

UNEP (2013) Global mercury assessment 2013: sources, emissions, releases and the environmental transport. Geneva

Velasco A, Arcega-Cabrera F, Oceguera-Vargas I et al (2016) Global Mercury Observatory System (GMOS): measurements of atmospheric mercury in Celestun, Yucatan, Mexico during 2012. Environ Sci Pollut R 23:17474–17483. https://doi.org/10.1007/s11356-016-6852-5

Chapter 16
Tackling Exposure to Chagas Disease in the Yucatan from a Human Ecology Perspective

Carlos N. Ibarra-Cerdeña, Adriana González-Martínez, Alba R. Valdez-Tah, Claudia Guadalupe Chi-Méndez, María Teresa Castillo-Burguete, and Janine M. Ramsey

Abstract Chagas disease is a zoonotic infectious disease, produced by the parasite *Trypanosoma cruzi*. Infection results principally from contact with parasite-contaminated bug feces, deposited during a triatomine vector's blood meal. The vectors are ubiquitous in conserved and human-transformed landscapes, since they inhabit sylvatic mammal nests, but persist and disperse among human communities. Exposure to infected vectors occurs both in conserved habitats and along human-transformation gradients. Several studies have focused on biological determinants, but only a handful of studies analyze sociocultural determinants of transmission. We explore cultural and social patterns within landscapes that affect human exposure in Mayan populations of Yucatan state. Primary factors contributing to vector exposure are: the lack of knowledge regarding the consequences of insect blood feeding habits, houses susceptible to bug infestation, farming and land-use practices, hunting,

C. N. Ibarra-Cerdeña (✉) · M. T. Castillo-Burguete
Departamento de Ecología Humana, Cinvestav-Mérida, Mérida, Yucatán, México
e-mail: cibarra@cinvestav.mx; maria.castillo@cinvestav.mx

A. González-Martínez
Departamento de Ecología Humana, Cinvestav-Mérida, Mérida, Yucatán, México

Facultad de Ciencias Biológicas, Universidad Autónoma de Nuevo León, Monterrey, México

A. R. Valdez-Tah
Department of Anthropology, University of California, Irvine, Irvine, CA, USA

C. G. Chi-Méndez
Departamento de Ecología Humana, Cinvestav-Mérida, Mérida, Yucatán, México

Departamento de Ciencias de la Salud y Ecología Humana, Universidad de Guadalajara, Centro Universitario de la Costa Sur, Autlán de Navarro, Jalisco, México
e-mail: claudia.chi@academicos.udg.mx

J. M. Ramsey
Centro Regional de Investigación en Salud Pública, Instituto Nacional de Salud Pública, Tapachula, Chiapas, México
e-mail: jramsey@insp.mx

H. Azcorra, F. Dickinson (eds.), *Culture, Environment and Health in the Yucatan Peninsula*, https://doi.org/10.1007/978-3-030-27001-8_16

and rural migration. Gender affects the differential contribution of all of these factors. Exposure is aggravated by deficient infection diagnosis and treatment options, which raise risk for all transmission modes (vector-borne, congenital, blood transfusion and organ transplant, or oral).

16.1 Introduction

Most emerging human diseases are zoonotic (i.e., diseases from pathogens that are maintained in wildlife, and spread to humans) (Taylor et al. 2001). In fact, zoonotic infectious diseases have been important concerns to mankind since the beginning of livestock domestication 10,000 years ago (Gebreyes et al. 2014). Such a remarkable historical impact of pathogens on human history contrasts with our current understanding of spatiotemporal dynamics of disease epidemics, and how to reduce disease burden sustainably in underprivileged populations (Zeppelini et al. 2016). Moreover, recent methodological developments shed light on the enormous variety of zoonotic hazards for humans that could emerge from current patterns of global change (Olival et al. 2017; Carroll et al. 2018). Likewise, there is a global increase in zoonotic disease exposure prevalence related to socioeconomic, environmental, and ecological factors (Jones et al. 2008). Optimism by public health sectors by the mid-twentieth century, which predicted a positive impact to eradicate certain infectious diseases after the initiation of massive vaccine and antibiotic administration, have been obscured by this new evidence (Fauci 2001).

Notably, the appearance of drug-resistant pathogen strains or insecticide-resistant disease vectors reveals the evolutionary properties of these organisms, and the dynamic relationships between microorganisms, their hosts, and the environment, that challenge our efforts to control them (Morens et al. 2004). As such, more integrative strategies are needed that consider pathogens and their vectors not as static entities and human populations not as uniformly vulnerable to infection or disease risk. This approach relies on a human ecology perspective which integrates in a multidisciplinary or transdisciplinary model, those determinants that individually have been insufficient to unravel and predict the emergence of infectious diseases (i.e., social, clinical, microbiological, and ecological sciences among others) (Parkes et al. 2005). Chagas disease (CD) is an important example, due to its universal distribution in the American continent, highlighted by the World Health Organization (WHO) and other countries participating in the 2030 Millenium Goals, due to its chronicity and impact on disability-adjusted life years (DALYs) five times superior to that of malaria and twice that for dengue (WHO 2014). The cost burden is principally on the poorest and least attended populations (Ramsey et al. 2014; Sánchez-González et al. 2016). CD continues to be ignored by public health policy agendas in many endemic countries (Tarleton et al. 2014), including Mexico (Carabarin-Lima et al. 2013; Manne-Goehler et al. 2014). It is considered a Neglected Tropical Disease (NTD) by WHO and the international community, together with 19 other communicable diseases affecting up to one billion individuals, and costing billions of dollars in 149 countries (http://www.who.int/neglected_diseases/5_strategies/en/). World Health Organization's (WHO) London Declaration on Neglected Tropical Diseases proposed the elimination of intradomicile vector transmission, as well as the interruption of transfusional transmission and the control of congenital transmission, for the three most important forms of CD transmission in Latin America, for the 2020 goals. Following concerns raising questions regarding the feasibility of these goals (Tarleton et al. 2014), new goals have been set for 2030.

The current analysis focuses on our experience and those of different colleagues investigating several aspects of CD in the Yucatan peninsula, where most at-risk populations live in underprivileged conditions. Three main factors of human vulnerability are known for *T. cruzi* infection and CD: (1) lack of sufficient knowledge about disease risk due to vector contact, (2) greater exposure to vectors in anthropogenic landscapes occurs during their peak of dispersal (i.e., warm and end dry season, usually from March to June), and (3) the lack of access to infection diagnosis and multiple barriers for treatment. We believe, however, that a major proportion of disease risk could be mitigated by reducing vulnerability through formal information and health promotion programs which target the danger of vector contact during all activities (inside houses or outdoors), and continued permanent population-based surveillance to reinforce local knowledge building, which would trigger a bottom-up response motivating institutional compliance. Thus, public demand needs to drive operational and budgetary institutional responses, to provide appropriate diagnostic, treatment, and follow-up care and evidence-based prevention. Key actions to ensure information and action campaigns are discussed.

16.2 Chagas Disease

American trypanosomiasis and CD are caused by the protozoan parasite *Trypanosoma cruzi,* which can be transmitted via multiple modes: by an insect vector, blood transfusion, organ transplant, mother to fetus congenital, and oral contamination. The zoonotic infection in most mammal reservoir species is autochthonous and endemic in all countries of the American continent between parallels 40 North and South, and absent only in Canada (Rassi et al. 2010). CD is the parasitic infection with highest disease burden and a major cause of morbidity and mortality in the region (Hotez et al. 2008). The disease affects all population groups, although clear inequities in disease prevention and control programs which target acute urban viral disease priorities, and in healthcare inequities due to a lack of institutional efficacy for undetected, undiagnosed, and untreated infected individuals (due to ethnic, rural, educational, and economic determinants). The vast majority of *T. cruzi* infections go undiagnosed because of nonspecific mild symptoms in the acute phase, lack of access to medical care, and lack of program-driven actions aimed at CD (Rassi et al. 2010; Viotti et al. 2014). CD is therefore considered a Neglected Tropical Disease for which improvement in diagnosis and treatment currently requires research and development efforts with nonprofit interests, as well as political will (Rocha-Gaso et al. 2017). In Mexico, it is a significant public health concern due to the estimated two million infections (1.5% of blood donations infected), which represents more than 650,000 in chronic phase, and 5–6% of the country's mortality (Ramsey et al. 2014, 2015). Yet, less than 0.5% of infections have access to treatment (Manne et al. 2013). The average prevalence of CD in the Yucatan Peninsula (Campeche, Quintana Roo, and Yucatán) detected in blood banks between 2006 and 2017 is 1.33% (95% confidence intervals = 1–1.78%), based on a sample size of 103,686 potential donors. This is at least two times higher than the national average for the same period, 0.51% (95% confidence intervals = 0.49–0.53%), an estimation based on more than 450,000 potential donors (data compiled and reviewed by Arnal et al. 2018). Although this region presents the higher prevalence of blood bank positive tests in Mexico (Garcia-Montalvo et al. 2011), there is a disproportioned low number of diagnostics in clinical practice (Méndez-Domínguez et al. 2017).

Even if principal disease burden of CD affects vector-endemic tropical countries, the ease of traveling and current migratory trends now affect all continents (Coura and Viñas, 2010). In order to face the epidemiological challenges of CD due to the increasing complexity of all *T. cruzi* transmission routes in endemic and non-endemic countries (Tarleton et al. 2014), access to diagnosis and early treatment is the most cost-effective way to reduce the CD burden (Ramsey et al. 2014). Unfortunately, multiple barriers exist to improve and provide adequate approaches for the Mexican population, related to systemic and administrative obstacles, or poorly designed or integrated strategies which have not considered or incorporated pertinent and existing evidence. Information provision to exposed populations is frequently erroneous for disease awareness, or inadequate in most cases related to seeking or relying (mistrust) on medical attention after reporting bug bites, or even when diagnosis is provided following transfusion blood donation, but no follow-up to treat these patients occurs. Lack of disease awareness and its relationship to bug bites obviates the right to be informed that the individual may carry the parasite and develop severe disease, and to receive treatment.

Two principal clinical stages are recognized, the first being the acute phase which may last a few months after parasite inoculation or acquisition, development of specific antibodies developed around 3–4 weeks after challenge, and may be asymptomatic (>60%), or entail more acute symptoms (acute myocarditis, or meningoencephalitic symptoms in children). This acute phase is followed by the chronic, initially asymptomatic (chronic asymptomatic), in which the individual may remain for life, feeling healthy and with no suspected organ alterations (WHO 2002). Between 25 and 40% of chronic individuals progress to a chronic symptomatic stage, that reduces their life expectancy (Rassi et al. 2010). The chronic symptomatic phase develops cardiac insufficiency with progressive dilated cardiomyopathy in over 90% of cases, and in the remaining 10%, megavisceral syndromes, principally esophageal and colon, which may develop alone or along with cardiomyopathies (Prata 2001). Chronic phase symptoms develop 10–20 years after initial infection, although information from novel biomarkers should in the near future provide more robust evidence for metabolic and immune processes causing a broad range of pathology (Rassi et al. 2010). Importantly, disease severity is not necessarily related to observed clinical stage, and to date there are no symptoms or tests that can inform regarding prognosis. However, despite late recognition of symptoms, untreated patients experience severe fatigue and perhaps achalasia, causing a decrease in physical activity and appropriate nutrition, defining CD as a disease with highest social and economic impacts, associated with high disability-adjusted life indicators (Lozano et al. 2012). *T. cruzi* infection can be treated (i.e., nifurtimox or benznidazole), however, with favorable prognosis when diagnosis and treatment occurs during the acute phase, while efficacy declines as disease progresses (Ramsey et al. 2014). Therefore, treatment success also depends on shortening the time between diagnosis and clinical care (Morillo et al. 2015).

Despite the relative high number of entomological publications concerned with the CD vector, there are very few published studies on clinical manifestations found in patients from the Yucatan Peninsula. Early reports are from 1975 when four cases were diagnosed as acute CD (Zavala-Velázquez 2003). In 1992, it was reported the first study that used electrocardiographic (ECC) analysis, finding 6 cases out of 416. The cases include one 9-year-old girl in the asymptomatic chronic phase with normal ECC and five symptomatic cases with different conditions of ECC (Barrera Pérez Mario et al. 1992). In 2011, 91 patients were screened in hospitals from Campeche, where 14 tested T. cruzi positive (Alducín-Téllez et al. 2011). One of the most complete pediatric clinical cases reported in Mexico correspond with a 7-year-old child from Yucatán with Chagasic cardiomyopathy. In this case, its physician found a "chagoma" (an area with erythematous and indurated skin that result of the allergic reaction of bug bite), and suspected that a triatomine could bit him. Serological tests determined the infection and clinical studies confirmed positive signs of cardiac damages with cardiomegaly (Méndez-Dominguez et al. 2017).

Disease prevention has relied mostly on eliminating domiciled vectors using pyrethroid insecticides, where invasive species are principally if not uniquely domesticated (Schofield et al. 2006). However, insecticide resistance in *Triatoma infestans* and *Rhodnius prolixus*, primary domesticated invasive vectors in South America, has been described (Mougabure-Cueto and Inés 2015). Given maintenance of sylvatic populations of all other vectors, and extradomicile transmission by the vector, indoor spraying could be ineffective long term for many vector species (i.e., those dispersing from sylvatic habitats or peripheral zones of villages) (Ramsey et al. 2003; Waleckx et al. 2015b).

16.3 *Trypanosoma cruzi* Transmission and Hosts

The etiological agent of CD is *Trypanosoma cruzi* (Kinestoplastida: Trypanosomatidae), a parasite that alternates between a vertebrate host (i.e., terrestrial mammals) and an invertebrate host (i.e., the vector; Hemiptera, Reduviidae: Triatominae) (Brener 1973). The parasite cycles between intracellular or extracellular (insect) replicative forms and extracellular infective forms (*metacyclic trypomastigotes* via a process called metacyclogenesis), although even *epimastigotes* can infect mammalian cells in vitro. Blood-circulating *trypomastigotes* ingested by the invertebrate host (i.e., the vector triatomine) convert into epimastigotes in the insect midgut and multiply repeatedly along the digestive tube. Finally, in the hindgut, these forms adhere to the rectal cuticle and transform into the *metacyclic trypomastigotes,* which are released in the feces while feeding on the blood source. These are the infective cells that penetrate mammalian hosts and eventually convert into amastigotes (i.e., in a process called amastigogenesis), which replicate in the mammal cells. The replicative amastigote forms complete several cycles before they cause cytolysis (i.e., breakdown of cells by the destruction of the membrane) and released into blood as *trypomastigotes* that will be absorbed by the triatomine or immediately invade new cells, thus fulfilling the cycle (Goldenberg and Avila 2011). The circulating trypomastigotes also invade cells in tissues producing amastigotes, which perpetuates the mammal cycles described above.

The parasite has an extensive genetic diversity with at least six discrete typing units (DTUs) and a potential seventh clade (Tcbat), which are robust units of analysis for molecular epidemiological studies (Tibayrenc 2003). Each DTU has low host species specificity within its geographical range (Izeta-Alberdi et al. 2016), and in Mexico, these have been detected in at least 64 mammal species, although many more species could be involved in the transmission (Stephens et al. 2009). Rodents and bats are the dominant host orders (>50% of host species) (Ibarra-Cerdeña et al. 2017), and there is evidence that these hosts play an important role in the parasite dispersal within landscapes and between patch classes (i.e., crops, forest, and villages) (López-Cancino et al. 2015). Medium-size mammals are also important reservoirs transmitting different *T. cruzi* lineages (Izeta-Alberdi et al. 2016) and carnivore mammals may amplify the parasite due to predation in the ecological community, or act as bioaccumulators of genetic diversity, due to numerous predation on different infected prey (Rocha et al. 2013).

The invertebrate hosts of *Trypanosoma cruzi* (i.e., the vectors) belong to the subfamily Triatominae, currently composed of 151 described species characterized by obligate hematophagy (149 existing and two fossil) (Justi and Galvão 2016). The blood-sucking habit and specific morphological adaptations for blood feeding are distinctive attributes of triatomines (i.e., head a mouthparts), which made them a recognizable group despite considerable overall variation in shape and color (Jurberg and Galvão 2006). In Mexico, there are 32 species (+ 8 subspecies) of triatomines capable of *T. cruzi* vector transmission, although more commonly transmission to humans occurs due to contact with one of 19 species in advanced process of adaptation to human modified habitats (i.e., to establish populations that can be sustained without insect immigration) (Ibarra-Cerdeña et al. 2009; Ramsey et al. 2015). Most Mexican triatomines are generalists for their blood source, and are associated with terrestrial, arboreal, rock outcropping, and cave mammal burrows or roosts (i.e., bat caves) (Ibarra-Cerdeña et al. 2009; Georgieva et al. 2017). Triatomines can be found in almost all regions of Mexico and are known to tolerate anthropogenic disturbance such as land-cover change, land-use change, and urbanization (Ramsey et al. 2000, 2005, 2012, 2015; Ibarra-Cerdeña et al. 2009; López-Cancino et al. 2015).

As outlined above, the ecological community of *T. cruzi* reservoirs in Mexico (vertebrate and invertebrate) includes 400 potential species of terrestrial mammals and approximately 32 independent lineages of vector species. This translates into exceptional opportunity for diverse vector–mammal interactions through which parasite dispersal can reach human populations (Ibarra-Cerdeña et al. 2017). Particularly relevant to this dispersal is the fact that many mammal and vector species already coexist with humans along their economic, cultural, occupational, and social activities in all areas, due to their synanthropic habits (i.e., tolerance to human influence) (Ramsey et al. 2012; Lopez-Cancino et al. 2015; Ibarra-Cerdeña et al. 2017).

16.4 The *Triatoma dimidiata* Complex in the Yucatan

As stated previously, Mexico has relatively high triatomine diversity compared with other biodiverse countries [i.e., 12 species in Costa Rica (Ayala Landa 2017), 15 species in Ecuador (Abad-Franch et al. 2001), 24 species in Colombia (Guhl et al. 2007), and 62 species in Brazil (Gurgel-Gonçalves et al. 2012)]. However, this species richness is not evenly distributed geographically (Ibarra-Cerdeña et al. 2009), since the primary vectors in domestic habitats south and east of the Tehuantepec Isthmus in Mexico belong to the *Triatoma dimidiata* complex (Fig. 16.1) (Ramsey et al. 2015) that includes broadly distributed haplogroups and at least one previously

Fig. 16.1 A female *Triatoma dimidiata* collected in the Yucatan and its eggs. (Photo credits: Kayu Vilchis and Adriana González-Martínez)

described sister species, *T. hegneri*, all considered to have important roles as CD vectors in Mexico, all countries in Central America, Colombia, Venezuela, Ecuador, and Northern Peru (Dorn et al. 2007; Pech-May et al. 2019). This range encompasses a wide spectrum of ecological conditions, climate zones, landscape modification gradients, and tolerance to high levels of anthropization (i.e., urbanization) (Dorn et al. 2007; Ibarra-Cerdeña et al. 2009; de La Vega and Schilman 2018). Blood meal sources of the complex are also diverse, leading to the conclusion that this can be one of the more generalist species groups of Mexican triatomines (Ibarra-Cerdeña et al. 2017). In fact, single individuals of this species have been reported to have as many as 13 vertebrate host species (Dumonteil et al. 2018), including mixed blood meals with human and domestic animals, combined with wildlife hosts (Dumonteil et al. 2018; López-Cancino et al. 2015). Its ability to colonize houses in most regions across its geographic range has led the complex to be classified as an invasive vector (Schofield et al. 2006; Crowl et al. 2008). The ability or potential to establish permanent populations in human communities may not be ubiquitous within the complex, as two distinct scenarios have been reported in the Yucatan peninsula: dispersal into houses seasonally (Dumonteil et al. 2007; Ramirez-Sierra et al. 2010; Waleckx et al. 2015b), as opposed to year-round colonization (i.e., presence of nymph stages as evidence of reproduction) (Lopez-Cancino et al. 2015). Low seasonality is observed in both rural and urban areas (i.e., Merida), where bug sightings occur over the year (albeit they are certainly more frequent during the dry season). Future population genetic studies will need to provide evidence for either or both strategies, as data may be biased due to the complex's resistance to many collecting methods. Knowledge regarding gene flow of *T. dimidiata* in rural communities will be essential to properly design control and monitoring strategies to effectively reduce and sustain low risk for human–vector contact in Mexican houses. It is clear due to vector life history traits, tolerance to anthropization, and differential *T. cruzi* infection that traditional strategies developed in South American countries to control *Triatoma infestans* and other domesticated vectors will be insufficient (Noireau et al. 2005; Schofield et al. 2006). Seasonal dispersal from adjacent habitats (i.e., sylvatic patches, secondary forests, grasslands) into rural human communities is a challenging attribute for vector control because residual insecticides have short half-lives (3–6 months), requiring at least 3 yrs. of 2 applications/yr., and a consolidation phase at least of 2 years of 1 application/yr., and perhaps even longer. Surveillance would be population-based, requiring effective primary healthcare support to maintain vector program links with all communities (Hashimoto et al. 2015; Ramsey et al. 2003; Waleckx et al. 2015b). Less-expensive and sustainable measures that avoid vector–human contact such as window and door insect screens have certain efficacy to prevent bugs from entering houses (Waleckx et al. 2015a). Nevertheless, the risk of vector contact is not limited to the intradomicile of houses in rural areas, where most individuals are farmers, who often hunt at night for subsistence, or have additional income by contracts for nighttime property security surveillance (Valdez-Tah et al. 2015; Gonzalez-Martínez et al. unpublished). In fact, men interviewed in the Yucatan have mentioned that they have received the landing of up to 80 triatomines in a night while waiting in tree-blinds for prey (Fig. 16.2).

Fig. 16.2 A hunter from a rural community in the Yucatan, hanging on a hammock attached to the tree. Hunters wait overnight until a prey is sighted. Note that clothing allows skin contact with bugs. (Photo credits: Kayu Vilchis)

16.5 Vector–Human Contact Interruption

Vector transmission of *Trypanosoma cruzi* is the major source of parasite acquisition in humans. Likewise, controlling domestic vector transmission could reduce between 46 and 83% of the relative number of new acute Chagas cases, according to a mathematical model calibrated in Yucatan scenarios (Lee et al. 2018). Yet, the role of non-domestic vector transmission (i.e., enzooty) has been neglected. While there are a few reports showing the presence of human blood in vectors collected from the forest (Ramsey et al. 2012; Stevens et al. 2014; López-Cancino et al. 2015) and human practices leading to exposition away from home (Valdez-Tah et al. 2015), infection risk of farmers has not been measured. Moreover, gender is a determinant for exposure to vectors, due to cultural and economic activities in rural communities of the Yucatan Peninsula. For instance, in an evaluation of the relationship between gender and spatial location in the landscape, most infected men report spending several hours each day and night in their fields, while few women will have similar presence in extra-community areas (Valdez-Tah et al. 2015; Gonzalez-Martinez et al. unpublished). Evaluating the relationship between gender and exposure in rural communities is important to determine the real vulnerability of the population in rural landscapes. Therefore, sustained reduction or interruption of *T. cruzi* vector transmission will be dependent upon incorporating these two key elements (i.e., the relationship between gender and occupation-habitat) (Triana et al. 2016).

While an extensive use of bed nets to reduce contact and ensure insecticide exposure would reduce triatomine populations in houses, any robust strategy must use complementary methods to establish permanent surveillance and barriers to vector amplification in modified habitats. This means that in a single village, several types of mechanisms for vector elimination will be necessary and dependent on house permeability (Fig. 16.3), presence of both pets and livestock (Fig. 16.4), and Mayan home gardens, where synanthropic mammal hosts and vectors interact.

16.6 Social Knowledge of Vector-Borne Transmission

A human ecology multidisciplinary approach for *T. cruzi* transmission in the Yucatan involves an approach which includes social knowledge related to both risk factors and human vulnerability (Valdez-Tah et al. 2015). Certainly, local communities do not only behave to the biological hazard of a zoonosis, but they rely on social beliefs, values, and different kinds of knowledge and representations from personal experiences, gender roles, healthcare practices, and their use of the natural environment. Although there is a general lack of specific knowledge related to *T. cruzi* transmission (Rosecrans et al. 2014; Valdez-Tah et al. 2015), rural communities elaborate their own ethno-ecology, generating local theories about insects, animals, and disease throughout the landscape, and according to seasons. According to social

Fig. 16.3 A house in a rural community of the Yucatan. New constructions are usually coupled with traditional houses. Red arrows show permeability of traditional house to insect invasion. Inside the room there are hammocks hanging where people rest and sleep at night during the warm season (i.e., the dispersal season of Chagas vectors). (Photo credits: Carlos Ibarra-Cerdeña)

Fig. 16.4 Farm animals living in close proximity to humans in rural communities of the Yucatan. These animals are typical blood sources of Chagas disease vectors. Mixed blood sources composed of farm and domestic animals with humans are common in *Triatoma dimidiata*. (Photos credit: Adriana González-Martínez)

representation theory, whereas educational interventions might increase knowledge about health risk derived from vector contact, people usually generate their own synthesis with interwoven information from personal and popular concepts and the biomedical sciences (Ventura-Garcia et al. 2013). Thus, the recognition of social knowledge not only contributes to better understand the space-time dynamics of transmission, but also to design intercultural interventions adapted to local settings.

While investigating social knowledge about *T. cruzi* transmission, we should keep in mind that it needs to be embedded, and should be understood, within a larger structural "risk environment." In Mexico, the absence of an appropriate and effective CD public health policy translates into the lack of an authorized voice—as it should be health officers—addressing the disease as a health concern through their institutional response and actions. Structural factors have obviated public conversation regarding Chagas, and therefore there is urgency for truly owned social knowledge that could prompt a social dialogue. Such processes could only occur in parallel with the *living experience* of the disease, particularly that at-risk populations and transfusion blood donors need to be targeted by physician and healthcare personnel training and massive health diagnosis campaigns, in order to receive treatment for *T. cruzi* infection. In addition, sociopolitical inequalities impede access to healthcare equally across the territory, imposing additional financial burden among rural populations (Ramsey et al. 2014). Travel distance for healthcare, daily wage loss, out-pocket cost for transportation, and related expenses may be key factors when seeking healthcare in large cities (Loyola-Sanchez et al. 2016).

16.7 Houses Hospitable to Bug Infestation

Domestic risk for triatomine bites, as was described previously, is dependent on the spatial and seasonal patterns of biological and ecological determinants of *T. cruzi* (Lopez-Cancino et al. 2015). Human practices and the houses' physical condition may provide improved environmental stability, nesting conditions, resources, and year-round resource availability for reservoirs (grain, food leftovers, livestock feed) (invasive sylvatic or livestock), essential not only to provide improved bug survival (fitness), but also for parasite transmission. Understanding the practices that favor the presence of triatomines implies an in-depth knowledge of the processes that create vulnerability. On the one hand, there are more direct and local factors of housing, such as livelihood, lifestyles, and the sociocultural context: cleaning methods and dusting of furniture, furnishings and their placement, adornments and internal additions to the house, composition and management of peridomestic areas, storage of different products, the accumulation of materials and the disposal of waste (organic and inorganic), the cleanliness of the peridomestic area and inside the house, the size of the property and its location with respect to the habitats that make up the landscape, materials used in the construction, style and structure of the house (number of bedrooms, pieces and location of the kitchen) (Ramsey et al. 2003;

Bustamante et al. 2009). On the other hand, the former factors might be affected by others such as land tenure, socioeconomic status, and access to credit systems (Gomes et al. 2013). A constant migratory mobility of the residents might also lead to a poor maintenance of the structure and internal spaces of the house.

Despite greatest interaction with humans in the dry season (reduced abundance of sylvatic hosts), and a decrease in the use of contact barriers (bed nets) and insect control (insecticides) in the same season, human practices will continue to provide improved resources, maintain triatomine interchange between habitats year-round, and sustain continuous risk over all landscape fragments (Ramirez-Sierra et al. 2010). Community-based housing improvements targeting walls do not have to be costly, rather they need to focus on creating an unstable microhabitat, which repels bugs (constant cleaning or sweeping), or impedes their movement across ceilings, walls, or floors. These recommendations for house improvement must consider measures to block the access to vectors for doors and windows (i.e., insect screens) as well as bed nets modified to be coupled in hammocks (the preferred sleeping mode in Yucatan). These modified bed nets must be used by people spending nights outdoors were vector presence is known.

Acknowledgements We thank the support for this contribution provided by PRODEP to CNIC and TCB Project CINVESTAV-CA-7, Red Patrimonio Biocultural-Node 24 for support of CNCI, and CONACyT FOSSIS # 261006 for support of JMR, AGM received a PhD Scholarship and CGC a MSc Scholarship from CONACyT.

References

Abad-Franch F, Paucar CA, Carpio CC et al (2001) Biogeography of Triatominae (Hemiptera: Reduviidae) in Ecuador: implications for the design of control strategies. Mem Inst Oswaldo Cruz 96:611–620

Alducin-Téllez C, Rueda-Villegas E, Medina-Yerbes I et al (2011) Prevalencia de serología positiva para trypanosoma cruzi en pacientes con diagnóstico clínico de miocardiopatía dilatada en el Estado de Campeche, México. Arch Cardiol Mex 81:204–207

Arnal, A., Waleckx, E., Herrera, C., et al (2018). Estimating the current burden of Chagas disease in Mexico: a systematic review of epidemiological surveys from 2006 to 2017. BioRxiv, 423145

Ayala Landa JM (2017) Los triatominos de Costa Rica (Heteroptera, Reduviidae, Triatominae). Arq Entomolóxicos 18:189–215

Barrera-Pérez Mario A, Rodríguez-Félix María E, del Guzmán-Marín Eugenia S, Zavala-Velázquez Jorge E (1992) Prevalencia de la Enfermedad de Chagas en el Estado de Yucatán. Rev Biomed 3:133–139

Brener Z (1973) Biology of Trypanosoma cruzi. Annu Rev Microbiol 27:347–382

Bustamante DM, Monroy C, Pineda S et al (2009) Risk factors for intradomiciliary infestation by the Chagas disease vector Triatoma dimidiata in Jutiapa, Guatemala. Cad Saude Publica 25:83–92

Carabarin-Lima A, González-Vázquez MC, Rodríguez-Morales O et al (2013) Chagas disease (American trypanosomiasis) in Mexico: an update. Acta Trop 127:126–135. https://doi.org/10.1016/j.actatropica.2013.04.007

Carroll D, Daszak P, Wolfe ND et al (2018) The global Virome project. Science 359:872–874. https://doi.org/10.1126/science.aap7463

Coura JR, Viñas PA (2010) Chagas disease: a Latin American health problem becoming a world health problem. Nature 465:S6–S7. https://doi.org/10.1016/j.actatropica.2009.11.003

Crowl TA, Crist TO, Parmenter RR et al (2008) The spread of invasive species and infectious disease as drivers of ecosystem change. Front Ecol Environ 6:238–246. https://doi.org/10.1890/070151

de La Vega GJ, Schilman PE (2018) Ecological and physiological thermal niches to understand distribution of Chagas disease vectors in Latin America. Med Vet Entomol 32:1–13. https://doi.org/10.1111/mve.12262

Dorn PL, Monroy C, Curtis A (2007) Triatoma dimidiata (Latreille, 1811): a review of its diversity across its geographic range and the relationship among populations. Infect Genet Evol 7:343–352. https://doi.org/10.1016/j.meegid.2006.10.001

Dumonteil E, Tripet F, Ramirez-Sierra MJ et al (2007) Assessment of Triatoma dimidiata dispersal in the Yucatan Peninsula of Mexico by morphometry and microsatellite markers. Am J Trop Med Hyg 76:930–937

Dumonteil E, Ramirez-Sierra M-J, Pérez-Carrillo S et al (2018) Detailed ecological associations of triatomines revealed by metabarcoding and next-generation sequencing: implications for triatomine behavior and Trypanosoma cruzi transmission cycles. Sci Rep 8:4140. https://doi.org/10.1038/s41598-018-22455-x

Fauci AS (2001) Infectious diseases: considerations for the 21st century. Clin Infect Dis 32:675–685. https://doi.org/10.1086/319235

Gebreyes WA, Dupouy-Camet J, Newport MJ et al (2014) The global one health paradigm: challenges and opportunities for tackling infectious diseases at the human, animal, and environment interface in low- resource settings. PLoS Negl Trop Dis 8:e3257. https://doi.org/10.1371/journal.pntd.0003257

Georgieva AY, Gordon ERL, Weirauch C (2017) Sylvatic host associations of Triatominae and implications for Chagas disease reservoirs: a review and new host records based on archival specimens. PeerJ 5:e3826. https://doi.org/10.7717/peerj.3826

Goldenberg S, Avila AR (2011) Aspects of Trypanosoma cruzi stage differentiation. In: Rollinson D, Hay SI (eds) Advances in parasitology, vol 75. Academic Press, London, pp 285–305

Guhl F, Aguilera G, Pinto N, Vergara D (2007) Actualización de la distribución geográfica y eco-epidemiología de la fauna de triatominos (Reduviidae: Triatominae) en Colombia. Biomedica 27:143–162

Gurgel-Gonçalves R, Galvão C, Costa J, Peterson AT (2012) Geographic distribution of Chagas disease vectors in Brazil based on ecological niche modeling. J Trop Med 2012:705326. https://doi.org/10.1155/2012/705326

García-Montalvo, B. (2011). Trypanosoma cruzi antibodies in blood donors in Yucatan state, Mexico. Revista Médica del Instituto Mexicano del Seguro Social, 49(4), 367–372.

Gomes, T. F., Freitas, F. S., Bezerra, C. M., Lima, M. M., & Carvalho-Costa, F. A. (2013). Reasons for persistence of dwelling vulnerability to Chagas disease (American trypanosomiasis): a qualitative study in northeastern Brazil. World Health & Population, 14(3), 14–21.

Hashimoto K, Zúniga C, Nakamura J, Hanada K (2015) Integrating an infectious disease programme into the primary health care service: a retrospective analysis of Chagas disease community-based surveillance in Honduras. BMC Health Serv Res 15:116. https://doi.org/10.1186/s12913-015-0785-4

Hotez PJ, Bottazzi ME, Franco-Paredes C et al (2008) The neglected tropical diseases of Latin America and the Caribbean: a review of disease burden and distribution and a roadmap for control and elimination. PLoS Negl Trop Dis 2:e300. https://doi.org/10.1371/journal.pntd.0000300

Ibarra-Cerdeña CN, Sánchez-Cordero V, Peterson AT, Ramsey JM (2009) Ecology of North American Triatominae. Acta Trop 110:178–186. https://doi.org/10.1016/j.actatropica.2008.11.012

Ibarra-Cerdeña CN, Valiente-Banuet L, Sánchez-Cordero V et al (2017) Trypanosoma cruzi reservoir—triatomine vector co-occurrence networks reveal meta-community effects by synanthropic mammals on geographic dispersal. PeerJ 5:e3152. https://doi.org/10.7717/peerj.3152

Izeta-Alberdi A, Ibarra-Cerdeña CN, Moo-Llanes DA, Ramsey JM (2016) Geographical, landscape and host associations of Trypanosoma cruzi DTUs and lineages. Parasit Vectors 9:631. https://doi.org/10.1186/s13071-016-1918-2

Jones KE, Patel NG, M a L et al (2008) Global trends in emerging infectious diseases. Nature 451:990–993. https://doi.org/10.1038/nature06536

Jurberg J, Galvão C (2006) Biology, ecology, and systematics of Triatominae (Heteroptera, Reduviidae), vectors of Chagas disease, and implications for human health. Biol Linz 50:1096–1116. https://doi.org/10.4137/EHI.S16003

Justi SA, Galvão C (2016) The evolutionary origin of diversity in Chagas disease vectors. Trends Parasitol 33:42–52

Lee BY, Bartsch SM, Skrip L et al (2018) Are the London Declaration's 2020 goals sufficient to control Chagas disease? Modeling scenarios for the Yucatan Peninsula. PLoS Negl Trop Dis 12:e0006337

López-Cancino SA, Tun-Ku E, De la Cruz-Felix HK et al (2015) Landscape ecology of Trypanosoma cruzi in the southern Yucatan Peninsula. Acta Trop 151:58–72. https://doi.org/10.1016/j.actatropica.2015.07.021

Lozano R, Naghavi M, Foreman K et al (2012) Global and regional mortality from 235 causes of death for 20 age groups in 1990 and 2010: a systematic analysis for the Global Burden of Disease Study 2010. Lancet 380:2095–2128. https://doi.org/10.1016/S0140-6736(12)61728-0

Loyola-Sanchez, A., Richardson, J., Wilkins, S., Lavis, J. N., Wilson, M. G., Alvarez-Nemegyei, J., & Pelaez-Ballestas, I. (2016). Barriers to accessing the culturally sensitive healthcare that could decrease the disabling effects of arthritis in a rural Mayan community: a qualitative inquiry. Clinical Rheumatology, 35(5), 1287–1298.

Manne JM, Snively CS, Ramsey JM et al (2013) Barriers to treatment access for Chagas disease in Mexico. PLoS Negl Trop Dis 7:e2488. https://doi.org/10.1371/journal.pntd.0002488

Manne-Goehler J, Ramsey JM, Salgado MO et al (2014) Increasing access to treatment for Chagas disease: the case of Morelos. Mexico Am J Trop Med Hyg 91:1125–1127. https://doi.org/10.4269/ajtmh.14-0357

Méndez-Domínguez N, Chi-Méndez C, Canto-Losa J et al (2017) Cardiopatía Chagásica En Un Escolar: Reporte de Caso. Rev Chil Pediatr 88:583–589

Morens DM, Folkers GK, Fauci AS (2004) The challenge of emerging and re-emerging infectious diseases. Nature 430:242–249. https://doi.org/10.1038/nature02759

Morillo CA, Marin-Neto JA, Avezum A et al (2015) Randomized trial of Benznidazole for chronic Chagas' cardiomyopathy. N Engl J Med 373:1295–1306. https://doi.org/10.1056/NEJMoa1507574

Mougabure-Cueto G, Inés M (2015) Insecticide resistance in vector Chagas disease: evolution, mechanisms and management. Acta Trop 149:70–85. https://doi.org/10.1016/j.actatropica.2015.05.014

Noireau F, Cortez MGR, Monteiro FA et al (2005) Can wild Triatoma infestans foci in Bolivia jeopardize Chagas disease control efforts? Franc. Trends Parasitol 21:2003–2006. https://doi.org/10.1016/j.pt.2004.10.009

Olival KJ, Hosseini PR, Zambrana-Torrelio C et al (2017) Host and viral traits predict zoonotic spillover from mammals. Nature 546:646–650. https://doi.org/10.1038/nature22975

Parkes MW, Bienen L, Breilh J et al (2005) All hands on deck: transdisciplinary approaches to emerging infectious disease. EcoHealth 2:258–272. https://doi.org/10.1007/s10393-005-8387-y

Pech-May A, Mazariegos-Hidalgo CJ, Izeta-Alberdi A et al (2019) Genetic variation and phylogeography of the Triatoma dimidiata complex evidence a potential center of origin and recent divergence of haplogroups having differential Trypanosoma cruzi and DTU infections. PLoS Negl Trop Dis 13:e0007044

Prata A (2001) Clinical and epidemiological aspects of Chagas disease. Lancet Infect Dis 1:92–100. https://doi.org/10.1016/S1473-3099(01)00065-2

Ramirez-Sierra MJ, Herrera-Aguilar M, Gourbière S, Dumonteil E (2010) Patterns of house infestation dynamics by non-domiciliated Triatoma dimidiata reveal a spatial gradient of infestation

in rural villages and potential insect manipulation by Trypanosoma cruzi. Tropical Med Int Health 15:77–86. https://doi.org/10.1111/j.1365-3156.2009.02422.x
Ramsey JM, Ordonez R, Cruz-Celis A et al (2000) Distribution of domestic Triatominae and stratification of Chagas disease transmission in Oaxaca, Mexico. Med Vet Entomol 14:19–30
Ramsey JM, Cruz-celis A, Salgado L et al (2003) Efficacy of Pyrethroid insecticides against domestic and Peridomestic populations of Triatoma pallidipennis and Triatoma barberi (Reduviidae : Triatominae) vectors of Chagas ' disease in Mexico. J Med Entomol 40:912–920
Ramsey JM, Alvear AL, Ordoñez R et al (2005) Risk factors associated with house infestation by the Chagas disease vector Triatoma pallidipennis in Cuernavaca metropolitan area, Mexico. Med Vet Entomol 19:219–228. https://doi.org/10.1111/j.0269-283X.2005.00563.x
Ramsey JM, Gutierrez-Cabrera AE, Salgado-Ramirez L et al (2012) Ecological connectivity of Trypanosoma cruzi reservoirs and Triatoma pallidipennis Hosts in an anthropogenic landscape with endemic Chagas disease. PLoS One 7:e46013. https://doi.org/10.1371/journal.pone.0046013
Ramsey JM, Elizondo-Cano M, Sanchez-González G et al (2014) Opportunity cost for early treatment of Chagas disease in Mexico. PLoS Negl Trop Dis 8:e2776. https://doi.org/10.1371/journal.pntd.0002776
Ramsey JM, Peterson AT, Carmona-Castro O et al (2015) Atlas of Mexican Triatominae (Reduviidae: Hemiptera) and vector transmission of Chagas disease. Mem Inst Oswaldo Cruz 110:339–352. https://doi.org/10.1590/0074-02760140404
Rassi AJ, Rassi A, Marin-Neto JA (2010) Chagas disease. Lancet 375:1388–1402. https://doi.org/10.1016/S0140-6736(10)60061-X
Rocha FL, Roque ALR, de Lima JS et al (2013) Trypanosoma cruzi infection in Neotropical wild carnivores (Mammalia: Carnivora): at the top of the T. cruzi transmission chain. PLoS One 8:e67463. https://doi.org/10.1371/journal.pone.0067463
Rocha-Gaso MI, Villarreal-Gómez LJ, Beyssen D et al (2017) Biosensors to diagnose Chagas disease: a brief review. Sensors 17:1–12. https://doi.org/10.3390/s17112629
Rosecrans K, Cruz-Martin G, King A, Dumonteil E (2014) Opportunities for improved chagas disease vector control based on knowledge, attitudes and practices of communities in the Yucatan Peninsula, Mexico. PLoS Negl Trop Dis 8:e2763. https://doi.org/10.1371/journal.pntd.0002763
Sánchez-González G, Figueroa-Lara A, Elizondo-Cano M et al (2016) Cost-effectiveness of blood donation screening for Trypanosoma cruzi in Mexico. PLoS Negl Trop Dis 10:e0004528. https://doi.org/10.1371/journal.pntd.0004528
Schofield CJ, Jannin J, Salvatella R (2006) The future of Chagas disease control. Trends Parasitol 22:583–588. https://doi.org/10.1016/j.pt.2006.09.011
Stephens CR, Gimenez HJ, Gonzalez C et al (2009) Using biotic interaction networks for prediction in biodiversity and emerging diseases. PLoS One 4:e5725. https://doi.org/10.1371/journal.pone.0005725
Stevens L, Monroy MC, Rodas AG, Dorn PL (2014) Hunting, swimming, and worshiping: human cultural practices illuminate the blood meal sources of cave dwelling Chagas vectors (Triatoma dimidiata) in Guatemala and Belize. PLoS Negl Trop Dis 8:e3047. https://doi.org/10.1371/journal.pntd.0003047
Tarleton RL, Gürtler RE, Urbina JA et al (2014) Chagas disease and the London declaration on neglected tropical diseases. PLoS Negl Trop Dis 8:e3219. https://doi.org/10.1371/journal.pntd.0003219
Taylor LH, Latham SM, woolhouse MEJ (2001) Risk factors for human disease emergence. Philos Trans R Soc B Biol Sci 356:983–989. https://doi.org/10.1098/rstb.2001.0888
Tibayrenc M (2003) Kinetoplastid biology and disease genetic subdivisions within Trypanosoma cruzi (discrete typing units) and their relevance for molecular epidemiology and experimental evolution. Kinetoplastid Biol Dis 2:1–6

Triana DRR, Mertens F, Zúniga CV et al (2016) The role of gender in Chagas disease prevention and control in Honduras: an analysis of communication and collaboration networks. EcoHealth 13:535–548. https://doi.org/10.1007/s10393-016-1141-9

Valdez-Tah A, Huicochea-Gómez L, Ortega-Canto J et al (2015) Social representations and practices towards Triatomines and Chagas disease in Calakmul, México. PLoS One 10:e0132830. https://doi.org/10.1371/journal.pone.0132830

Ventura-Garcia L, Roura M, Pell C et al (2013) Socio-cultural aspects of Chagas disease: a systematic review of qualitative research. PLoS Negl Trop Dis 7:e2410. https://doi.org/10.1371/journal.pntd.0002410

Viotti R, Grijalva MJ, Guhl F et al (2014) Towards a paradigm shift in the treatment of chronic Chagas disease. Antimicrob Agents Chemother 58:635–639. https://doi.org/10.1128/AAC.01662-13

Waleckx E, Camara-Mejia J, Ramirez-Sierra MJ et al (2015a) Una intervención innovadora de ecosalud para el control vectorial de la enfermedad de Chagas en Yucatán, Mexico. Rev Biomed 26:75–86

Waleckx E, Gourbière S, Dumonteil E (2015b) Intrusive versus domiciliated triatomines and the challenge of adapting vector control practices against Chagas disease. Mem Inst Oswaldo Cruz 110:324–338. https://doi.org/10.1590/0074-02760140409

WHO (2002) Control of Chagas disease. Second Report of the WHO Expert Committee World Health Organization

WHO (2014) Global burden of disease estimates for 2000–2012. WHO, Geneva

Zavala-Velázquez J (2003) La enfermedad de Chagas en el Estado de Yucatán, México (1940–2002). Rev Bioméd 14:34–43

Zeppelini CG, Maria A, Almeida P, De Cordeiro-estrela P (2016) Zoonoses as ecological entities: a case review of plague. PLoS Negl Trop Dis 10:1–10. https://doi.org/10.1371/journal.pntd.0004949

Chapter 17
Conclusions

Federico Dickinson and Hugo Azcorra

Our purpose in this book has been to provide the reader with a clear vision, from a human ecology perspective, of the research available on the interrelation between ecosystems, socioeconomic and cultural systems, and the biology of human populations on the Yucatan Peninsula, from ancient times to the present.

The Peninsula is a dynamic region, and this dynamic is determined by the interaction between social, demographic, economic, political, and environmental processes, as described by Luis Ramírez in Chap. 2. Mayan groups are numerous in the region, but its overall population currently consists of different ethnic and social groups. This is particularly apparent on the Caribbean coast in cities such as Cancún and Playa del Carmen, which are home to immigrants from all over Mexico, Latin America, and even Europe and North America. But diversity can also be found in Calakmul, where farmers from various states in Mexico have colonized the rainforest.

More than half the Peninsula's population now lives in urban environments. These are artificial ecosystems created, maintained, and transformed by human beings. On our species' evolutionary scale (not to mention our genus's), to be born, grow up, work, and reproduce in an urban environment is novel. Much research is still needed on humans' biological responses to the urban environment in the region (Fig. 17.1).

The aforementioned dynamic has serious environment implications, including but not limited to general deforestation of the Peninsula, severe pollution of groundwater and marine environments, collapsing fisheries, and social conflict over land ownership. These undermine the Peninsula's environmental capacity to sustain human populations at current, highly differential, consumption levels.

F. Dickinson (✉)
Departamento de Ecología Humana, Cinvestav-Mérida, Mérida, Yucatán, México
e-mail: federico.dickinson@cinvestav.mx

H. Azcorra
Departamento de Ecología Humana, Cinvestav-Mérida, Mérida, Yucatán, México

Centro de Investigaciones Silvio Zavala, Universidad Modelo, Mérida, Yucatán, México

H. Azcorra, F. Dickinson (eds.), *Culture, Environment and Health in the Yucatan Peninsula*, https://doi.org/10.1007/978-3-030-27001-8_17

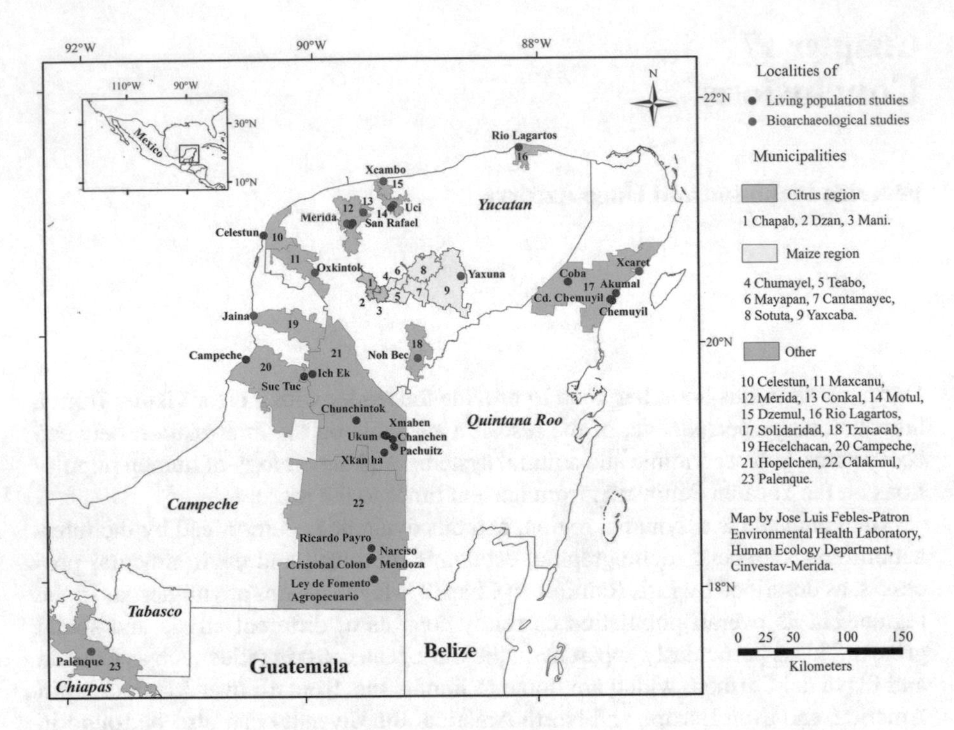

Fig. 17.1 Map of the Yucatan Peninsula showing the locations of studies reported in this volume

The human ecology perspective that articulates the contributions to this book (Chap. 1, Figs. 1.1 and 1.2) paints a partial but rather disheartening panorama of the biological condition of the Peninsula's human populations. Children and adolescents are growing up under adverse conditions that compromise their development and future health and well-being. Simultaneously, adult women and men, be they rural or urban, experience strong stressors throughout their adult life. The larger part of the Peninsula's Maya population has suffered literally centuries of historic trauma. If their biological conditions are to substantially improve, they will require various generations of notably better living conditions. These include access to justice, health and education services far superior to those currently available, higher income, a balanced and sufficient diet, and less discrimination (Chap. 5).

We believe a society's main resource is its individuals. If these are not in good health, it can seriously limit a society's possibility of attaining adequate levels of economic and cultural productivity, autonomy, sovereignty, well-being (*sensu* White 2008; White and Jha 2018), and, why not, happiness.

17.1 A Long Road to Travel

Human biology research on the Yucatan Peninsula from the human ecology perspective is scarce and far from covering the region (Map 17.1). Entire areas have not been addressed with this approach. Of particular note is the almost complete absence of knowledge on the human biology of the Caribbean coast populations. Developed as a series of large-scale tourism centers, deficient urban planning together with the arrival of hundreds of thousands of labor immigrants has generated huge socioeconomic and well-being disparities that are surely expressed in biological terms. Another case is the oil boom town of Ciudad del Carmen, in the state of Campeche, which has experienced decades of intense oil industry development. The social, demographic, and environmental impacts of this development up to the early twenty-first century have been generally described by Baños Ramírez (2007) and Barrera Peniche (2007), but to our knowledge no similar studies have been done of its effects on human biological condition.

Even less work has been done from a human ecology perspective on how the well-being of the Peninsula's human populations are effected by endemic diseases such as zika, chikungunya, and dengue as well as acquired diseases such as metabolic syndrome (plus associated disorders like diabetes, hypertension, and obesity). We think it is especially important to apply the human ecology perspective to study the potential effects of vector-transmitted diseases on the unborn, which has been done in diverse populations and regions worldwide (WHO, 1999).

In the state of Yucatan, research is needed on the lifestyle changes in rural communities caused by the cultural, economic, and behavioral influence of emigrants residing in the United States and returnees (Chap. 6). These changes imply alterations in diet and the appearance of increase in illegal drug use, but also modify multiple facets of community life such as intergenerational relations and female reproductive health (Chan Mex 2015). These in turn increase stressors whose effects on the biological condition of individuals and groups are as yet unknown.

Our hope is that this book will help to motivate interdisciplinary research on the main human ecology problems on the Yucatan Peninsula as well as identify possible ways of resolving, or at least mitigating, these problems. We think this effort is best guided by the methodology proposed by García (2006), which has been applied successfully in concrete cases (Gracía et al. 1988; Tudela 1989), and by incorporating the recommendations of Goodman and Leatherman (1998 p 21–24) for human ecology research.

Interaction between human beings and the environment occurs on the individual level but also on the social level. Among its many aspects is social production. This needs to be recognized as occurring within a context of power that encompasses inequality between the groups within a society, which has biological consequences in humans. Understanding how this production is organized, and who makes decisions and controls the process is fundamental to transcending the immediate effects of this interaction on individuals. This political perspective necessarily broadens the sociopolitical and environmental context of analysis and demands that the concept of adaption be reconceptualized. This is an especially tricky term in human ecology given its multiple meanings.

In terms of human biology and ecology, adaptation needs to be understood as a process (Frisancho 1981), rather than a mere set of measurements (Goodman and Leatherman 1998), since these are not the final stages of biological processes but "points" or "moments" within the adaptation process. When we refer to adaptation we also need to ask ourselves what is its context, for whom is it adaptive, and at what level and moment (Goodman and Leatherman 1998, p 23). We also need to acknowledge that adaptation is never perfect, always a compromise and therefore implies "costs" and "benefits."

In closing, we think that it is absolutely necessary that human biology research done within the human ecology framework recognize that human biology occurs within a political field. It is immersed in power relationships in which individuals and groups do not necessarily have the same growth and development opportunities, or the same access to the means to meet their biological, emotional, and social needs. This is expressed, among other ways, in differing levels of well-being.

References

Baños Ramírez O (2007) Petróleo y política en Campeche. In: Ramírez Carrillo LA (ed), En la ruta del petróleo, impactos de una eventual explotación petrolera en Yucatán. In: El contexto regional: política, empresariado y petróleo en Campeche y Tabasco, 1st edn. Fundación Plan Estratégico de Mérida, Mérida, México, pp 83–472

Barrera Peniche JM (2007) La experiencia de Isla del Carmen. In: Ramírez Carrillo LA (ed) En la ruta del petróleo, impactos de una eventual explotación petrolera en Yucatán. In: El contexto regional: política, empresariado y petróleo en Campeche y Tabasco, 1st edn. Fundación Plan Estratégico de Mérida, Mérida, México, pp 473–502

Chan Mex TJ (2015) Salud sexual y reproductiva de mujeres yucatecas en contextos migratorios. Dissertation. Centro de Investigación y de Estudios Avanzados del Instituto Politécnico Nacional

Frisancho AR (1981) Human adaptation. A functional interpretation. University of Michigan Press, Ann Arbor, Michigan

García R (2006) Sistemas complejos. Conceptos, método y fundamentación epistemológica de la investigación interdisciplinaria, 1st edn. Gedisa, Barcelona

García R, Sanz SC, Baraona MD et al (1988) Deterioro ambiental y pobreza en la abundancia productiva. El caso de la Comarca Lagunera., 1st ed. IFIAS – Cinvestav, México, D. F

Goodman AH, Leatherman TL (1998) Traversing the chasm between biology and culture: An Introduction. In: Goodman AH, Leatherman TL (eds) Building a new biocultural synthesis. Political-economic perspectives on human biology, 1st edn. The University of Michigan Press, Ann Arbor, pp 3–41

Tudela F (ed) (1989) La modernización forzada del trópico: El caso de Tabasco. Proyecto Integrado del Golfo. El Colegio de México, México, D. F.

White SC (2008) But what is well-being? A framework for analysis in social and development policy and practice. Paper presented at the Conference on Regeneration and Wellbeing: Research into Practice. University of Bradford, Bradford, April 24th 2008.

White SC, Jha S (2018) Towards an interdisciplinary approach to wellbeing: Life histories and Self-Determination Theory in rural Zambia. Soc Sci Med 212:153–160

World Health Organization (1999) Situation report. Zika virus, Microcephaly, Guillain-Barré Syndrome. Available via DIALOG. http://origin.searo.who.int/bhutan/who-zika-28-7-16.pdf. Accessed 10 Mar 2019

Glossary

Acclimatization The process in which an individual organism adjusts to a change in its environment (such as a change in altitude, temperature, humidity, photoperiod, or pH), allowing it to maintain performance across a range of environmental conditions.

Achalasia Is a motility disorder of the esophagus characterized by loss of enteric neurons leading to absence of peristalsis and impaired relaxation of the lower esophageal sphincter.

Actiheart Compact device that records movement, which allows physical activity to be recorded in a synchronized manner with the heart rate. The Actiheart is worn on the chest and consists of two electrodes.

Adaptation A change in structure, function, or behavior by which a species or individual improves its chance of survival in a specific environment. Adaptations develop as the result of natural selection operating on random genetic variations that are capable of being passed from one generation to the next.

Adiposity A condition of being severely overweight or obese.

Aetiological The cause or set of causes of a disease or condition.

Age-specific fertility rates An estimator of fertility by age group in a population.

Aguada Spanish word for a water body.

Aljibe Spanish word for a tank, usually underground, built to collect rainwater.

Amino acids Any of the chemical substances found in plants and animals that combine to make protein.

Anemia A condition marked by a deficiency of red blood cells or of hemoglobin in the blood, resulting in pallor and weariness.

Anthropization The disturbance of landscapes and natural environments by human action.

Anthropogenic Of human origin or produced by humans.

Anthropogenic landscapes Areas where direct human alteration of ecological patterns and processes is profound, ongoing, and directed towards servicing the needs of human populations for food, fuel, fiber, timber, shelter, trade, and recreation.

H. Azcorra, F. Dickinson (eds.), *Culture, Environment and Health in the Yucatan Peninsula*, https://doi.org/10.1007/978-3-030-27001-8

Aquifer A water-bearing bed or stratum of permeable rock which can contain or transmit groundwater.

Arthralgia Pain in the joint.

Arthritis Painful inflammation and stiffness of the joints.

Arthroses A degenerative disease of a joint.

ASFR See age-specific fertility rates.

Ataxia The loss of full control of bodily movements.

Bioarchaeology The study of bones and other biological materials found in archaeological remains in order to provide information about human life or the environment in the past.

Bioimpedance analysis A method for analyzing the water content of the body through variations in the flow of an electric current through the tissues between different types of tissue.

Biomarkers Refers to a broad subcategory of medical signs. Is a characteristic that can be objectively measured and evaluated as an indicator of normal biological processes, pathogenic processes, or pharmacologic responses to a therapeutic intervention.

BMI See body mass index.

Body mass index (BMI) An indicator of body density as determined by the relationship of body weight (in kilograms) divided by height (in meters) squared. This index correlates with body fat (adipose tissue). Their relationship varies with age and gender.

Bubonic plague A serious, sometimes fatal, infection with the bacterial toxin *Yersinia pestis*, transmitted by fleas from infected rodents and characterized by high fever, weakness, and the formation of buboes, especially in the groin and armpits.

Calories Units used in measuring the amount of energy food provides when eaten and digested. A calorie is equal to the amount of heat needed to increase the temperature of one gram of water by one degree Celsius.

Cardiomyopathy Refers to diseases of the heart muscle. In most cases, cardiomyopathy causes the heart muscle to become enlarged, thick, or rigid.

Cardiopulmonary resuscitation The artificial substitution of heart and lung action as indicated for heart arrest resulting from electric shock, drowning, respiratory arrest, or other causes. The two major components of cardiopulmonary resuscitation are artificial ventilation (respiration, artificial) and closed-chest cardiac massage.

CD See Chagas disease.

Cenote From *ts'ono'ot*, a Maya word, underground water body.

Cerebral palsy A condition marked by impaired muscle coordination and/or other disabilities, typically caused by damage to the brain before or at birth.

Chagas disease Sickness caused by *Trypanosoma cruzi*, mainly transmitted by bloodsucking bugs distributed in all America, excepting Canada, causes damage to the heart and central nervous system.

Chaltun From the Maya language: flat, smooth rock of great size.

Chronic undernutrition State of nutritional deficiency caused by long-term nutritional inadequacies and/or recurrent infections, includes stunting (low height-for-age).

Circular migration Temporary and cyclical movement of a migrant worker between his/her home and usually an employment area. It could be international, cross-country, or rural-urban.

Classic Archaeological period of the Pre-Columbian Maya civilization, between 250-950 CE. This is the era which saw the consolidation of Maya civilization in which they perfected mathematics, astronomy, architecture, and the visual arts and refined and perfected the calendar.

Cofradía An association of religious people, organized to meet pious objectives, recognized by Catholic Church.

Cohort Set of individuals that share a defining event in a specified time period (birth date, marriage, graduation, etcetera).

Complex system The representation of a cut of the reality, conceived like an organized totality in which the constituent elements cannot be studied in isolation.

Concept Embryo.

Corporeity The quality of having a physical body or existence.

Cortisol Glucocorticoid $C_{21}H_{30}O_5$ produced by the adrenal cortex upon stimulation by ACTH that mediates various metabolic processes (such as gluconeogenesis), has anti-inflammatory and immunosuppressive properties, and whose levels in the blood may become elevated in response to physical or psychological stress.

Criba orbitalia Lesions characterized by pitting of the compact bone of the skull, which is usually associated with an increase in thickness of the adjacent diploic bone. The lesions can vary in size from less than 1 mm to large coalescing apertures, and in area from the orbital roof (criba orbitalia) to the skull vault, particularly the frontal, parietal, and occipital bones.

Cytoskeleton A microscopic network of protein filaments and tubules in the cytoplasm of many living cells, giving them shape and coherence.

DALYs See disability-adjusted life years.

Decompression illness Is caused by intravascular or extravascular bubbles that are formed as a result of reduction in environmental pressure (decompression). The term covers both arterial gas embolism, in which alveolar gas or venous gas emboli (via cardiac shunts or via pulmonary vessels) are introduced into the arterial circulation, and decompression sickness, which is caused by in situ bubble formation from dissolved inert gas.

Developmental induction A process whereby a stimulus or insult at a critical period of development has lasting or lifelong impact on the phenotype of the organism.

Developmental stress Constraining force(s) or influence(s) exerted on an organism during early stage of ontogeny resulting on short- and long-term negative consequences on well-being.

Diabetes A disease in which the body's ability to produce or respond to the hormone insulin is impaired, resulting in abnormal metabolism of carbohydrates.

Diagenesis The physical and chemical changes occurring during the conversion of sediment to sedimentary rock.

Disability-adjusted life years A measure of overall disease burden, expressed as the number of years lost due to ill health, disability, or early death.

Disease risk Any attribute, characteristic, or exposure of an individual that increases the likelihood of developing a disease or injury.

Disease vectors Invertebrates or nonhuman vertebrates which transmit infective organisms from one host to another.

Dizziness A sensation of spinning around and losing one's balance.

Dose–response Relating to, exhibiting, or designating a relationship between the size of a dose and a measure of the response to it.

Dual burden' households Is defined as a household in which at least one member is underweight and at least one member is overweight.

Dual-burden of nutrition Is defined as the coexistence of under- and overnutrition in the same population/group, the same household/family, or the same person.

Economic transition An economy that is changing from being one under government control to be a market economy.

Ecosystem All the living things in an area and the way they affect each other and the environment.

Ejido A system of communal land possession institutionalized after the Mexican Revolution of 1910.

Enamel The hard, white, shiny substance that form the covering of a tooth.

Enamel hypoplasia A quantitative defect that occurs due to the reduced thickness of enamel. The disorder is morphologically identified as one or more shallow horizontal lines on the crown of the tooth. Most hypoplastic defects are associated with systemic physiological stress such as starvation, infectious diseases, metabolic disorders, and physical or psychological trauma.

Encomendero A man that, during some time of the Spanish Colonial times in America, was entrusted with some number of Indian people who must work for him or pay him in kind in exchange of Catholic religious instruction and defense.

Encomienda The assignment, by Spanish crown, of some number of Indians to a Spaniard subject in compensation for his services.

Endemic (Of a disease or condition) regularly found among particular people or in a certain area.

Endemic diseases The constant presence of diseases or infectious agents within a given geographic area or population group. It may also refer to the usual prevalence of a given disease with such area or group.

Energy-dense food Provide a high number of calories per gram of food.

Epidemiological Is the study of how often diseases occur in different groups of people and why.

Epigenetic transmission The transmission of information from one generation of an organism to the next that influence the traits of offspring without alteration of the primary structure of DNA.

Essential nutrients Nutritional substances required for optimal health. These must be in the diet, because they are not formed metabolically within the body.

Ethnic group A community or population made of people who share a common cultural background or descent.

Ethnographic Of or relating to ethnography, that is rigorous explanations of cultural phenomena by comparing human cultures.

Extradomicile Out of domicile, the area surrounding the house.

Famine A situation in which there is not enough food for a great number of people, causing illness and death.

Fertility transition A major change in the general pattern of reproduction in a society.

Gender The different characteristics that begin to define a person as masculine or feminine—consists of several categories apart from the traditional binary ends of the male/female spectrum

Gene flow The alteration of the frequencies of alleles of particular **genes** in a population, resulting from interbreeding with organisms from another population having different frequencies.

Globalization The process of interaction and integration among people, companies, and governments worldwide.

Groundwater Water within the earth that supplies wells and springs.

Growth Quantitative increase in size or mass.

Guillain–Barre syndrome An acute inflammatory autoimmune neuritis caused by T cell-mediated cellular immune response directed towards peripheral myelin. Demyelination occurs in peripheral nerves and nerve roots.

Hacienda A large plantation with a dwelling house.

Hematophagy Is the practice of certain animals of feeding on blood.

Half-live Is the time required for a quantity to reduce to half its initial value. The term is commonly used in nuclear physics to describe how quickly unstable atoms undergo, or how long stable atoms survive, radioactive decay.

Haplogroups A group of single chromosomes, or single DNA strands, which share a common ancestor.

Height-for-age z-scores A z-score (aka, a standard score) indicates how many standard deviations an element is from the mean.

Hemorrhagic fevers, viral A group of viral diseases of diverse etiology but having many similar clinical characteristics; increased capillary permeability, leukopenia, and thrombocytopenia are common to all.

Henequen A Mexican agave (*Agave fourcroydes*) with large fleshy leaves, cultivated for fiber production.

Heterarchy The Late Classic Maya power was a heterarchy, meaning that Maya capitals used to coordinate authorities of different levels and characteristics in a segmental state of conical clans, integrated by corporative units. These units were collective groups with their own identity.

Hieroglyphics A system of writing that uses pictures instead of words.

High fidelity medical simulation A controlled learning environment that closely represents reality.

Historical trauma theory A scheme that proposes that populations subjected to long-term, mass trauma show higher rates of disease, even several generations have passed since the primary traumatic event occurred.

Holistic Characterized by the belief that the parts of something are intimately interconnected and explicable only by reference to the whole.

Homeostasis The ability or tendency of a living organism, cell, or group to keep the conditions inside it the same despite any changes in the conditions around it, or this state of internal balance.

Huipil From *huipilli*, a Nahuatl word. A typical dress of Maya women in the Peninsula of Yucatan.

Human biology An interdisciplinary approach to understanding the human being from biological, behavioral, social, and cultural perspectives.

Human ecology Human ecology is an interdisciplinary and transdisciplinary study of the relationship between humans and their natural, social, and built environments.

Hyperglycemia An excess of glucose in the bloodstream, often associated with diabetes mellitus.

IGF The insulin-like growth factors (IGFs) are proteins with high sequence similarity to insulin. IGFs are part of a complex system that cells use to communicate with their physiologic environment.

Intergenerational influences hypothesis Theory that propose that the biology and health status of individuals are shaped by the conditions experienced by their recent ancestors during their critical periods of growth.

Intradomicile Of or related to the area inside a house.

Junk foods Foods that are high in calories but low in nutritional content.

Karst A relief made by limestone, dolomite, or gypsum rocks. The karst is highly soluble to water.

Landscapes An area that is spatially heterogeneous in at least one factor of interest. For example, from a wildlife perspective, we might define landscape as an area of land containing a mosaic of habitat patches.

Leptokurtic distribution Statistic. Related to a frequency *distribution* or its graphical representation that is more heavily concentrated about the *mean* than a normal *distribution*.

Life history The series of changes undergone by an organism during its lifetime.

Longitudinal models A set of statistical analyses in which observations (outcomes/treatments/exposures) were collected at multiple follow-up or repeated times. Broadly, longitudinal analyses are used to characterize and predict the change of observations and estimate the influence of factors on the change and trajectory of outcomes.

Maya people Mesoamerican ethnic group with thousands of years of history. The location of Maya groups goes from southeastern part of Mexico to the Central America in the current nations of Guatemala, Belize, El Salvador, and Honduras.

Measles An infectious viral disease, causing fever and a red rash on the skin, typically occurring in childhood.

Megavisceral syndromes The enlargements of the colon, heart, or esophagus that are typical manifestations of infection with *T. cruzi*.

Meningoencephalitic Is the predominant lesion, with focal accumulations of a mixed inflammatory cell infiltrate in the meninges and brain.

Menopausal transition The time when the levels of hormones produced by the aging ovaries fluctuate, leading to irregular menstrual patterns (irregularity in the length of the period, the time between periods, and the level of flow).

Mesoamerica A relatively homogeneous pre-Columbian cultural area encompassing southern Mexico, and current Belize, El Salvador, and Guatemala and part of Costa Rica, Honduras, and Nicaragua.

Metabolic disease A medical condition in which the normal chemical processes that occur within a living organism in order to maintain life is altered.

Metabolic syndrome A cluster of biochemical and physiological abnormalities associated with the development of cardiovascular disease and type 2 diabetes.

Methylmercury Any of various toxic compounds of mercury containing the complex CH_3Hg- that often occur as pollutants which accumulate in living organisms (such as fish) especially in higher levels of a food chain.

Microcephaly A congenital abnormality in which the cerebrum is underdeveloped, the fontanels close prematurely, and, as a result, the head is small.

Microtubules A microscopic tubular structure present in numbers in the cytoplasm of cells, sometimes aggregating to form more complex structures.

Morbidity Illness.

Mortality Death.

Myocarditis Inflammation of the heart muscle.

Natural fertility Fertility pattern in which women reproduce throughout their reproductive life span.

New Spain It was an integral territorial entity of the Spanish Empire, established by Habsburg Spain during the Spanish colonization of the Americas, including what now is Mexico and some parts of Central America, United States of America, and even Canada.

Non-anthropic From nonhuman activities or processes.

Nutritional dual burden The coexistence of low height-for-age or stunting and overweight/obesity at individual or population level.

Nutritional status The sum or result of biological and sociocultural processes involved in the obtaining, assimilation, and metabolism of nutrients by the organism.

Nutritional transition Is the shift in dietary consumption and energy expenditure that coincides with economic, demographic, and epidemiological changes. Specifically, the term is used for the transition of developing countries from traditional diets high in cereal and fiber to more Western pattern diets high in sugars, fat, and animal-source food.

Obvenciones During the Colonial times in New Spain, a kind of religious tax paid by Indians.

Ontogeny The development or course of development especially of an individual organism.

Oportunidades Official Mexican program, launched at 2002, oriented to support families living in poverty to increase their feeding, health, and education capabilities by providing them economic resources and services (https://www.gob.mx/prospera/documentos/que-es-prospera) 1st April 2019. See *Prospera*.

***Oportunidades* Program** A government social assistance program in Mexico founded in 2002. *Oportunidades* was designed to target poverty by providing cash payments to families in exchange for regular school attendance, health clinic visits, and nutrition support.

Osteological age Age of a bone; it could be different from the chronological age, i.e., the number of years elapsed between the birth date and the time of observation.

Osteopenia Reduced bone mass of lesser severity than osteoporosis.

Osteoporosis A medical condition in which the bones become brittle and fragile from loss of tissue.

Overcrowding The presence of more people in a space that is comfortable and safe.

Paleodemography The field of enquiry that attempts to identify demographic parameters from past populations (usually skeletal samples) derived from archaeological contexts, and then to make interpretations regarding the health and well-being of those populations.

Paleodiet Chemical signatures in bone that may offer archaeologists a means to assess past human access to various food groups and the habitats where the foods were produced.

Paleopathology The study and application of methods and techniques for investigating diseases and related conditions from skeletal and soft tissue remains.

Palpitation A noticeably rapid, strong, or irregular heartbeat due to agitation, exertion, or illness.

Pellagra A disease caused by a dietary deficiency of nicotinic acid, characterized by burning or itching often followed by scaling of the skin, inflammation of the mouth, diarrhea, and mental disturbance.

Perception Awareness of something through the senses.

Peridomestic See extra domicile.

Periostitis Inflammation of the periosteum. The term implicitly refers to the soft-tissue membrane, not the bone itself, and to a very specific pathological response, inflammation.

Phenotype A set of observable characteristics of an individual resulting from the interaction of its genotype and environment.

Physiographic Related to the science of physical geography.

Platykurtic distribution Statistic. Related to a frequency *distribution or its graphical representation that is* less concentrated about the *mean* than the corresponding normal.

Plasticity The capacity of organisms with the same genotype to vary in developmental pattern, in phenotype, or in behavior according to varying environmental conditions.

Political ecology Is the study of the relationships between political, economic, and social factors with environmental issues and changes.

Porotic hyperostosis A pathological condition that affects bones of the cranial vault, and is characterized by localized areas of spongy or porous bone tissue.

Post-Classic Archaeological period of the Pre-Columbian Maya civilization, between 900 1521 CE.

Poverty A state or situation in which necessities of food, health, clothing, education, and recreation are not met at individual or population level.

Preclassic An archaeological period in Maya history stretches from the beginning of permanent village life c. 1000 BCE. until the advent of the Classic Period c. 250 CE,

Prevalence Is the proportion of a population who has a specific characteristic in a given time period.

Prospera Official Mexican program, launched at 1997, oriented to support families living in poverty to increase their feeding, health, and education capabilities providing them economic resources and services. In 2002, this program was renamed *Oportunidades* (https://www.gob.mx/prospera/documentos/que-es-prospera) 1st April 2019. See *Oportunidades*.

Pruritus An intense itching sensation that produces the urge to rub or scratch the skin to obtain relief.

Rapamycin (mTOR) A protein that helps control several cell functions, including cell division and survival, and binds to rapamycin and other drugs.

Repartimiento At New Spain (see above), during the Colony (sixteenth to nineteenth centuries), the assignment of Indians as free labor force to *encomenderos* (see *enconmendero*, above) or to the Spanish Crown.

Risk Measurement of a situation's danger using statistical methods.

Santeria A pantheistic Afro-Cuban religious cult developed from the beliefs and customs of the Yoruba people and incorporating some elements of the Catholic religion.

Sartenejas A Spanish word, for rounded hollows in calcareous rocks that keep rainwater.

SES See socioeconomic status

Saturated fat A type of fat containing a high proportion of fatty acid molecules without double bonds, considered to be less healthy in the diet than unsaturated fat.

Slash-and-burn A widely used method of growing food in which wild or forested land is clear-cut and any remaining vegetation burned. The resulting layer of ash provides the newly cleared land with a nutrient-rich layer to help fertilize crops.

Smallpox An acute contagious viral disease, with fever and pustules usually leaving permanent scars.

Social inequality Refers to the existence of unequal opportunities and rewards for different social positions or statuses within a group or society.

Socioeconomic status Is the social standing or class of an individual or group. It is often measured as a combination of education, income, and occupation.

Spongy hyperostosis Lesions characterized by pitting of the compact bone of the skull, which is usually associated with an increase in thickness of the adjacent diploic bone. The lesions can vary in size from less than 1 mm to large coalescing apertures, and in area of the skull vault, particularly the frontal, parietal, and occipital bones. See Criba orbitalia.

Stress In a medical or biological context stress is a physical, mental, or emotional factor that causes bodily or mental tension. ... Stress can cause or influence the course of many medical conditions including psychological conditions such as depression and anxiety.

Stunting A consequence of severe and long-lasting malnutrition in which a child fails to achieve the expected height for his or her age.

Subsistence agriculture A type of farming in which most of the produce (subsistence crop) is consumed by the farmer and his or her family, leaving little or nothing to be marketed.

Symphyses A place where two bones are closely joined, either forming an immovable joint or completely fused.

Symptoms A physical or mental feature, which is regarded as indicating a condition of disease, particularly such a feature that is apparent to the patient.

Synanthropic Related to a species of wild animals and plants that live near, and benefit from, an association with humans and the somewhat artificial habitats that humans create around them.

Syncope A transient loss of consciousness and postural tone caused by diminished blood flow to the brain (i.e., brain ischemia). Presyncope refers to the sensation of lightheadedness and loss of strength that precedes a syncopal event or accompanies an incomplete syncope.

Syncretic blends The attempted reconciliation or union of different or opposing principles, practices, or parties, as in philosophy or religion.

Taphonomy The study of the processes that affect organic remains as they become fossilized.

Teleological The explanation of phenomena in terms of the purpose they serve rather than of the cause by which they arise.

Theocracy A situation where the religious leaders assume a leading role in the state.

Thrifty genotype The idea that there are some people who have "thrifty genes" stems from the work of geneticist James Neel who argued that problems such as obesity and diabetes are manifestations of an adaptive evolutionary trait that would have been beneficial when human beings faced a reality that involved periods of famine.

Thrifty phenotype The thrifty phenotype hypothesis says that reduced fetal growth is strongly associated with a number of chronic conditions later in life. This increased susceptibility results from adaptations made by the fetus in an environment limited in its supply of nutrients.

Terra incognita Latin for unknown or unexplored territory.

TFR See total fertility rate.

Tortilla A thin, round, unleavened bread prepared from cornmeal or sometimes wheat flour, baked on a flat plate of iron, earthenware, or the like.

Total fertility rate A non-age-biased estimator of total fertility in a population.

Translactional Through maternal milk, by way of maternal milk.

Triatomine Name of the subfamily of the insect vectors of *Trypanosoma cruzi*, the etiological agent of Chagas disease.

Trypanosome A single-celled parasitic protozoan with a trailing flagellum, infesting the blood.

Tripanosomiasis Any tropical disease caused by trypanosomes and typically transmitted by biting insects, especially sleeping sickness and Chagas' disease.

Vector An organism, typically a biting insect or tick, that transmits a disease or parasite from one animal or plant to another.

Waterborne diseases Conditions caused by pathogenic microorganisms that are transmitted in water, whether for bathing, washing, or drinking water, or by eating food exposed to contaminated water.

Zoonotic Used to refer to a disease that can spread from animals to human.

Z-score The number of standard deviation units away from the mean a particular value of data lies.

Index

H. Azcorra, F. Dickinson (eds.), *Culture, Environment and Health in the Yucatan Peninsula*, https://doi.org/10.1007/978-3-030-27001-8

N

O

P

T

Printed in the United States
By Bookmasters